Processor Description Languages
Applications and Methodologies

The Morgan Kaufmann Series in Systems on Silicon

Series Editor: Wayne Wolf, Georgia Institute of Technology

Processor Description Languages
Applications and Methodologies

Prabhat Mishra
University of Florida

Nikil Dutt
University of California, Irvine

AMSTERDAM • BOSTON • HEIDELBERG • LONDON
NEW YORK • OXFORD • PARIS • SAN DIEGO
SAN FRANCISCO • SINGAPORE • SYDNEY • TOKYO

Morgan Kaufmann Publishers is an imprint of Elsevier

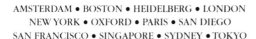

MORGAN KAUFMANN PUBLISHERS

Senior Acquisitions Editor	Charles Glaser
Publishing Services Manager	George Morrison
Assistant Editor	Gregory Chalson
Project Manager	Rajashree Satheesh Kumar
Designer	Joanne Blank
Typesetting	diacriTech
Cover Designer	Joanne Blank
Interior printer	RR Donnelley
Cover printer	Phoenix Color Corp.
Cover Image	istockphoto.com © Todd Harrison
Cover Design	Dick Hannus

Morgan Kaufmann Publishers is an imprint of Elsevier.
30 Corporate Drive, Suite 400, Burlington, MA 01803, USA

This book is printed on acid-free paper. ∞

Library of Congress Cataloging-in-Publication Data
Processor description languages: applications and methodologies/edited by Prabhat Mishra, Nikil Dutt.
    p. cm.
  Includes bibliographical references and index.
  ISBN 978-0-12-374287-2 (alk. paper)
1. Embedded computer systems. I. Mishra, Prabhat, 1973- II. Dutt Nikil.
  TK7895.E42P75 2008
  004.16–dc22

2008004949

ISBN: 978-0-12-374287-2
ISSN: 1875-9661

For information on all Morgan Kaufmann publications,
visit our website at *www.mkp.com* or *www.books.elsevier.com*

Printed and bound by CPI Group (UK) Ltd, Croydon, CR0 4YY
Transferred to Digital Print 2011

To Ahan and Tania - Prabhat
To Shefali, Shubir, and Sujata - Nikil

Contents

CHAPTER 6   EXPRESSION: An ADL for Software Toolkit Generation, Exploration, and Validation of Programmable SOC Architectures   133

Prabhat Mishra and Nikil Dutt

CHAPTER 10 **ADL++: Object-Oriented Specification of Complicated Instruction Sets and Microarchitectures** 247

Soner Önder

List of Contributors

Nupur Bhattacharyya Andrews
Tensilica

Nupur Bhattacharyya Andrews has been a primary developer of the Tensilica Instruction Extension (TIE) technology for many years. She has extensive experience in verification of microprocessors and systems at Tensilica and Silicon Graphics and in EDA tool design for DSP systems at Cadence Design Systems. She received her master of science degree in signal processing from Louisiana State University, and a bachelor's degree in electrical engineering from Jadavpur University, India.

Guido Araujo
University of Campinas, Brazil

Guido Araujo received his Ph.D. degree in electrical engineering from Princeton University in 1997. He worked as a consultant for Conexant Semiconductor Systems and Mindspeed Technologies from 1997 to 1999, and is currently a professor of computer science and engineering at the University of Campinas (UNICAMP), Sao Paulo, Brazil. His main research interests are code optimization, architecture simulation, and embedded system design. Araujo was corecipient of best paper awards at the DAC'97, SCOPES'03, and SBAC-PAD'04. He was also awarded the 2002 University of Campinas Zeferino Vaz Award for his contributions to computer science research and teaching.

Rodolfo Azevedo
University of Campinas, Brazil

Rodolfo Azevedo received his Ph.D. in computer science from University of Campinas (UNICAMP), Brazil, in 2002. He currently works as a professor in the Institute of Computing at UNICAMP. Azevedo was a corecipient of a best paper award at SBAC-PAD'04 for his work on the ArchC ADL. His main research interests are computer architecture, embedded systems, code compression, low power, ADLs, and system-level design, mainly using SystemC.

Nirmalya Bandyopadhyay
University of Florida, USA

Nirmalya Bandyopadhyay received his bachelor of engineering degree in computer science and technology from Bengal Engineering and Science University, Shibpur, India, in 2004. He spent 2 years in industry in the area of networking and EDA tool development. He was also associated with the Formal Verification group at the Indian Institute of Technology, Kharagpur. Currently, he is pursuing his Ph.D. degree from Department of Computer and Information Science and Engineering, University of Florida, Gainesville, USA. His research interest includes application of satisfiability in test case generation and multicore verification.

Kanad Basu
University of Florida, USA

Kanad Basu received his bachelor of electronics and teleCommunication engineering degree from Jadavpur University, Kolkata, India, in 2007. He is currently pursuing Ph.D. degree from the Computer and Information Science and Engineering department of the University of Florida. His research interests include embedded systems, nanosystem design and verification, and code compression. He has 7 publications in various international conferences in the areas of wireless networks and intelligent systems.

Anupam Chattopadhyay
RWTH Aachen University, Germany

Anupam Chattopadhyay received his bachelor of electronics and teleCommunication engineering from Jadavpur University, Kolkata, India, in 2000, and after a brief industrial stint, obtained his degree of master of engineering in embedded systems design from the University of Lugano, Switzerland, in 2002. He is currently pursuing his Ph.D. degree from the Institute for Integrated Signal Processing Systems (ISS), RWTH Aachen University, Germany. His research interests include automatic implementation of processors with LISA, architecture optimization techniques, and a toolflow for reconfigurable ASIPs.

Nikil Dutt
University of California, Irvine, USA

Nikil Dutt is a chancellor's professor of CS and EECS and is with the Center for Embedded Computer Systems at UC Irvine. He is a fellow of the IEEE and an ACM distinguished scientist. He has received numerous best paper awards at conferences and serves on the editorial boards of ACM TODAES (EiC), ACM TECS (AE), and IEEE T-VLSI (AE). His research interests are in embedded systems design automation, computer architecture, optimizing compilers, system specification techniques, and distributed embedded systems.

Daniel Gajski
University of California, Irvine, USA

Daniel Gajski is the Henry Samueli Endowed Chair in computer system design and Professor of EECS at UC Irvine. He directs the Center for Embedded Computer Systems and was instrumental in developing formalisms such as Y-chart numerous algorithms for high-level synthesis, the definition of CDFG and FSM with data (FSMD), and system-level languages such as SpecCharts and SpecC. He holds Dipl. Ing. and M.S. degrees in electrical engineering from the University of Zagreb, Croatia, and a doctoral degree in computer and information sciences from the University of Pennsylvania, Philadelphia.

Werner Geurts
Target Compiler Technologies

Werner Geurts is vice president of engineering of Target Compiler Technologies. Before cofounding Target in 1996, he was a researcher at IMEC, where he has been working on behavioral synthesis of data-path structures and on retargetable compilation. Werner Geurts has coauthored several papers in electronic design automation. He received master's degree in electrical engineering from the Hogeschool Antwerpen and K.U. Leuven, and a Ph.D. degree from K.U. Leuven, in 1985, 1988, and 1995, respectively.

Gert Goossens
Target Compiler Technologies

Gert Goossens is the CEO and a cofounder of Target Compiler Technologies, a provider of retargetable tools for the design of application-specific processors. Before cofounding Target in 1996, Gert Goossens was affiliated with the IMEC research center, where he headed research groups on behavioral synthesis and software compilation. Gert Goossens has authored or coauthored around 40 papers in electronic design automation. He received a masters and a Ph.D degree in electrical engineering from K.U. Leuven in 1984 and 1989, respectively.

Bita Gorjiara
University of California, Irvine, USA

Bita Gorjiara received her B.S. and M.S. from University of Tehran, and her Ph.D. in computer science and engineering from University of California, Irvine, in 2007. She is one of the key developers of NISC Technology, and her Ph.D. thesis focuses on efficient synthesis of NISC processors. Bita has received Hami Award of Excellence in Engineering and EECS Best Student Paper Award. Her research interests include custom processor design, embedded system design, high-level synthesis, and power optimization.

Frank Hannig
University of Erlangen-Nuremberg, Germany

Frank Hannig graduated in an interdisciplinary course in CS and EE from Paderborn University, where he later joined as a Ph.D. student. In 2003, he moved to the newly founded Department of Hardware/Software Co-Design at the University of Erlangen-Nuremberg. Here, Frank leads the architecture and compiler design group. His main research interests are the design of massively parallel architectures and mapping methodologies for domain-specific computing. Frank is a member of IEEE and reviewer for multiple journals and conferences including IEEE TVLSI, IEEE TSP, DAC, and DATE.

Masaharu Imai
Osaka University, Japan

Dr. Masaharu Imai received his Ph.D. degree in information science from Nagoya University in 1979. From 1979 through 1996, he has been with the Department of Information and Computer Sciences, Toyohashi University of Technology, Toyohashi, Japan. From 1996 to till now, he is with Osaka University, Osaka, Japan, where he is a professor at the Department of Information Systems Engineering, Graduate School of Information Science and Technology. He is one of the founders and CTO of ASIP Solutions, Inc.

Dmitrij Kissler
University of Erlangen-Nuremberg, Germany

Dmitrij Kissler received his master of science degree with honors in computer science from the University of Erlangen-Nuremberg, Germany, in April 2006. Since May 2006, he is working as a research scientist at the Department of Hardware/Software Co-Design at the University of Erlangen-Nuremberg. His main research interests include the design of massively parallel embedded hardware architectures, architecture/compiler coexploration, and efficient power modeling and estimation techniques.

Yuki Kobayashi
Osaka University, Japan

Yuki Kobayashi received his B.E. and MIST degrees from Osaka University, Japan, in 2003 and 2004, respectively. In 2007, he received his Ph.D. degree in information science and technology from Osaka University. Since April 2005, he is a research fellow at JSPS, Japan. He was researching at IMEC vzw, Belgium, for 9 months, from 2004 to 2006. His research interests include processor architecture and compiler technology. He is a member of IEEE.

Alexey Kupriyanov
University of Erlangen-Nuremberg, Germany

Alexey Kupriyanov received his master's degree in computer engineering from the Zaporozhye State Engineering Academy, Ukraine, in 2002 (with honors). In January 2003, he joined the research group of Prof. Jürgen Teich as a Ph.D. student in the Department of Hardware/Software Co-Design at the University of Erlangen-Nuremberg, Germany. His main research interests are the modeling and efficient simulation of multicore SoC architectures including systematic integration of application-specific instruction set processors and processor arrays, architecture description languages, and architecture/compiler codesign.

Dirk Lanneer
Target Compiler Technologies

Dirk Lanneer is the vice president of engineering of Target Compiler Technologies. Before cofounding Target, an IMEC spin-off, he worked from 1986 to 1996 as a researcher at IMEC, first on the "Cathedral 2nd" behavioural synthesis system, and later extending techniques from this research to retargetable compilation. Dirk Lanneer received his master's and Ph.D. degree in electrical engineering from the Katholieke Universiteit Leuven, Belgium, in 1986 and 1993, respectively.

Rainer Leupers
RWTH Aachen University, Germany

Rainer Leupers is a professor of software for systems on silicon at RWTH Aachen University. His research activities comprise software development tools, processor architectures, and electronic design automation for embedded systems, with emphasis on ASIP and MPSoC design tools. He is a cofounder of LISATek, an embedded processor design tools vendor, acquired by CoWare Inc. in 2003. In 2006, he edited the Morgan Kaufmann book Customizable Embedded Processors.

Sharad Malik
Princeton University, USA

Sharad Malik received his B.Tech. degree in electrical engineering from the Indian Institute of Technology, New Delhi, India in 1985 and M.S. and Ph.D. degrees in computer science from the University of California, Berkeley, in 1987 and 1990, respectively. Currently, he is a professor in the Department of Electrical Engineering, Princeton University. He is a fellow of the IEEE. His research in functional timing analysis and propositional satisfiability has been widely used in industrial electronic design automation tools.

Peter Marwedel
Technische Universität Dortmund, Germany

Peter Marwedel was born in Hamburg, Germany. He received his PhD. in physics from the University of Kiel, Germany, in 1974. From 1974 to 1989, he was a faculty member of the Institute for Computer Science and Applied Mathematics at the same university. He has been a professor at the University of Dortmund, Germany, since 1989. He is heading the embedded systems group at the CS department and is chairing ICD e.V., a local spin-off.

Heinrich Meyr
RWTH Aachen University, Germany

Heinrich Meyr is presently a professor of electrical engineering at RWTH Aachen University, Germany, and chief scientific officer of CoWare Inc., San Jose. He pursues a dual career as a researcher and an entrepreneur with over 30 years of professional experience. He is a fellow of IEEE and has received several best paper and professional awards. His present research activities include cross-disciplinary analysis and design of complex signal processing systems for communication applications.

Prabhat Mishra
University of Florida, USA

Prabhat Mishra is an assistant professor in the Department of Computer and Information Science and Engineering at University of Florida. He has a B.E. from Jadavpur University, M.Tech. from Indian Institute of Technology, Kharagpur, and Ph.D. from University of California, Irvine—all in computer science. He has published one book and many conference and journal articles. He has received several reearch and teaching awards including the NSF CAREER Award, CODES+ISSS Best Paper Award, EDAA Outstanding Dissertation Award, and International Educator of the Year Award. His research interests include functional verification and design automation of embedded systems.

Soner Önder
Michigan Technological University, USA

Dr. Soner Önder is an associate professor of computer science at Michigan Technological University. He got his Ph.D. in computer science in 1999 from the University of Pittsburgh. His main interests are microarchitecture techniques for instruction-level parallelism, compiler-hardware cooperation, and architecture description languages. His research has been supported by DARPA and NSF, and he is a recipient of National Science Foundation CAREER award (2004). Dr. Onder is a member of IEEE and ACM.

Johan Van Praet
Target Compiler Technologies

Johan Van Praet is the vice president of engineering of Target Compiler Technologies. Before cofounding Target in 1996, he worked out a processor modelling methodology for retargetable compilation and processor design at IMEC. He holds several patents for this research, and his Ph.D. thesis on this topic was awarded the IBM Belgium prize for computer science 1998. Johan Van Praet earned his master's and Ph.D. degree in electrical engineering from the K.U. Leuven in 1990 and 1997, respectively.

Subramanian Rajagopalan
Synopsys

Subramanian Rajagopalan received his B.Tech. in electrical engineering from Indian Institute of Technology, Madras, India, in 1998, and the M.A. and Ph.D. in electrical engineering from Princeton University, Princeton, in 2001 and 2004, respectively.

Since 2004, he has been with the Advanced Technology Group at Synopsys, India. His research interests include system-level design, parasitics extraction, low-power design, and analog/RF design automation.

Mehrdad Reshadi
University of California, Irvine, USA

Mehrdad Reshadi received his B.S. (Hons.) degree in computer engineering from Sharif University of Technology, Iran, in 1997, the M.S. degree in computer engineering from University of Tehran, Iran, in 2000, and Ph.D in computer science from University of California, Irvine, in 2007. His research interests include embedded system design automation, microprocessor modeling, and simulation, design synthesis, and specification-based compilation and simulation. He received the best paper award at International Conference on Hardware/Software Codesign and System Synthesis (CODES+ISSS) in 2003.

Sandro Rigo
University of Campinas, Brazil

Sandro Rigo received his Ph.D. in computer science from University of Campinas (UNICAMP), Brazil, in 2004. He currently works as a professor in the Institute of Computing at UNICAMP. Rigo was a corecepient a best paper award at SBAC-PAD'04 for his work on the ArchC ADL. He has conducted researches on garbage collection, code optimization, ADLs, and system-level design, mainly using SystemC. His current research interests focus on code optimization for parallel and multi-core architectures, ESL simulation and verification methodologies, and transactional memories.

Himanshu Sanghavi
Tensilica

Himanshu Sanghavi is the director of engineering at Tensilica, working on configurable microprocessor cores and tools for microprocessor design. Prior to Tensilica, he worked at Mediamatics (a division of National Semiconductor) on digital audio and video products, and at Intel on microprocessor, graphics, and video products. He holds an MS in electrical and computer engineering from The University of Texas at Austin, and a B.Tech. in electrical engineering from IIT Bombay, India.

Aviral Shrivastava
Arizona State University, USA

Aviral Shrivastava is an assistant professor in the Department of Computer Science and Engineering at the Arizona State University, where he has established and heads the Compiler and Microarchitecture Labs (CML). He received his Ph.D. and master's in information and computer science from University of California, Irvine. He recieved his bachelor degree in computer science and engineering from Indian Institute of Technology, Delhi. His research interests are intersection of compilers and computer architecture, with a particular interest in embedded systems. Dr. Shrivastava is a lifetime member of ACM, and serves on organizing and program committees of several premier embedded system conferences, including ISLPED, CODES+ISSS, CASES, and LCTES.

Yoshinori Takeuchi
Osaka University, Japan

Yoshinori Takeuchi received his B.E., M.E., and Dr. of Eng. degrees from Tokyo Institute of Technology in 1987, 1989, and 1992, respectively. From 1992 through 1996, he was a research associate with Department of Engineering, Tokyo University of Agriculture and Technology. From 1996, he has been with the Osaka University. He was a visiting professor at the University of California, Irvine, from 2006 to 2007. He is currently an associate professor of Graduate School of Information Science and Technology at Osaka University. His research interests include system-level design, VLSI design, and VLSI CAD. He is a member of IEICE of Japan, IPSJ, ACM, SP, CAS, and SSC Society of IEEE.

Jürgen Teich
University of Erlangen-Nuremberg, Germany

Jürgen Teich is an appointed full professor in the Department of Computer Science of the Friedrich-Alexander University Erlangen-Nuremberg, holding a chair in hardware/software co-design since 2003. Dr. Teich has been a member of multiple program committees of international conferences and workshops and program chair for CODES+ISSS 2007. He is a senior member of the IEEE. Since 2004, he acts also as a reviewer for the German Science Foundation (DFG) for the area of computer architecture and embedded systems. Prof. Teich is supervising more than 20 Ph.D. students currently.

Wei Qin
Boston University, USA

Wei Qin received his B.S. and M.S. degrees in electronics engineering from Fudan University in China, and his Ph.D. in electrical engineering from Princeton University. Since 2004, he has been an assistant professor at the ECE Department of Boston University. His research interests include modeling, synthesis, and verification of programmable processors; tools, methodologies, and architectures for heterogeneous multiprocessor embedded systems; and design languages for electronic systems. His research is supported by Intel and Freescale Semiconductors.

Preface

Today, computing devices are seamlessly weaved into the fabric of our everyday lives. We encounter two types of such computing devices: desktop-based computing devices and embedded computer systems. Desktop-based computing systems encompass traditional computers, including personal computers, notebook computers, workstations, and servers. Embedded computer systems are ubiquitous—they run the computing devices hidden inside a vast array of everyday products and appliances such as cell phones, toys, handheld PDAs, cameras, and microwave ovens. Cars are full of them, as are airplanes, satellites, and advanced military and medical equipments. As applications grow increasingly complex, so do the complexities of the embedded computing devices. Both types of computing devices use programmable components such as processors, coprocessors, and memories to execute the application programs. The complexity of these components is increasing at an exponential rate due to technological advances, as well as due to demand for realization of ever more complex applications in communication, multimedia, networking, and entertainment. Shrinking time-to-market coupled with short product lifetimes creates a critical need for design automation of increasingly sophisticated and complex programmable components.

Architecture Description Languages (ADLs), also known as processor description languages, have been successfully used in industry as well as in academic research for design and verification of application-specific instruction-set processors (ASIP). ADLs allow precise modeling and specification of processors. The ADL specification can be used to enable various design automation tasks including exploration, simulation, compilation, synthesis, test generation, and validation of architectures. Each automation task creates unique opportunities. For example, the exploration task is used to figure out the best possible processor architecture for the given set of application programs under various design constraints such as area, power, and performance. Manual or semi-automatic approaches are time consuming and error prone. In many realistic scenarios, it is not possible to explore all feasible design alternatives using manual or semi-automatic approaches. ADL-based specification and design automation reduces overall design and verification effort, and thereby enables generation of low-cost, efficient, and reliable systems.

This book presents a comprehensive understanding of ADLs and ADL-driven design automation methodologies. There has been a plethora of research in this area in the last 15 years. Unfortunately, there is no comprehensive text available to help designers and researchers efficiently integrate ADL-driven design automation steps into their current methodologies. For example, some of the ADLs were developed 10–15 years ago. They have gone through various transformations to suit the needs of contemporary design processes, as well as of the processors themselves. Although each ADL or its modification has appeared as a conference or journal paper, the modified form of the complete ADL or the up-to-date methodologies are not accessible in a comprehensive and unified format. The reader of this book will get a comprehensive understanding of representative processor description languages and how they can be used for design automation of embedded systems and system-on-chip (SOC) architectures. The first two chapters cover the basics of generic ADLs and ADL-driven design automation techniques. The remaining chapters cover specific contemporary ADLs and their supported tools, techniques, and methodologies.

AUDIENCE

This book is designed for readers interested in design and verification of application-specific processors and system-on-chip designs. This includes readers in the following three broad categories:

- The instructors planning to teach a senior-level undergraduate course or graduate-level course on design automation of embedded processors.

- Architects, managers, and design/verification engineers involved in early exploration, analysis, rapid prototyping, and validation of processor cores, coprocessors, and memory architectures.

- Researchers in academia as well as in industry interested in the area of high-level design and validation of embedded systems.

We hope you enjoy reading this book and find the information useful for your purpose.

Prabhat Mishra, University of Florida
Nikil Dutt, University of California, Irvine
January 1, 2008

About the Editors

Prabhat Mishra is an assistant professor in the Department of Computer and Information Science and Engineering at the University of Florida. He received his B.E. from Jadavpur University, Kolkata, India, in 1994, M.Tech. from the Indian Institute of Technology, Kharagpur, India, in 1996, and Ph.D. from the University of California, Irvine, in 2004—all in computer science. Prior to his current position, he spent several years in industry working in the areas of design and verification of microprocessors and embedded systems. His research interests are in the area of VLSI CAD, functional verification, and design automation of embedded and nanosystems. He is a coauthor of the book *Functional Verification of Programmable Embedded Architectures*, Springer, 2005. He has published more than 40 research articles in premier journals and conferences. His research has been recognized by several awards including the NSF CAREER Award from National Science Foundation in 2008, CODES+ISSS Best Paper Award in 2003, and EDAA Outstanding Dissertation Award from the European Design Automation Association in 2005. He has also received the International Educator of the Year Award from the College of Engineering for his significant international research and teaching contributions. He currently serves as information director of ACM Transactions on Design Automation of Electronic Systems (TODAES), as a technical program committee member of various reputed conferences (DATE, CODES+ISSS, ISCAS, VLSI Design, I-SPAN, and EUC), and as a reviewer for many ACM/IEEE journals, conferences, and workshops. Dr. Mishra is a member of ACM and IEEE.

Nikil Dutt is a chancellor's professor at the University of California, Irvine, with academic appointments in the CS and EECS departments. He received his B.E. (Hons) in mechanical engineering from the Birla Institute of Technology and Science, Pilani, India, in 1980, M.S. in computer science from the Pennsylvania State University in 1983, and Ph.D. in Computer Science from the University of Illinois at Urbana–Champaign in 1989. He is affiliated with the following Centers at UCI: Center for Embedded Computer Systems (CECS), California Institute for Telecommunications and Information Technology (Calit2), the Center for Pervasive Communications and Computing (CPCC), and the Laboratory for Ubiquitous Computing and Interaction (LUCI). His interests are in embedded systems, electronic design automation, computer architecture, optimizing compilers, system specification techniques, and distributed systems. He is a coauthor of six books: *Memory Issues in Embedded Systems-on-Chip: Optimizations and Exploration*, Kluwer Academic Publishers, 1999; *Memory Architecture Exploration for Programmable Embedded Systems*, Kluwer Academic Publishers, 2003; *SPARK: A Parallelizing Approach to the High-Level Synthesis of Digital Circuits*, Kluwer Academic Publishers, 2004; *Functional Verification of Programmable Embedded Architectures: A Top-Down Approach*, Springer-Verlag, 2005; and *On-Chip Communication Architectures: System on Chip Interconnect*, Morgan Kaufman, 2008. His research has been recognized by Best

Paper Awards at the following conferences: CHDL'89, CHDL'91, VLSI Design 2003, CODES+ISSS 2003, CNCC 2006, and ASPDAC 2006; and Best Paper Award Nominations at: WASP 2004, DAC 2005, and VLSI Design 2006. He has also received a number of departmental and campus awards for excellence in teaching at UC Irvine. He currently serves as Editor-in-Chief of ACM Transactions on Design Automation of Electronic Systems (TODAES) and as Associate Editor of ACM Transactions on Embedded Computer Systems (TECS) and of IEEE Transactions on VLSI Systems (TVLSI). He was an ACM SIGDA Distinguished Lecturer during 2001–2002, and an IEEE Computer Society Distinguished Visitor for 2003–2005. He has served on the steering, organizing, and program committees of several premier CAD and Embedded System conferences and workshops, including ASPDAC, DATE, ICCAD, CODES+ISSS, CASES, ISLPED, and LCTES. He is a fellow of the IEEE, an ACM Distinguished Scientist, an IFIP Silver Core awardee, and serves on, or has served on the advisory boards of ACM SIGBED, ACM SIGDA, and IFIP WG 10.5.

Introduction to Architecture Description Languages

Prabhat Mishra and Nikil Dutt

Embedded systems present a tremendous opportunity to customize designs by exploiting the application behavior. Rapid exploration and evaluation of candidate architectures (consisting of embedded processors, coprocessors, and memories) are necessary due to time-to-market pressure and short product lifetimes. Manual or semi-automatic design and verification of architectures is a time-consuming process. This can be done only by a set of skilled designers. Furthermore, the interaction among the different teams, such as specification developers, hardware designers, verification engineers, and simulator developers makes rapid exploration and evaluation infeasible. As a result, system architects rarely have tools or the time to explore architectural alternatives to find the best-in-class solution for the target applications. This situation is very expensive both in terms of time and engineering resources, and has a substantial impact on time-to-market. Without automation and a unified development environment, the design process is prone to error and may lead to inconsistencies between hardware and software representations. The solution is to use a golden specification to capture the architecture and generate the required executable models to enable design automation of embedded processors.

Modeling plays a central role in design automation of application-specific instruction-set processors (ASIPs). It is necessary to develop a specification language that can model complex processors at a higher level of abstraction and also enable automatic analysis and generation of efficient prototypes. The language should be powerful enough to capture the high-level description of the programmable architectures. On the other hand, the language should be simple enough to allow correlation of the information between the specification and the architecture manual. Specifications widely used today are still written informally in a natural language. Since natural language specifications are not amenable to automated analysis, there are possibilities of ambiguity, incompleteness, and contradiction: all problems that can lead to different interpretations of the specification. Clearly, formal specification languages are suitable for analysis and verification. Some have become popular because they are suitable as input languages for powerful verification tools such as model checkers. Such specifications are popular among verification engineers with

1

expertise in formal languages. However, these specifications may not be acceptable to designers and other tool developers, since they may not have the background or inclination to use formal languages. Indeed, it is often the case that the architecture manual—although possibly ambiguous and inconsistent—is often used as the de facto "golden reference model". Therefore, the ideal "Architecture Description Language" (ADL) specification should have both formal (unambiguous) semantics as well as an easy correlation with the architecture manual.

1.1 WHAT IS AN ARCHITECTURE DESCRIPTION LANGUAGE?

The phrase "Architecture Description Language" (ADL) has been used in the context of designing both software and hardware architectures. Software ADLs are used for representing and analyzing software architectures [1,2]. They capture the behavioral specifications of the components and their interactions that comprise the software architecture. However, hardware ADLs capture the structure (hardware components and their connectivity) and the behavior (instruction-set) of processor architectures. The term "**Processor Description Language**", also known as machine description language, refers to hardware ADLs. This book uses the term ADL to imply hardware ADLs.

The ADLs have been successfully used as a specification language for processor development. Fig. 1.1 shows the ADL-driven design automation methodology for embedded processors. The ADL specification is used to generate various executable models including simulator, compiler, and hardware implementation. The generated models enable various design automation tasks including exploration, simulation,

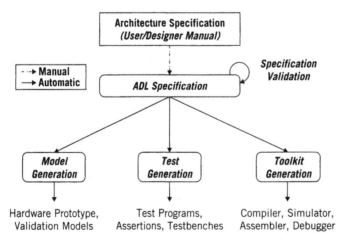

FIGURE 1.1

ADL-driven design automation of embedded processors.

compilation, synthesis, test generation, and validation. Each of these design automation activities both reduces the overall design effort and improves the quality of the final implementation. For example, the exploration is used to figure out the best possible processor architecture for a given set of application programs under various design constraints such as area, power, and performance. Chapter 2 describes various ADL-driven methodologies for development of efficient and reliable embedded processors.

1.2 ADLs AND OTHER LANGUAGES

How do ADLs differ from programming languages, hardware description languages, modeling languages, and the like? This section attempts to answer this question. However, it is not always possible to answer the question: given a language for describing an architecture, what are the criteria for deciding whether it is an ADL or not? Specifications widely in use today are still written informally in natural languages such as English. Since natural language specifications are not amenable to automated analysis, there are possibilities of ambiguity, incompleteness, and contradiction: all problems that can lead to different interpretations of the specification. Clearly, formal specification languages are suitable for analysis and verification. Some have become popular because they are input languages for powerful verification tools such as a model checker. Such specifications are popular among verification engineers with expertise in formal languages. However, these specifications are not easily accepted by designers and other tool developers. An ADL specification should have formal (unambiguous) semantics as well as easy correlation with the architecture manual.

In principle, ADLs differ from programming languages because the latter bind all architectural abstractions to specific point solutions, whereas ADLs intentionally suppress or vary such binding. In practice, the architecture is embodied and recoverable from code by reverse engineering methods. For example, it might be possible to analyze a piece of code written in C and figure out whether it corresponds to *Fetch* unit or not. Many languages provide architecture-level views of the system. For example, C++ offers the ability to describe the structure of a processor by instantiating objects for the components of the architecture. However, C++ offers little or no architecture-level analytical capabilities. Therefore, it is difficult to describe architecture at a level of abstraction suitable for early analysis and exploration. More importantly, traditional programming languages are not natural choice for describing architectures due to their inability to capture hardware features such as parallelism and synchronization. However, some variations of SystemC (such as ArchC described in Chapter 11) can be used as system ADLs.

The ADLs differ from modeling languages (such as UML) because the latter are more concerned with the behaviors of the whole rather than the parts, whereas ADLs concentrate on the representation of components. In practice, many modeling languages allow the representation of cooperating components and can represent

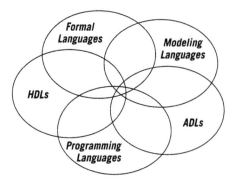

FIGURE 1.2

ADLs versus non-ADLs.

architectures reasonably well. However, the lack of an abstraction would make it harder to describe the instruction-set of the architecture. Traditional Hardware Description Languages (HDL), such as VHDL and Verilog, do not have sufficient abstraction to describe architectures and explore them at the system level. It is possible to perform reverse-engineering to extract the structure of the architecture from the HDL description. However, it is hard to extract the instruction-set behavior of the architecture. In practice, some variants of HDLs work reasonably well as ADLs for specific classes of embedded processors.

There is no clear line between ADLs and non-ADLs. In principle, programming languages, modeling languages, and hardware description languages have aspects in common with ADLs, as shown in Fig. 1.2. Languages can, however, be discriminated from one another according to how much architectural information they can capture and analyze. Languages that were born as ADLs show a clear advantage in this area over languages built for some other purpose and later co-opted to represent architectures.

1.3 CLASSIFICATION OF CONTEMPORARY ADLs

The concept of using machine description languages for specification of architectures has been around for a long time. Early ADLs such as ISPS [3] were used for simulation, evaluation, and synthesis of computers and other digital systems. This section classifies contemporary ADLs in the context of designing customizable and configurable embedded processors. There are many comprehensive ADL surveys available in the literature, including ADLs for retargetable compilation [4], SOC design [5], and embedded processor development [6]. Fig. 1.3 shows the classification of ADLs based on two aspects: *content* and *objective*. The content-oriented classification is based on the nature of the information an ADL can capture, whereas the objective-oriented classification is based on the purpose of an ADL. Contemporary ADLs

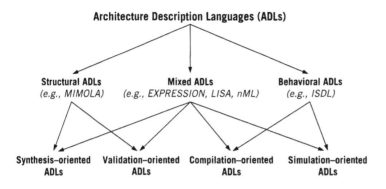

FIGURE 1.3

Taxonomy of ADLs.

can be classified into four categories based on the objective: simulation-oriented, synthesis-oriented, compilation-oriented, and validation-oriented. It is not always possible to establish a one-to-one correspondence between content-based and objective-based classification.

1.3.1 Content-based Classification of ADLs

The ADLs can be classified into three categories based on the nature of the information: structural, behavioral, and mixed. The structural ADLs capture the structure in terms of architectural components and their connectivity. The behavioral ADLs capture the instruction-set behavior of the processor architecture. The mixed ADLs capture both structure and behavior of the architecture.

Structural ADLs

There are two important aspects to consider in designing an ADL: level of abstraction versus generality. It is very difficult to find an abstraction to capture the features of different types of processors. A common way to obtain generality is to lower the abstraction level. Register transfer level (RT-level) is a popular abstraction level—low enough for detailed behavior modeling of digital systems, and high enough to hide gate-level implementation details. Early ADLs such as MIMOLA [7] are based on RT-level descriptions. Chapter 3 describes the MIMOLA language and the supported methodologies in detail. Structural ADLs are suitable for hardware synthesis and cycle-accurate simulation, but they are unfit for retargetable compiler generation.

Behavioral ADLs

The difficulty of instruction-set extraction can be avoided by abstracting behavioral information from the structural details. Behavioral ADLs such as ISDL [8] explicitly specifies the instruction semantics and ignore detailed hardware structures. Typically, there is a one-to-one correspondence between a behavioral ADL and an instruction-set reference manual. Chapter 14 describes the ISDL language. Behavioral ADLs are appropriate for generating a compiler and simulator for instruction-set architectures, but may not be well suited for generating cycle-accurate simulators or hardware implementations of the architecture.

Mixed ADLs

Mixed languages such as nML, LISA, EXPRESSION, ADL++, and HMDES capture both structural and behavioral details of the architecture. Chapter 6 describes the EXPRESSION language in detail. The LISA language and supported methodologies are described in Chapter 5. Chapter 14 describes the HMDES language and its variations. The nML language and its associated methodologies are described in Chapter 4. Mixed ADLs combine the benefits of both structural ADLs and behavioral ADLs. As a result, mixed ADLs are suitable for various design automation activities including retargetable software toolkit (including compiler and simulator) generation, exploration, architecture synthesis, and functional validation.

1.3.2 Objective-based Classification of ADLs

Contemporary ADLs can be classified into four categories based on the objective: compilation-oriented, simulation-oriented, synthesis-oriented, and validation-oriented. In this section we briefly describe the ADLs based on the objective-based classification. We primarily discuss the required capabilities in an ADL to perform the intended objective.

Compilation-oriented ADLs

The goal of such an ADL is to enable automatic generation of retargetable compilers. A compiler is classified as retargetable if it can be adapted to generate code for different target processors with a significant reuse of the compiler source code. Retargetability is typically achieved by providing target machine information in an ADL as input to the compiler along with the program corresponding to the application. Therefore, behavioral ADLs as well as mixed ADLs are suitable for compiler generation. There is a balance between the information captured in an ADL and the information necessary for compiler optimizations. Certain ADLs (e.g., AVIV [9] using ISDL, CHESS [10] using nML, and Elcor [11] using HMDES) explicitly capture all the necessary details such as instruction-set and resource conflict information. Recognizing that the architecture information needed by the compiler is not always in a form that may be well suited for other tools (such as synthesis) or does not permit concise specification, some research has focussed on extraction of such information from a more amenable specification. Examples include the MSSQ and RECORD compiler using MIMOLA [7], retargetable BURG-based C-compiler using MAML [12], compiler optimizers using MADL [13], retargetable C compiler based on LISA [14], and the EXPRESS compiler using EXPRESSION [15].

Simulation-oriented ADLs

Simulation can be performed at various abstraction levels. At the highest level of abstraction, functional simulation (instruction-set simulation) of the processor can be performed by modeling only the instruction-set. Behavioral ADLs can enable generation of functional simulators. The cycle-accurate and phase-accurate simulation models yield more detailed timing information since they are at lower levels of abstraction. Structural ADLs as well as mixed ADLs are good candidates for cycle-accurate simulator generation. Retargetability (i.e., the ability to simulate a wide variety of target processors) is especially important in the context of customizable processor design. Simulators with limited retargetability are very fast but may not be useful in all aspects of the design process. Such simulators (e.g., HPL-PD [11] using HMDES) typically incorporate a fixed architecture template and allow only limited retargetability in the form of parameters such as number of registers and ALUs. Based on the simulation model, simulators can be classified into three types: interpretive, compiled, and mixed. Interpretive simulators (e.g., GENSIM/XSIM [16] using ISDL) offer flexibility but are slow due to the fetch, decode, and execution models for each instruction. Compilation based approaches (e.g., [17] using LISA) reduce the runtime overhead by translating each target instruction into a series of host machine instructions, which manipulate the simulated machine state. Recently proposed techniques (JIT-CCS [18] using LISA and IS-CS [19] using EXPRESSION) combine the flexibility of interpretive simulation with the speed of compiled simulation. MAML (described in Chapter 12) supports mixed compiled/interpretive simulation for uniprocessor as well as multiprocessor architectures.

Synthesis-oriented ADLs

Structural ADLs such as MIMOLA are suitable for hardware generation. Some of the behavioral languages (such as ISDL) are also used for hardware generation. For example, the HDL generator HGEN [16] is based on ISDL description. Mixed ADLs such as nML, LISA, MAML, and EXPRESSION capture both structure and behavior of the processor and enable HDL generation [20, 21]. The synthesizable HDL generation approach based on the LISA language produces an HDL model of the architecture. The designer has the choice to generate a VHDL, Verilog, or SystemC representation of the target architecture [20]. Similarly, the synthesis tool GO [22] uses an nML description to generate synthesizable RTL models in VHDL or Verilog.

Validation-oriented ADLs

The ADLs have been successfully used in both academia as well as in the industry to enable test generation for functional validation of embedded processors. Traditionally, structural ADLs such as MIMOLA [7] are suitable for test generation. Mixed ADLs also enable test generation such as specification-coverage-based test generation using EXPRESSION [23–25], retargetable test program generation using nML [22], and automated test generation using LISA processor models [26]. ADLs have been used in the context of functional verification of embedded processors using a top-down validation methodology [27]. The first step in the methodology is to verify the ADL specification to ensure the correctness of the specified architecture [28]. The validated ADL specification can be used as a golden reference model for various validation tasks including property checking, test generation, and equivalence checking. For example, the generated hardware model (reference) can be used to perform both property checking and equivalence checking of the implementation using EXPRESSION ADL [29].

1.4 ADLs: PAST, PRESENT, AND FUTURE

The ADLs have been successfully used in academic research as well as industry for processor development. The early ADLs were either structure-oriented (MIMOLA), or behavior-oriented (ISDL). As a result, each class of ADLs is suitable for specific tasks. For example, structure-oriented ADLs are suitable for hardware synthesis, but may be unfit for compiler generation. Similarly, behavior-oriented ADLs are appropriate for generating compiler and simulator for instruction-set architectures, but may be unsuited for generating cycle-accurate simulators or hardware implementations of the architecture. The later ADLs (LISA, HMDES, and EXPRESSION) adopted the mixed approach where the language captures both structure and behavior of the architecture.

The ADLs designed for a specific domain (such as DSP or VLIW) or for a specific purpose (such as simulation or compilation) can be compact and may enable automatic generation of efficient (in terms of area, power, and performance) tools and hardware prototypes. However, it is difficult to design an ADL for a wide variety of architectures to perform different tasks using the same specification. Generic ADLs require the support of powerful methodologies to generate high-quality results compared to domain- and task-specific ADLs. A majority of the ADLs were initially designed to serve a specific purpose. For example, nML was designed to capture instruction-set behavior to enable instruction-set simulation and compilation. Similarly, LISA and RADL were designed for simulation of processor architectures. Likewise, HMDES and EXPRESSION were designed mainly for generating retargetable compilers.

Contemporary ADLs have gone through various transformations with the new features and methodologies to perform the required design automation tasks. For example, nML is extended by Target Compiler Technologies to perform hardware synthesis and test generation [22]. Similarly, the LISA language has been used for hardware generation [30, 31], compiler generation [14], instruction encoding synthesis [32], and JTAG interface generation [33]. Likewise, EXPRESSION has been used for hardware generation [21], instruction-set synthesis [34], test generation [24, 25], and specification validation [28, 35].

In the future, the existing ADLs will go through changes in two dimensions. First, ADLs will specify not only processor, memory, and co-processor architectures but also other components of the system-on-chip architectures including peripherals and external interfaces. Second, ADLs will be used for software toolkit generation, hardware synthesis, test generation, instruction-set synthesis, and validation of microprocessors. Furthermore, multiprocessor SOCs will be captured and various attendant tasks will be addressed[1]. The tasks include support for formal analysis, generation of real-time operating systems (RTOS), exploration of communication architectures, and support for interface synthesis. The emerging ADLs will have these features.

1.5 BOOK ORGANIZATION

The rest of the book is organized as follows. The first two chapters introduce the need for ADLs and ADL-driven methodologies for design automation of embedded processors and system-on-chip designs. The remaining chapters describe various contemporary ADLs and their associated methodologies for processor development.

- Chapter 1 introduces the need for ADLs and classifies contemporary ADLs in terms of their capabilities in capturing today's embedded processors and enabling various design automation steps.

- Chapter 2 describes the required ADL-driven methodologies for compiler and simulator generation, architecture synthesis, design space exploration, and functional validation of embedded processors.

- Chapter 3 presents MIMOLA, which is a fully synthesizable language. This chapter also describes the associated methodologies for architecture synthesis, compilation, and test generation.

- Chapter 4 describes the nML language which is a mixed ADL. This chapter also describes the associated methodologies for retargetable compilation, retargetable instruction-set simulation, test generation, and architecture synthesis for ASIP design.

- Chapter 5 presents LISA, which is a uniform ADL for embedded processor modeling, and generation of software toolsuite and implementation. This chapter also describes all the associated methodologies.

- Chapter 6 describes EXPRESSION, which is an ADL for software toolkit generation, exploration, and validation of programmable SOC architectures.

- Chapter 7 presents ASIP Meister framework. This chapter describes how to generate the compiler and simulator, as well as the hardware implementation for ASIP design.

---

[1]Chapter 12 describes the MAML language which is an early attempt in ADL-driven design and validation of multiprocessor architectures.

- Chapter 8 describes the TIE language and its use for creating application specific instruction extensions for the Xtensa microprocessor core.

- Chapter 9 presents the MADL language, which is based on a formal and flexible concurrency model. This chapter also describes associated development tools including simulator and compiler.

- Chapter 10 describes ADL++, which allows object-oriented specification of complicated instruction sets and micro-architectures.

- Chapter 11 presents ArchC language, which is based on SystemC. This chapter also describes the ADL-driven generation of fast microprocessor simulators.

- Chapter 12 describes the MAML language, which is an ADL for modeling and simulation of single and multiprocessor architectures. This chapter also describes associated methodologies for the simulator and implementation generation.

- Chapter 13 presents GNR, which is a formal language for specification, compilation, and synthesis of custom-embedded processors.

- Chapter 14 describes HMDES, ISDL, and other contemporary ADLs. It also describes the associated design automation methodologies.

We recommend that you read Chapter 2 next, before deciding how to proceed with reading the rest of this book. For instance, to cover ADLs based on their content (i.e., structure, behavior, etc.), you can read Chapter 3 (MIMOLA) for structural ADLs, Chapter 14 (ISDL) for behavioral ADLs, and so on. Alternatively, you can choose a particular chapter based on a specific scenario—Chapter 10 for how to exploit object-oriented paradigm for processor specification, Chapter 11 for how to use SystemC for SOC exploration, Chapter 12 for how to design and verify multiprocessors, and so on. Once you read Chapters 1 and 2, you can read the remaining chapters in any order since they are completely independent. We hope you will enjoy reading the chapters and find many useful concepts in this book.

REFERENCES

[1] N. Medvidovic and R. Taylor. A framework for classifying and comparing architecture description languages. In M. Jazayeri and H. Schauer, editors, *Proceedings of the 6th European Conference held jointly with the 5th ACM SIGSOFT International Symposium on Foundations of Software Engineering*, pages 60–76. Springer-Verlag, 1997.

[2] Paul C. Clements. A survey of architecture description languages. In *Proceedings of International Workshop on Software Specification and Design (IWSSD)*, pages 16–25, 1996.

[3] M. R. Barbacci. Instruction set processor specifications (ISPS): The notation and its applications. *IEEE Transactions on Computers*, C-30(1):24–40, January 1981.

[4] W. Qin and S. Malik. Architecture description languages for retargetable compilation. In *The Compiler Design Handbook: Optimizations & Machine Code Generation*. CRC Press, 2002.

[5] H. Tomiyama, A. Halambi, P. Grun, N. Dutt, and A. Nicolau. Architecture description languages for systems-on-chip design. In *Proceedings of Asia Pacific Conference on Chip Design Language*, pages 109–116, 1999.

[6] P. Mishra and N. Dutt. Architecture description languages for programmable embedded systems. *IEE Proceedings on Computers and Digital Techniques*, 152(3):285–297, May 2005.

[7] R. Leupers and P. Marwedel. Retargetable code generation based on structural processor descriptions. *Design Automation for Embedded Systems*, 3(1):75–108, 1998.

[8] G. Hadjiyiannis, S. Hanono, and S. Devadas. ISDL: An instruction set description language for retargetability. In *Proceedings of Design Automation Conference (DAC)*, pages 299–302, 1997.

[9] S. Hanono and S. Devadas. Instruction selection, resource allocation, and scheduling in the AVIV retargetable code generator. In *Proceedings of Design Automation Conference (DAC)*, pages 510–515, 1998.

[10] D. Lanneer, J. Praet, A. Kifli, K. Schoofs, W. Geurts, F. Thoen, and G. Goossens. CHESS: Retargetable code generation for embedded DSP processors. In P. Marwedel and G. Goossens, editors, *Code Generation for Embedded Processors*, pages 85–102. Kluwer Academic Publishers, 1995.

[11] The MDES User Manual. *http://www.trimaran.org*, 1997.

[12] D. Fischer, J. Teich, M. Thies, and R. Weper. BUILDABONG: a framework for architecture/compiler co-exploration for ASIPs. *Journal for Circuits, Systems, and Computers, Special Issue: Application Specific Hardware Design*, pages 353–375, 2003.

[13] W. Qin, S. Rajagopalan, and S. Malik. A formal concurrency model based architecture description language for synthesis of software development tools. In *Proceedings of ACM Conference on Languages, Compilers, and Tools for Embedded Systems (LCTES)*, pages 47–56, 2004.

[14] M. Hohenauer, H. Scharwaechter, K. Karuri, O. Wahlen, T. Kogel, R. Leupers, G. Ascheid, H. Meyr, G. Braun, and H. Someren. A methodology and tool suite for c compiler generation from ADL processor models. In *Proceedings of Design Automation and Test in Europe (DATE)*, pages 1276–1283, 2004.

[15] A. Halambi, P. Grun, V. Ganesh, A. Khare, N. Dutt, and A. Nicolau. EXPRESSION: A language for architecture exploration through compiler/simulator retargetability. In *Proceedings of Design Automation and Test in Europe (DATE)*, pages 485–490, 1999.

[16] G. Hadjiyiannis, P. Russo, and S. Devadas. A methodology for accurate performance evaluation in architecture exploration. In *Proceedings of Design Automation Conference (DAC)*, pages 927–932, 1999.

[17] S. Pees, A. Hoffmann, and H. Meyr. Retargetable compiled simulation of embedded processors using a machine description language. *ACM Transactions on Design Automation of Electronic Systems*, 5(4):815–834, October 2000.

[18] A. Nohl, G. Braun, O. Schliebusch, R. Leupers, H. Meyr, and A. Hoffmann. A universal technique for fast and flexible instruction-set architecture simulation. In *Proceedings of Design Automation Conference (DAC)*, pages 22–27, 2002.

[19] M. Reshadi, P. Mishra, and N. Dutt. Instruction set compiled simulation: A technique for fast and flexible instruction set simulation. In *Proceedings of Design Automation Conference (DAC)*, pages 758–763, 2003.

[20] O. Schliebusch, A. Chattopadhyay, M. Steinert, G. Braun, A. Nohl, R. Leupers, G. Ascheid, and H. Meyr. RTL processor synthesis for architecture exploration and implementation. In *Proceedings of Design Automation and Test in Europe (DATE)*, pages 156-160, 2004.

[21] P. Mishra, A. Kejariwal, and N. Dutt. Synthesis-driven exploration of pipelined embedded processors. In *Proceedings of International Conference on VLSI Design*, pages 921-926 2004.

[22] http://www.retarget.com. *Target Compiler Technologies*, January 2008.

[23] H. Koo and P. Mishra. Functional test generation using property decompositions for validation of pipelined processors. In *Proceedings of Design Automation and Test in Europe (DATE)*, pages 1240-1245, 2006.

[24] P. Mishra and N. Dutt. Graph-based functional test program generation for pipelined processors. In *Proceedings of Design Automation and Test in Europe (DATE)*, pages 182-187, 2004.

[25] P. Mishra and N. Dutt. Functional coverage driven test generation for validation of pipelined processors. In *Proceedings of Design Automation and Test in Europe (DATE)*, pages 678-683, 2005.

[26] O. Luthje. A methodology for automated test generation for LISA processor models. In *Proceedings of Synthesis and System Integration of Mixed Technologies (SASIMI)*, pages 266-273, 2004.

[27] P. Mishra and N. Dutt. *Functional Verification of Programmable Embedded Architectures: A Top-Down Approach*. Springer, 2005.

[28] P. Mishra and N. Dutt. Automatic modeling and validation of pipeline specifications. *ACM Transactions on Embedded Computing Systems (TECS)*, 3(1):114-139, 2004.

[29] P. Mishra, N. Dutt, N. Krishnamurthy, and M. Abadir. A top-down methodology for validation of microprocessors. *IEEE Design & Test of Computers*, 21(2):122-131, 2004.

[30] A. Hoffmann, O. Schliebusch, A. Nohl, G. Braun, O. Wahlen, and H. Meyr. A methodology for the design of application specific instruction set processors (ASIP) using the machine description language LISA. In *Proceedings of International Conference on Computer-Aided Design (ICCAD)*, pages 625-630, 2001.

[31] O. Schliebusch, A. Hoffmann, A. Nohl, G. Braun, and H. Meyr. Architecture implementation using the machine description language LISA. In *Proceedings of Asia South Pacific Design Automation Conference (ASPDAC)/International Conference on VLSI Design*, pages 239-244, 2002.

[32] A. Nohl, V. Greive, G. Braun, A. Hoffmann, R. Leupers, O. Schliebusch, and H. Meyr. Instruction encoding synthesis for architecture exploration using hierarchical processor models. In *Proceedings of Design Automation Conference (DAC)*, pages 262-267, 2003.

[33] O. Schliebusch, D. Kammler, A. Chattopadhyay, R. Leupers, G. Ascheid, and H. Meyr. Automatic generation of JTAG interface and debug mechanism for ASIPs. In *Proceedings of GSPx Conference (www.gspx.com)*, 2004.

[34] P. Biswas and N. Dutt. Reducing code size for heterogeneous-connectivity-based VLIW DSPs through synthesis of instruction set extensions. In *Proceedings of Compilers, Architectures, Synthesis for Embedded Systems (CASES)*, pages 104-112, 2003.

[35] P. Mishra, N. Dutt, and H. Tomiyama. Towards automatic validation of dynamic behavior in pipelined processor specifications. *Kluwer Design Automation for Embedded Systems (DAES)*, 8(2-3):249-265, June-September 2003.

ADL-driven Methodologies for Design Automation of Embedded Processors

2

Prabhat Mishra and Aviral Shrivastava

This chapter describes various ADL-driven methodologies for the development of efficient and reliable embedded processors. The ADLs designed for a specific domain (such as DSP or VLIW) or for a specific purpose (such as simulation or compilation) can be compact, and it is possible to automatically generate efficient tools and hardwares. However, it is difficult to design an ADL for a wide variety of architectures to perform different tasks using the same specification. Generic ADLs require the support of powerful methodologies to generate high-quality results compared to domain-specific/task-specific ADLs. This chapter presents a comprehensive overview of all the supported methodologies. It describes the fundamental challenges and required techniques to support powerful methodologies in the presence of generic ADLs. Chapters 3–14 present specific methodologies supported by the respective ADLs.

This chapter is organized as follows. Section 2.1 outlines ADL-driven design space exploration using generated software tools and hardware prototypes. Section 2.2 presents retargetable compiler generation approaches. Section 2.3 describes ADL-driven simulator generation techniques. ADL-driven implementation (hardware models) generation is discussed in Section 2.4. Section 2.5 describes various techniques for verifying the specification as well as the implementation. Finally, Section 2.6 concludes this chapter.

2.1 DESIGN SPACE EXPLORATION

Embedded systems present a tremendous opportunity to customize designs by exploiting the application behavior. Rapid exploration and evaluation of candidate architectures are necessary due to time-to-market pressure and short product lifetimes. The ADLs are used to specify processor and memory architectures and generate software toolkit including compiler, simulator, assembler, profiler, and debugger. Fig. 2.1 shows a traditional ADL-based design space exploration flow.

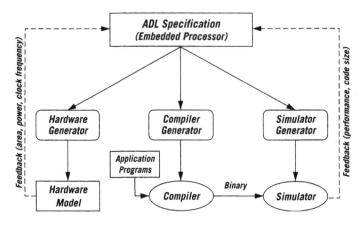

FIGURE 2.1

ADL-driven design space exploration.

The application programs are compiled and simulated, and the feedback is used to modify the ADL specification with the goal of finding the best possible architecture for a given set of application programs under various design constraints such as area, power, and performance. An ADL-driven software toolkit generation enables performance-driven exploration. The simulator produces profiling data and thus may answer questions concerning the instruction set, the performance of an algorithm, and the required size of memory and registers. However, the required silicon area, clock frequency, and power consumption can only be determined by generating a synthesizable HDL (hardware) model.

One of the main purposes of an ADL is to support automatic generation of a high-quality software toolkit, including at least an ILP (instruction-level parallelism) compiler and a cycle-accurate simulator. However, such tools require detailed information about the processor, typically in a form that is not concise and easily specifiable. Therefore, it becomes necessary to develop procedures to automatically generate such tool-specific information from the ADL specification. For example, reservation tables (RTs) are used in many ILP compilers to describe resource conflicts. However, manual description of RTs on a per-instruction basis is cumbersome and error-prone. Instead, it is easier to specify the pipeline and datapath resources in an abstract manner, and generate RTs on a per-instruction basis [1]. Sections 2.2 and 2.3 describe some of the challenges in automatic generation of software tools (focusing on compilers and simulators) and survey some of the approaches adopted by current tools. Section 2.4 describes ADL-driven hardware generation approaches.

2.2 RETARGETABLE COMPILER GENERATION

The advent of System-on-Chip (SOC) technology resulted in a paradigm shift for the design process of embedded systems employing programmable processors with custom hardware. Traditionally, embedded systems developers performed limited exploration of the design space using standard processor and memory architectures. Furthermore, software development was usually done using existing processors (with supported integrated software development environments) or done manually using processor-specific low-level languages (assembly). This was feasible because the software content in such systems was low, and also because the processor architecture was fairly simple (e.g., no instruction-level parallelism features) and well defined (e.g., no parameterizable components).

In the SOC domain, system-level design libraries increasingly consist of Intellectual Property (IP) blocks such as processor cores that span a spectrum of architectural styles, ranging from traditional DSPs and superscalar RISC, to VLIWs and hybrid ASIPs. These processor cores typically allow customization through parameterization of features (such as number of functional units, operation latencies, etc.). Furthermore, SOC technologies permit the incorporation of novel on-chip memory organizations (including the use of on-chip DRAM, frame buffers, streaming buffers, and partitioned register files). Together, these features allow exploration of a wide range of processor-memory organizations in order to customize the design for a specific embedded application.

An important and noticeable trend in the embedded SOC domain is the increasing migration of system functionality from hardware to software, resulting in a high degree of software content for newer SOC designs. This trend, combined with shrinking time-to-market cycles, has resulted in intense pressure to migrate the software development to a high-level language-based environment (such as C, C++, Java) in order to reduce the time spent in system design. To effectively explore the processor-memory design space and develop software in a high-level language, the designer requires a high-quality software toolkit (primarily a highly optimizing compiler and cycle-accurate simulator). Compilers for embedded systems have been the focus of several research efforts.

The compilation process can be broadly broken into two steps: analysis and synthesis [2]. During analysis, the program (in high-level language) is converted into an intermediate representation (IR) that contains all the desired information such as control and data dependences. During synthesis, the IR is transformed and optimized in order to generate efficient target-specific code. The synthesis step is more complex and typically includes the following phases: instruction selection, scheduling, resource allocation, code optimizations/transformations, and code generation. The effectiveness of each phase depends on the target architecture and the application. Additionally, a further problem during the synthesis step is that the optimal ordering between these phases (and other optimizations) is highly dependent on the target architecture and the application program. For example, the ordering between the memory assignment optimizations and instruction scheduling passes is very critical for memory-intensive applications. As a result, traditionally, compilers have been painstakingly hand-tuned to a particular architecture (or architecture class) and application domain(s). However, stringent time-to-market constraints for SOC designs no longer make it feasible to manually generate compilers tuned to particular architectures.

A promising approach to automatic compiler generation is the "retargetable compiler" approach. A compiler is classified as retargetable if it can be adapted to generate code for different target processors with significant reuse of the compiler source code. Retargetability is typically achieved by providing target machine information (in an ADL) as input to the compiler along with the program corresponding to the application. The origins of retargetable compiler technology can be traced back to UNCOL (Universal Computer-Oriented Language) project [3] where the key idea of separate front-ends and back-ends was suggested to reuse the middle-end of the compiler. According to this, all source code is converted to a common intermediate representation (CIR), on which the majority of code transformations/optimizations are performed. Once optimized, machine-specific back-end is used to generate the executable. This basic style is still extensively used in popular compilers like SUIF (Stanford University Intermediate Format), and GCC (GNU Compiler Collection). In the context of SOC design environments, the application specification remains the same, therefore, there is just one front-end; however, there will be processor-specific back-ends. Further research in this area has extended the concept of compiler code reuse. Several important compiler phases like instruction scheduling and register allocation must be performed in the back-end, and there is tremendous opportunity to reuse compiler code in them.

In the remainder of this section, we will classify ADL-based retargetable compilers in several ways to provide insight into the capabilities and differences of the various ADL-based retargetable compilers.

2.2.1 Retargetability Based on ADL Content

The first classification is based on the level of detail in which the processor architecture is described in the ADL.

Parameter-based retargetability

In this scheme, several microarchitectural parameters, including operation latencies, number of functional units, number of registers, and delay slots, are provided in the ADL. Retargetability in the compiler is provided by means of some compiler phases being parameterized on these microarchitectural parameters. This is the simplest form of retargetability, popularly used to retarget the register allocator by the number of general purpose registers and instruction scheduler by using operation latencies.

Structure-based retargetability

In this scheme, the structure of the processor is described as a netlist in the ADL. The compiler optimizes the code for the processor structure. This kind of retargetability is popularly used to describe the processor pipeline structure and perform detailed fine-grain instruction scheduling by modeling and avoiding resource and data hazards.

Behavior-based retargetability

Retargeting the instruction selection requires the description of the semantics, or the behavior of instructions of the target processor in terms of instructions of the CIR (Common Intermediate Representation). In the general form of this approach, the mapping for instruction trees of CIR to instruction trees of the target processor is specified in the ADL, and the compiler effectively tiles the application in CIR using input tree patterns and replaces them with the equivalent target instruction tree patterns.

2.2.2 Retargetability Based on Compiler Phases

Another insightful way of classifying ADL-based retargetable compilers is according to the compiler phases which are retargetable. When compiling for a different processor, at least the instruction selection and the register allocation phases of the compiler must be modified; these must be retargetable in a "retargetable compiler".

Retargetable instruction selection

The minimum information required to retarget instruction selection is the specification of the mapping of each CIR instruction in terms of the target instructions in the ADL. While this minimum information is sufficient, better instruction selection can be achieved if the ADL contains several mappings of CIR instructions to target instructions. The target instruction mappings can vary depending on the power, performance, code-size requirements, and the compiler can choose the mapping depending on its optimization metric. Several retargetable compilers use the IBurg [4] library to perform these complex mappings.

Retargetable register allocation

While simple retargetable register allocators are able to retarget if the number of general purpose registers is changed, in reality this is rarely the case. Register allocation is typically very closely tied to instruction selection. Very often, operands of instructions can only be mapped to a specific set of registers. The registers are therefore specified in groups of register classes, and the mapping of operands of an instruction to the register classes is specified in the ADL. Each variable in the application then gets associated with several register classes, depending on the appropriate instructions. Then the variable can be assigned to a register in the intersection of all the classes to which it belongs.

Retargetable instruction scheduling

At the basic level, the job of an instruction scheduler is to predict pipeline hazards, and find a legitimate reordering of instructions (that does not break data dependencies) to minimize them. The simplest form of retargetable instruction selection is when the instruction selection takes the operation latency of each operation, and generates a schedule using them. However, due to multiple pipelines, the delay of an operation may be dependent on the dependent operation itself.

Some retargetable compilers including GCC allow users to specify an automaton of instruction sequences that will minimize/avoid pipeline resource hazards. Instead of explicitly specifying the automatons in the ADL, some retargetable compilers deduce reservation tables from the ADL, which describes the processor pipeline, and the flow of instructions through it. Reservation tables of instructions can be combined to detect all resource hazards in a schedule. In the most general case, operation tables can be extracted from the ADL if the pipeline structure of the processor and the flow of instructions and its operands are specified in the ADL.

2.2.3 Retargetability Based on Architectural Abstractions

The third and very insightful differentiating classification of various existing ADL-based retargetable compilers is based on the architectural abstractions that the compiler can provide retargetability for. Fig. 2.2 shows processor design abstractions for instruction set architecture, processor pipeline design, processor-memory interface, and the memory design. Existing retargetable compilers differ in the abstraction for which they provide retargetability.

ISA retargetability

Functional retargetability can be achieved by compiling for a new instruction set. Retargetability toward microarchitectural features is only required for optimizing the compiler. Instruction set retargetability requires retargetable instruction selection and register allocation. There are various examples of such retargetable compilers including AVIV [5] using ISDL, CHESS [6] using nML and Elcor [7]. One very interesting ISA retargetability is the ability of the EXPRESS [8] compiler to generate good code for "dual instruction set architectures". Dual instruction set architectures typically have two instruction sets—one is the normal 32-bit instruction set, and the other is a narrow 16-bit wide instruction set. While the 32-bit ISA is a complete instruction set (IS), the narrow 16-bit ISA has only a compressed version of some of the most frequently used instructions. The idea is that if the whole application can be expressed only using the narrow instructions, then it would lead to a 50% reduction in the code size; however there are several challenges to achieve it. First,

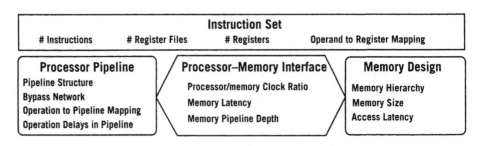

FIGURE 2.2

Processor design abstractions for compiler retargetability.

not all instructions have mapping to the narrow IS, second, the narrow instructions can access only a fraction of the register file, and therefore indiscriminate conversion will lead to an increase in spilling, causing an increase in code size. Advanced compiler techniques of using a register pressure-based heuristic to determine the "regions of code to convert" provide consistently high degrees of code compression. The ADL-based retargetable compilers can perform this profitable conversion after describing the narrow instruction set and its mapping from the normal 32-bit instruction set in the ADL.

Processor pipeline retargetability

Processor pipeline retargetability implies that the compiler should be able to generate a good-quality code, even if the processor pipeline is modified. Compilers like GCC require the user to define instruction automatons, which include pipeline information, for example, instruction dependencies in them. While this information is directly usable by the compiler to perform instruction scheduling, it is not very convenient for the user. An SOC (System-on-Chip) developer has to understand the pipeline structure and the flow of operations in them, and then translate that information into instruction automatons. Very often, this is time consuming and an error-prone effort.

Advanced ADL-based retargetable compilers automatically derive the information needed by the compiler from a structural description of the processor pipeline. The processor pipeline is specified as a DAG of pipeline stages, and the flow of operations in the pipeline is indicated as an attribute of each pipeline stage. From this information, retargetable compilers can automatically generate Reservation Tables (RTs), even for multi-cycle and pipelined instructions. The RTs of instructions can be combined to detect all resource hazards, and therefore avoid resource hazards during instruction scheduling. There are various compilers in this category including ASIP Meister based on Cosy [9], LISA [10], MAML [11], MADL [12], GNR [13], and TIE [14].

Further developments in the ADL-based retargetable compiler technology allow the compilers to detect not only the resource hazards, but also data hazards, by automatically generating Operation Tables, or OTs [15], from the structural description of the processor pipeline and operation binding to the pipeline stages in the ADL description. Operation tables like RTs specify not only the resources that the operation uses in each cycle, but also what the instruction does with data in each cycle of its execution. Operation Tables can be used to accurately detect all resource and data hazards in a pipeline, even in the presence of partial bypassing in the processor.

Memory hierarchy retargetability

The significance of the memory hierarchy on both the runtime and the power consumption of the processor makes it extremely important for any retargetable compilers to know about the memory hierarchy, and optimize for it. Research on ADL-based, memory aware retargetable compilers can be divided into two parts—first is for cache-based systems, and second is for scratch-pad-based systems. For cache-based systems, initial research looked at code and data placement in the memory to reduce conflict misses. The cache parameters like block size, associativity, and cache size are specified in the ADL, and the retargetable compiler places the data and code in the memory to reduce conflict misses.

The ADL-based retargetable compiler research has also looked at compiling for other cache hierarchies, for example, horizontally partitioned caches (HPCs). There are multiple (typically two) caches at the same level of memory hierarchy in this architecture, and data has to be exclusively divided among the two caches. By partitioning high and low temporal locality data, interference can be reduced. Recent research has shown that HPCs are very effective in reducing the energy consumption of the memory subsystem when one of the cache is small. The hierarchy of the caches is specified as a DAG (Directed Acyclic Graph), and the design parameters of each cache are specified. ADL-based retargetable compilers use this information to partition and map the data in the two caches. There are various compilers in this category including MIMOLA [16] and EXPRESS [17].

Significant research advances in memory-aware retargetable compilers has been made for scratch-pad-based systems. Scratch pads are (typically small) on-die memories, which unlike caches are mapped to a separate address space. Consequently, unlike caches, the use of scratch pad is explicit in the assembly. An assembly writer (a programmer, or a compiler) has to explicitly move data onto the scratch pad, use it from there, and then write it back to the memory, if needed. Techniques have been developed for using the scratch pad for global static data, heap data, and the local stack data. While the main benefit of using scratch pad is in the reduced power consumption due to the absence of miss-management hardware, the access times are much faster for scratch pads. The ADL-based retargetable compilers take the scratch pad size as an input, and are able to generate code to manage the scratch pad to gain power and performance improvements.

Processor-memory interface retargetability

With increasing memory latency, cache misses are becoming ever more important, and they pose interesting optimization challenges. The question is: what can we do while the processor is stalled, waiting for data from memory. For example, how do we minimize the energy consumption of the processor while it is stalled. If we wanted to switch to a low-power mode, it is not possible, because even to switch to the closest low-power mode, for example, in the Intel XScale processor, it takes more than 180 processor cycles while the memory latency or cache miss latency is only 30 cycles. Compiler techniques have been proposed to aggregate several stalls and create a large stall during which the processor can be profitably switched to low-power mode. Processor memory interface parameters are specified in the ADL description, and the processor can perform code transformations to enable the aggregation. The EXPRESS [18] compiler using EXPRESSION ADL supports retargetability based on processor-memory interface details.

The ADL-based retargetable compilers are able to automatically construct data structures containing information they need for their phases from the structural and behavioral description of the processor. Using that, they are now able to optimize for several architectural and microarchitectural features to generate good code; however, a lot remains to be done. The microarchitectural exploration space is limited only by human creativity, and many features open new doors for compiler optimization. As compared to a native compiler, a retargetable compiler fundamentally should have only a compilation-time penalty; however, in practice, because of the microarchitectural nuances and idiosyncrasies, a native compiler typically performs better than a retargetable compiler. However, this improvement of a native compiler comes more as an evolution, as compiler designers become more and more aware of microarchitectural complexities, and implement pointed heuristics to achieve code quality, as the maturing of a compiler takes a long time. Retargetable compilers are invaluable as the first cut compiler for a new processor, and are immensely useful in early design space exploration of processor architectures.

2.3 RETARGETABLE SIMULATOR GENERATION

Simulators are critical components of the exploration and software design toolkit for the system designer. They can be used to perform diverse tasks such as verifying the functionality and/or timing behavior of the system (including hardware and software), and generating quantitative measurements (e.g., power consumption), which can be used to aid the design process. Simulation of the processor system can be performed at various abstraction levels. At the highest level of abstraction, a functional simulation of the processor can be performed by modeling only the instruction set (IS). Such simulators are termed instruction-set simulators (ISS) or instruction-level simulators (ILS). At lower levels of abstraction are the cycle-accurate and phase-accurate simulation models that yield more detailed timing information. Simulators can be further classified based on whether they provide bit-accurate models, pin-accurate models, exact pipeline models, or structural models of the processor.

Typically, simulators at higher levels of abstraction (e.g., ISS, ILS) are faster, but gather less information as compared to those at lower levels of abstraction (e.g., cycle-accurate, phase-accurate). Retargetability (i.e., the ability to simulate a wide variety of target processors) is especially important in the arena of embedded SOC design with emphasis on the exploration and co-design of hardware and software. Simulators with limited retargetability are very fast but may not be useful in all aspects of the design process. Such simulators typically incorporate a fixed architecture template and allow only limited retargetability in the form of parameters such as number of registers and ALUs. Examples of such simulators are numerous in the industry and include the HPL-PD [7] simulator using the MDES ADL. The model of simulation adopted has a significant impact on the simulation speed and flexibility of the simulator. Based on the simulation model, simulators can be classified into three types: interpretive, compiled, and mixed.

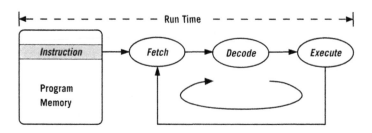

FIGURE 2.3

Interpretive simulation.

2.3.1 **Interpretive Simulation**

Such simulators are based on an interpretive model of the processor's instruction set. Interpretive simulators store the state of the target processor in host memory. It then follows a fetch, decode, and execute model: instructions are fetched from memory, decoded, and then executed in serial order as shown in Fig. 2.3. Advantages of this model include ease of implementation, flexibility, and the ability to collect varied processor state information. However, it suffers from significant performance degradation as compared to the other approaches, primarily due to the tremendous overhead in fetching, decoding, and dispatching instructions. Almost all commercially available simulators are interpretive. Examples of research interpretive retargetable simulators include SIMPRESS [19] using EXPRESSION, and GENSIM/XSIM [20] using ISDL.

2.3.2 **Compiled Simulation**

Compilation-based approaches reduce the runtime overhead by translating each target instruction into a series of host machine instructions which manipulate the simulated machine state, as shown in Fig. 2.4. Such translation can be done either at compile time (static compiled simulation) where the fetch-decode-dispatch overhead is completely eliminated, or at load time (dynamic compiled simulation), which amortizes the overhead over repeated execution of code. Simulators based on the static compilation model are presented by Zhu et al. [21] and Pees et al. [22]. Examples of dynamic compiled code simulators include the Shade simulator [23], and the Embra simulator [24].

2.3.3 **Mixed Approaches**

Traditional interpretive simulation is flexible but slow. Instruction decoding is a time-consuming process in a software simulation. Compiled simulation performs compile time decoding of application programs to improve the simulation performance. However, all compiled simulators rely on the assumption that the complete program code is known before the simulation starts and is further more runtime static. Due to the restrictiveness of the compiled technique, interpretive simulators

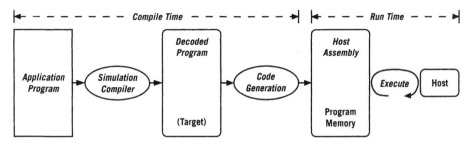

FIGURE 2.4

Compiled simulation.

are typically used in embedded systems design flow. Two recently proposed simulation techniques (JIT-CCS [25] and IS-CS [26]) combine the flexibility of interpretive simulation with the speed of the compiled simulation.

The *just-in-time cache compiled simulation* (JIT-CCS) technique compiles an instruction during runtime, *just-in-time* before the instruction is going to be executed. Subsequently, the extracted information is stored in a simulation cache for direct reuse in a repeated execution of the program address. The simulator recognizes if the program code of a previously executed address has changed, and initiates a recompilation. The *instruction set compiled simulation* (IS-CS) technique performs time-consuming instruction decoding during compile time. In case an instruction is modified at runtime, the instruction is re-decoded prior to execution. It also uses an *instruction abstraction* technique to generate aggressively optimized decoded instructions that further improve simulation performance [26, 27].

2.4 ARCHITECTURE SYNTHESIS

There are two major approaches in the literature for synthesizable HDL generation. The first one is a parameterized processor-core-based approach. These cores are bound to a single processor template whose architecture and tools can be modified to a certain degree. The second approach is based on processor specification languages.

2.4.1 Implementation Generation Using Processor Templates

Examples of processor-template-based approaches are Xtensa [14], Jazz [28], and PEAS-I [29, 30]. Xtensa [14] is a scalable RISC processor core. Configuration options include the width of the register set, caches, and memories. New functional units and instructions can be added using the Tensilica Instruction Language (TIE). A synthesizable hardware model along with software toolkit can be generated for this class of architectures. Improv's Jazz [28] processor is supported by a flexible design methodology to customize the computational resources and instruction set of the processor. It allows modifications of data width, number of registers, depth of hardware task queue, and addition of custom functionality in Verilog. PEAS-I [29, 30] is a GUI-based hardware/software codesign framework. It generates HDL code along with a software toolkit. It has support for several architecture types and a library of configurable resources.

2.4.2 ADL-driven Implementation Generation

Fig. 2.1 shows a typical framework of processor description language-driven HDL generation and exploration. The generated hardware models are also used for implementation validation, as described in Section 2.5. Structure-centric ADLs such as MIMOLA are suitable for hardware generation. Some of the behavioral languages (such as ISDL) are also used for hardware generation. For example, the HDL generator HGEN [20] uses ISDL description. Mixed languages such as nML, LISA, and EXPRESSION capture both the structure and behavior of the processor. The synthesizable HDL generation approach based on LISA language [31] produces an HDL model of the architecture. The designer has the choice to generate a VHDL, Verilog, or SystemC representation of the target architecture [31]. Similarly, the synthesis tool GO [32] uses nML description to generate synthesizable RTL models in VHDL or Verilog. The HDL generation methodology presented by Mishra et al. [33] combines the advantages of the processor-template-based environments and the language-based specifications using EXPRESSION ADL. The MAML language allows RTL generation based on highly parameterizable templates written in VHDL. Itoh et al. (PEAS-III [34,35]) have proposed a micro-operation-description-based synthesizable HDL generation. It can handle processor models with a hardware interlock mechanism and multi-cycle operations [34].

2.5 TOP-DOWN VALIDATION

Validation of microprocessors is one of the most complex and important tasks in the current System-on-Chip (SOC) design methodology. Fig. 2.5 shows a traditional architecture validation flow. The architect prepares an informal specification of the microprocessor in the form of a natural language such as English. The logic designer implements the modules in the register-transfer level (RTL). The *RTL design* is validated using a combination of simulation-based techniques and formal methods. One of the most important problems in today's processor design validation is the lack of a golden reference model that can be used for verifying the design at different levels of abstraction. Thus, many existing validation techniques employ a *bottom-up approach* to pipeline verification, where the functionality of an existing pipelined processor is, in essence, reverse-engineered from its RT-level implementation.

Mishra et al. [36] have presented an ADL-driven validation technique that is complementary to these bottom-up approaches. It leverages the system architect's

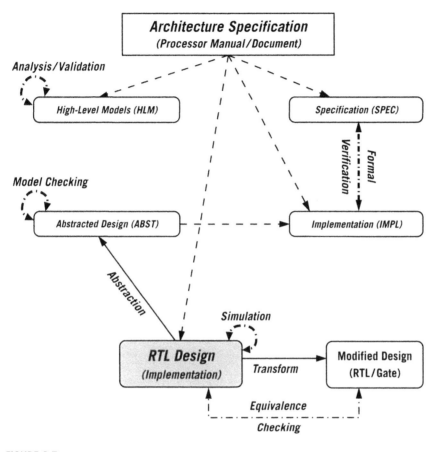

FIGURE 2.5

Traditional bottom-up validation flow.

knowledge about the behavior of the processor/memory architectures through ADL constructs, thereby allowing a powerful *top-down approach* to microprocessor validation. Fig. 2.6 shows an ADL-driven top-down validation methodology. This methodology has two important steps: validation of ADL specification, and ADL-driven validation of microprocessors.

2.5.1 Validation of ADL Specification

One of the most important requirements in a top-down validation methodology is to ensure that the specification (reference model) is golden. This section presents techniques to validate the static and dynamic behaviors of the architecture specified in an ADL. It is necessary to validate the ADL specification to ensure the correctness

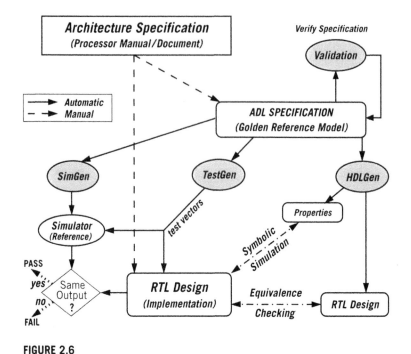

FIGURE 2.6

Top-down validation flow.

of both the architecture specified and the generated executable models including a software toolkit and hardware implementation. The benefits of validation are two fold. First, the process of any specification is error-prone and thus verification techniques can be used to check for the correctness and consistency of the specification. Second, changes made to the processor during exploration may result in incorrect execution of the system and verification techniques can be used to ensure correctness of the modified architecture.

One of the major challenges in validating the ADL specification is to verify the pipeline behavior in the presence of hazards and multiple exceptions. There are many important properties that need to be verified to validate the pipeline behavior. For example, it is necessary to verify that each operation in the instruction set can execute correctly in the processor pipeline. It is also necessary to ensure that execution of each operation is completed in a finite amount of time. Similarly, it is important to verify the execution style of the architecture.

Typical validation scenario requires two models: specification (assumed to be correct) and implementation (needs to be verified). A set of tests can be applied on both specification (or its simulatable model) and implementation, and the corresponding outputs can be compared. However, in case of specification validation, we have only one model that needs to be verified. As a result, property checking (or model checking) is very suitable for specification validation. Property

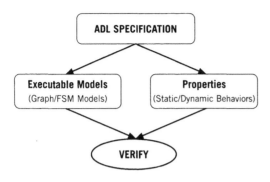

FIGURE 2.7

Validation of ADL specification.

checking ensures that a specification model satisfies a set of properties (or intended behaviors). Fig. 2.7 shows a property checking-based flow for specification validation. The ADL specification cannot be directly used for property checking. Therefore, it is necessary to generate executable formal models (such as graph or FSM models) from the specification. It is also required to generate a set of properties (behaviors) from the specification. The properties can be applied on the generated models using a model checking framework. The generated models and properties need to be suitable for the model checking framework. For example, a property can be a function that operates on a graph model [37]. In case a model checking-based framework such as SMV (symbolic model verifier) [38] is used, the model as well as the properties need to be specified in a specific SMV/temporal language.

Chapter 6 presents two property checking approaches for specification validation using EXPRESSION ADL. The first approach verifies a set of static behaviors/properties including connectedness, false pipeline and data-transfer paths, and completeness using a graph-based model of the architecture [37]. The second approach generates an FSM model from the specification to verify dynamic behaviors such as determinism and in-order execution in the presence of hazards and multiple exceptions [39]. The validated ADL specification can be used as a golden reference model for top-down validation of programmable architectures consisting of processor cores, coprocessors, and memory subsystems.

2.5.2 Implementation Validation

The ADL-driven validation approach has been demonstrated in two directions: simulation-based validation using directed test generation, and design validation using a combination of equivalence checking and symbolic simulation.

Simulation-based validation using directed test generation

Existing processor validation techniques employ a combination of simulation-based techniques and formal methods. Simulation is the most widely used form of processor validation. Use of ADL specification improves the overall validation effort, since both simulator and directed tests can be automatically generated from the ADL specification, as shown in Fig. 2.6. The generated tests can be applied on the hardware implementation as well as on the generated cycle-accurate simulator (reference model), and the outputs can be compared to check the correctness of the implementation. Since ADL specification can be used to generate simulation models at different levels of abstraction, the same validation methodology can be used for verifying simulators. For example, an instruction-set simulator (reference model) can be used to verify a cycle-accurate simulator (implementation).

Various types of test programs are used during simulation: random, constrained-random, and directed tests. The directed test vectors are generated based on certain coverage metrics such as pipeline coverage, functional coverage, and so on. Directed tests are very promising in reducing the validation time and effort, since a significantly less number of directed tests are required compared to random tests to obtain the same coverage goal. Test generation for functional validation of processors has been demonstrated using MIMOLA [16], EXPRESSION [40], nML [32], and LISA [41]. The basic idea is to generate the required behaviors and constraints from the ADL specification to direct the test generation process. For example, the model checking-based approach is used in EXPRESSION framework [40, 42]. This approach generates a graph-based model of the pipelined processor. Based on the graph coverage, a set of properties/behaviors are generated from the specification. Finally, the negated version of the properties are applied on the design using a model checker. The model checker produces a set of counterexamples which are converted into test programs consisting of instruction sequences.

Property/model checking

Fig. 2.8 shows an ADL-driven property checking methodology. The basic idea is to generate the properties based on a certain coverage metric such as pipeline coverage, functional coverage, etc. These properties and RTL implementation cannot be directly applied to a model checker due to capacity restrictions. Instead, an abstracted version of the RTL implementation is used to reduce space complexity in model checking. However, the process of abstraction may introduce errors (false negative) or may suppress errors (false positive) present in the RTL implementation. Symbolic simulation can be used to apply properties directly on the RTL implementation [43]. Symbolic simulation combines traditional simulation with formal symbolic manipulation [44]. Each symbolic value represents a signal value for different operating conditions, parameterized in terms of a set of symbolic Boolean variables. By this encoding, a single symbolic simulation run can cover many conditions that would require multiple runs of a traditional simulator.

Equivalence checking

Equivalence checking is a branch of static verification that employs formal techniques to prove that two versions of a design either are or are not functionally equivalent. Fig. 2.9 shows an ADL-driven approach that uses the generated hardware

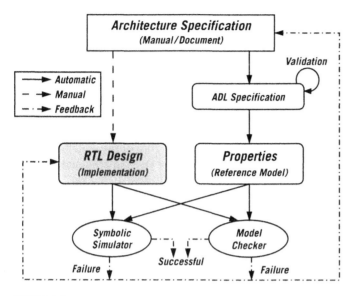

FIGURE 2.8

ADL-driven property checking.

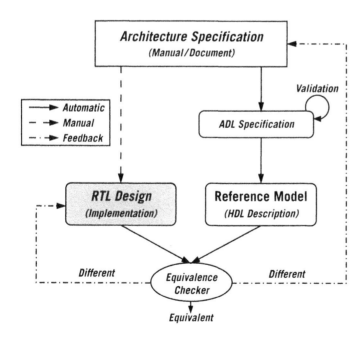

FIGURE 2.9

ADL-driven equivalence checking.

prototype as a reference model for equivalence checking with the RTL implementation [43]. An equivalence checker will try to match the compare points between the designs. The unmatched compare points need to be mapped manually. The tool tries to establish equivalence for each matched compare point. In case of failure, the failing compare points are analyzed to verify whether they are actual failures or not. The feedback is used to perform additional setup (in case of a false negative), or to modify the implementation (*RTL design*).

The ADL-driven hardware generation and validation of design implementation using equivalence checking has one limitation: the structure of the generated hardware model (reference) needs to be similar to that of the implementation. This requirement is primarily due the limitation of the equivalence checkers available today. Equivalence checking is not possible using these tools if the reference and implementation designs are large and drastically different. Property checking can be useful in such scenarios to ensure that both designs satisfy a set of properties. However, property checking does not guarantee equivalence between two designs. As a result, it is also necessary to use other complementary validation techniques (such as simulation) to verify the implementation.

2.6 CONCLUSIONS

Design of embedded systems presents a tremendous opportunity to customize the implementation by exploiting the application behavior. Architecture Description Languages have been used successfully to capture a wide variety of architectures, and automatically generate software toolkits including the compiler, simulator, assembler, and debugger. The generated tools allow efficient exploration of design alternatives to determine the best possible architecture under various constraints such as area, power, and performance. The ADL specification is also used to generate hardware models as well as functional test patterns to enable top-down validation of embedded processors. The ADL-driven methodologies reduce time-to-market and enable generation of cost-effective and reliable embedded systems.

REFERENCES

[1] P. Grun, A. Halambi, N. Dutt, and A. Nicolau. RTGEN: An algorithm for automatic generation of reservation tables from architectural descriptions. *IEEE Transactions on Very Large Scale Integration (VLSI) Systems*, 11(4):731–737, August 2003.

[2] A. Aho, R. Sethi, and J. Ullman. Compilers: Principles, techniques and tools. Addition-Wesley, 1986.

[3] M. Conway. Proposal for an UNCOL. *Communications of ACM*, 1(10):5–8, 1958.

[4] C. Fraser, R. Henry, and T. Proebsting. BURG: Fast optimal instruction selection and tree parsing. *SIGPLAN Notes*, 27(4):68–76, 1992.

[5] S. Hanono and S. Devadas. Instruction selection, resource allocation, and scheduling in the AVIV retargetable code generator. In *Proc. of Design Automation Conference (DAC)*, pages 510-515, 1998.

[6] W. Geurts, G. Goossens, D. Lanneer, and J. Praet. Design of application-specific instruction-set processors for multi-media, using a retargetable compilation flow. In *Proc. of the International Signal Processing Conference (GSPx)*, Santa Clara, October 2005.

[7] The MDES User Manual. *http://www.trimaran.org*, 1997.

[8] A. Shrivastava, P. Biswas, A. Halambi, N. Dutt, and A. Nicolau. Compilation framework for code size reduction using reduced bit-width ISAs (rISAs). *ACM Transactions on Design Automation of Electronic Systems (TODAES)*, 11(1):123-146, January 2006.

[9] Associated Compiler Experts. *http://www.ace.nl*, January 2008.

[10] CoWare LISATek Products. *http://www.coware.com*, January 2008.

[11] F. Hannig and J. Teich. Resource constrained and speculative scheduling of an algorithm class with run-time dependent conditionals. In *Proc. of International Conference on Application-Specific Systems, Architectures and Processors (ASAP)*, pages 17-27, 2004.

[12] W. Qin, S. Rajagopalan, and S. Malik. A formal concurrency model based architecture description language for synthesis of software development tools. In *Proc. of ACM Conference on Languages, Compilers, and Tools for Embedded Systems (LCTES)*, pages 47-56, 2004.

[13] M. Reshadi and D. Gajski. A cycle-accurate compilation algorithm for custom pipelined datapaths. In *Proc. of the 3rd IEEE/ACM/IFIP International Conference on Hardware/Software Codesign and System Synthesis (CODES+ISSS)*, pages 21-26, 2005.

[14] Tensilica Inc. *http://www.tensilica.com*, January 2008.

[15] A. Shrivastava, E. Earlie, N. Dutt, and A. Nicolau. Operation tables for scheduling in the presence of incomplete bypassing. In *Proc. of the IEEE/ACM/IFIP International Conference on Hardware/Software Codesign and System Synthesis (CODES+ISSS)*, pages 194-199, 2004.

[16] R. Leupers and P. Marwedel. Retargetable code generation based on structural processor descriptions. *Design Automation for Embedded Systems*, 3(1):75-108, 1998.

[17] P. Grun, N. Dutt, and A. Nicolau. Memory aware compilation through accurate timing extraction. In *Proc. of Design Automation Conference (DAC)*, pages 316-321, 2000.

[18] A. Shrivastava, E. Earlie, N. Dutt, and A. Nicolau. Aggregating processor free time for energy reduction. In *Proc. of the 3rd IEEE/ACM/IFIP international conference on Hardware/software codesign and system synthesis (CODES+ISSS)*, pages 154-159, 2005.

[19] A. Khare, N. Savoiu, A. Halambi, P. Grun, N. Dutt, and A. Nicolau. V-SAT: A visual specification and analysis tool for system-on-chip exploration. In *Proc. of EUROMICRO Conference*, pages 1196-1203, 1999.

[20] G. Hadjiyiannis, P. Russo, and S. Devadas. A methodology for accurate performance evaluation in architecture exploration. In *Proc. of Design Automation Conference (DAC)*, pages 927-932, 1999.

[21] J. Zhu and D. Gajski. A retargetable, ultra-fast, instruction set simulator. In *Proc. of Design Automation and Test in Europe (DATE)*, 1999.

[22] S. Pees, A. Hoffmann, and H. Meyr. Retargetable compiled simulation of embedded processors using a machine description language. *ACM Transactions on Design Automation of Electronic Systems*, 5(4):815-834, October 2000.

[23] R. Cmelik and D. Keppel. Shade: A fast instruction-set simulator for execution profiling. *ACM SIGMETRICS Performance Evaluation Review*, 22(1):128-137, May 1994.

[24] E. Witchel and M. Rosenblum. Embra: Fast and flexible machine simulation. In *Proc. of ACM SIGMETRICS International Conference on Measurement and Modeling of Computer Systems*, pages 68-79, 1996.

[25] A. Nohl, G. Braun, O. Schliebusch, R. Leupers, H. Meyr, and A. Hoffmann. A universal technique for fast and flexible instruction-set architecture simulation. In *Proc. of Design Automation Conference (DAC)*, pages 22-27, 2002.

[26] M. Reshadi, P. Mishra, and N. Dutt. Instruction set compiled simulation: A technique for fast and flexible instruction set simulation. In *Proc. of Design Automation Conference (DAC)*, pages 758-763, 2003.

[27] M. Reshadi, N. Bansal, P. Mishra, and N. Dutt. An efficient retargetable framework for instruction-set simulation. In *Proc. of International Symposium on Hardware/Software Codesign and System Synthesis (CODES+ISSS)*, pages 13-18, 2003.

[28] http://www.improvsys.com. *Improv Inc*, January 2008.

[29] J. Sato, N. Hikichi, A. Shiomi, and M. Imai. Effectiveness of a HW/SW codesign system PEAS-I in the CPU core design. In *Proc. of Asia Pacific Conference on Hardware Description Languages (APCHDL)*, pages 259-262, 1994.

[30] N. Binh, M. Imai, A. Shiomi, and N. Hikichi. A hardware/software partitioning algorithm for pipelined instruction set processor. In *Proc. of European Design Automation Conference (EURO-DAC)*, pages 176-181, 1995.

[31] O. Schliebusch, A. Chattopadhyay, M. Steinert, G. Braun, A. Nohl, R. Leupers, G. Ascheid, and H. Meyr. RTL processor synthesis for architecture exploration and implementation. In *Proc. of Design Automation and Test in Europe (DATE)*, pages 156-160, 2004.

[32] http://www.retarget.com. *Target Compiler Technologies*, January 2008.

[33] P. Mishra, A. Kejariwal, and N. Dutt. Synthesis-driven exploration of pipelined embedded processors. In *Proc. of International Conference on VLSI Design*, 2004.

[34] M. Itoh, S. Higaki, Y. Takeuchi, A. Kitajima, M. Imai, J. Sato, and A. Shiomi. PEAS-III: An ASIP design environment. In *Proc. of International Conference on Computer Design (ICCD)*, page 430, 2000.

[35] M. Itoh, Y. Takeuchi, M. Imai, and A. Shiomi. Synthesizable HDL generation for pipelined processors from a micro-operation description. *IEICE Transactions on Fundamentals of Electronics, Communications and Computer Sciences*, E83-A(3):394-400, March 2000.

[36] P. Mishra. *Specification-driven Validation of Programmable Embedded Systems*. PhD thesis, University of California, Irvine, March 2004.

[37] P. Mishra and N. Dutt. Automatic modeling and validation of pipeline specifications. *ACM Transactions on Embedded Computing Systems (TECS)*, 3(1):114-139, 2004.

[38] http://www.cs.cmu.edu/~modelcheck. *Symbolic Model Verifier*, January 2008.

[39] P. Mishra, N. Dutt, and H. Tomiyama. Towards automatic validation of dynamic behavior in pipelined processor specifications. *Kluwer Design Automation for Embedded Systems (DAES)*, 8(2-3):249-265, June-September 2003.

[40] P. Mishra and N. Dutt. Graph-based functional test program generation for pipelined processors. In *Proc. of Design Automation and Test in Europe (DATE)*, pages 182-187, 2004.

[41] A. Chattopadhyay, A. Sinha, D. Żhang, R. Leupers, G. Ascheid, and H. Meyr. ADL-driven test pattern generation for functional verification of embedded processors. In *IEEE European Test Symposium*, 2007.

[42] H. Koo and P. Mishra. Functional test generation using property decompositions for validation of pipelined processors. In *Proc. of Design Automation and Test in Europe (DATE)*, pages 1240-1245, 2006.

[43] P. Mishra, N. Dutt, N. Krishnamurthy, and M. Abadir. A top-down methodology for validation of microprocessors. *IEEE Design & Test of Computers*, 21(2):122-131, 2004.

[44] R. Bryant. Symbolic simulation—Techniques and applications. In *Proc. of Design Automation Conference (DAC)*, pages 517-521, 1990.

MIMOLA—A Fully Synthesizable Language

3

Peter Marwedel

3.1 INTRODUCTION

3.1.1 Origin of the Language

In the early seventies of the last century, Gerhard Zimmermann headed a group of researchers at the radio astronomy observatory of the University in Kiel, Germany. The observatory was used to measure the radio signals from the sun. The goal was to enhance our understanding of processes in the sun by monitoring the signals at a resolution of about 2 milliseconds at various frequencies and analyzing the results. Initially, results were recorded on paper, most of the time at a lower resolution. The number of paper rolls increased over time and there was hardly any hope that the collected data could ever be analyzed.

Therefore, Zimmermann started a project aiming at recording the data digitally on magnetic tape. Recording every sample was still unfeasible, so data reduction techniques had to be applied. Zimmermann proposed an algorithm for approximating recorded data. The algorithm was too complex to be executed in real time on the available computer. A special purpose processor, fast enough to execute the algorithm in real time, was needed. The following question arose: how do we systematically design processors such that they execute a given algorithm at the required speed? This question led to the (co-) invention of high-level synthesis. Zimmermann proposed using a very wide instruction format and instructions that were directly interpreted by the hardware. Machines of this type were later called VLIW machines, but this term did not exist in the seventies. Therefore, Zimmermann's ideas also included some of the first work on VLIW machines.

In order to synthesize (a term also not used in the seventies) such machines from an algorithm, a notation for a bit-true representation of the algorithm and its mapping to hardware resources was needed. Therefore, Zimmermann proposed a first version of a language comprising the required features. He called this language MIMOLA (Machine Independent Microprogramming Language) [1]. The first version of the language was based on postfix-notation and was difficult to read.

35

Around 1975, the author of this chapter joined the project. Together, they developed the required synthesis techniques and enhanced the language. The language was later turned into an infix-language and became more readable. In the years that followed, the syntax of MIMOLA was influenced to a large extent by other computer languages, predominantly by PASCAL.

A major set of innovative tools was built around the language. The first set of tools, called MSS1 (MIMOLA Software System 1), was used by Honeywell for application studies in the early 1980s. The design of a second set of tools, called MSS2, was started during the same period. MSS2 has been used by some academic design groups until the mid-1990s. Important milestones of the MIMOLA project are listed in Table 3.1.

3.1.2 Purpose of the Language

The semantics of available hardware description languages like VHDL and SystemC was initially defined for simulation. Synthesis came in as an afterthought. Some language elements cannot be efficiently synthesized, and for some language elements, there is a difference between simulation and synthesis semantics. As can be expected and as confirmed by designers, much design time is currently wasted for considering the differences between synthesis and simulation semantics. MIMOLA was designed for synthesis, but is also simulatable. Hence, the time for considering the different semantics is saved.

In architectural synthesis, the main input consists of the behavior to be implemented in the form of a program (see Fig. 3.1(a)). For MIMOLA, this program is described in a PASCAL-like syntax. Additional inputs include information about structural elements (available library components and possibly predefined (partial) structures). Finally, there can be hints for linking behavior and structure.

Programs are also input to retargetable code generation (see Fig. 3.1(b)). In contrast to architectural synthesis, the structure is fixed and the compiler has to generate binary code for the target structure. This is the code for the lowest

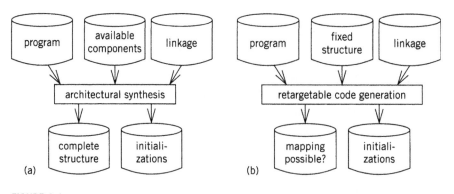

FIGURE 3.1

(a) Architectural synthesis. (b) Retargetable code generation.

Table 3.1   History of the MIMOLA project.

1976	G. Zimmermann proposes high-level synthesis [1]. Target architectures are of a VLIW type.
1977	G. Zimmermann publishes the MIMOLA language definition [2]. P. Marwedel writes software for MIMOLA.
1979	First international publications [3, 4].
1980	Zimmermann moves to Honeywell, Minneapolis and starts using MSS1. At Kiel, researchers lead by P. Marwedel start to work on a new version of the language, based on PASCAL. Also, work on retargetable compilation starts.
1981/84	Work on retargetable compilation is published [5, 6].
1986	Work on a second generation high-level synthesis tool is published [7].
	First paper on test program generation [8].
1987/89	Work on second generation compilation is published [9, 10].
1987	Handicaps resulting from the use of a mainframe providing just a flat file system come to an end. The software is ported to workstations. The resulting stable software version is called Version 3.45. Work on MIMOLA 4.0 starts.
1990	Members of the design team move from Kiel to Dortmund, Germany.
1993	Retargetable code generation becomes a hot topic. This stimulates publishing details of such code generation from MIMOLA [11]. This topic leads to the work of companies such as Tensilica.
1995	The CHIPS project funded by the European Union leads to the first workshop on code generation for embedded processors. The resulting book, published together with Gert Goossens of IMEC [12], becomes a key source of information for this area. The workshop is the first in the series of SCOPES workshops.
	First paper on the second generation test program generation published [13].
1997	MIMOLA is used as a hardware description language for the third generation retargetable RECORD compiler [14]. Further use of MIMOLA comes to an end.

programmable level, which can either be machine code or microcode. The MSS2 also includes simulators for simulating both at the behavioral and at the structural level. Finally, tools for generating self-test programs have also been designed [8, 13, 15, 16]. This variety of tools allows for smooth transitions between different design tasks [17].

3.1.3 Related Work: State of the Art in the Early Years

What are the contributions that the MIMOLA project makes to the state of the art? How does the project extend the knowledge on system design methods? These questions require a brief summary of the state of the art during the time the MIMOLA project made its contributions. Obviously, it is not possible to provide a complete set of references in the diverse areas in which the MIMOLA project advances the state of the art. What was the situation in the late seventies/early eighties?

The situation can be elaborated as follows:

- Design automation tools are mainly focusing on the geometrical level. Conferences are dominated by presentations on automatic placement and routing tools. Schematic entry is the prevailing technique for specifying a design. First logic synthesis tools are being developed. Darringer proposes to take a new look at logic synthesis [18]. Concurrent to our group at Kiel, a group of researchers at Carnegie-Mellon University starts to look at synthesis for a level higher than logic synthesis [19, 20].

- Some hardware description languages have been developed. IFIP supports the design of a "consensus language." This leads to the design of CONLAN [21]. See Mermet et al. [22] for a survey of European languages designed in those days. These languages are mostly academic niche topics and are essentially designed for simulation. Designing a language for synthesis is an exception.

- Concerning processor architectures, the most recent trend is the introduction of RISC processors by John Hennessy [23] and others. Dataflow computers emerge as a new and initially promising direction [24]. VLIW processors are hardly considered.

- Retargetable compilers hardly exist. There is Baba's microcode compiler [25]. The thesis of Cattell [26] provides insights into the generation of code generators. Glanville and Graham propose a table-driven code generator [27]. A survey by Ganapathi et al. summarizes the state of the art in retargetable compilation [28].

- Testing processors is mainly based on using large and expensive testers. However, Abraham et al. propose to run test programs on processors [29, 30].

3.1.4 Outline of This Chapter

In the following sections we summarize the contributions made during our work on the MIMOLA language and related tools. The outline of this chapter is as follows: Section 3.2 comprises of a description of the MIMOLA language. We cover the salient language features, the description of programs, the description of system structures, and mechanisms for linking programs and structures. Section 3.3 contains an overview over the capabilities of the tools that were developed around the MIMOLA language. Finally, Section 3.4 comprises the conclusion of this chapter. This chapter extends earlier papers [17, 22] on the language and MSS2. We are going to use MIMOLA Version 3.45 [31], the version for which synthesis, the first two code generators, and the first test generation tool were written.

3.2 SALIENT FEATURES OF THE MIMOLA LANGUAGE

3.2.1 Overall Structure of Description

A language that supports the mentioned tools has to have description capabilities for (partial or complete) structures, for the required system behavior (programs) and for links between these two. In the following section we describe how MIMOLA provides these capabilities. In MIMOLA, *design entities* are called **modules**. The description of modules comprises sections for describing types, programs, structures, and linkage information. Fig. 3.2 demonstrates the overall syntax of MIMOLA descriptions (leaving out some possible repetitions).

3.2.2 Declarations, Data Types, and Operations

Type declarations include the declaration of data types used in the description. Many simulation-oriented languages support various data types. Mapping these to data types supported in hardware typically requires a number of type-converting functions. Simulators can either actually evaluate these functions or treat these functions as intrinsics. For MIMOLA, it was decided to avoid complications resulting

```
module identifier module_interface;
    [<constant and type declarations>]
    [<program>]
    [<structure>]
    [<linkage>]
end;
```

FIGURE 3.2

Global syntax of MIMOLA descriptions.

from functions having a different interpretation in simulation and synthesis. It was decided to support bit vectors of arbitrary length as the only predefined data type. Due to this, descriptions closely resemble the actual hardware structure. Vectors of these values can be used throughout the language. These vectors have descending index ranges and are denoted by "(upper-index : lower-index)." For example: (15:0) is a bit vector of length 16. In contrast to VHDL, all data elements are described in terms of such vectors. There is no special data type for single bits: a single bit is just a bit vector with a single element.

Arguments and results of arithmetic functions are also always bit vectors. MIMOLA does not include arithmetic functions with various combinations of argument types. This is important in order to avoid combinatorial explosion of the number of predefined functions.

Elements of MIMOLA bit vectors can have any of the four values "0," "1," "X," and "Z." "X" and "Z" can only be used in limited contexts. "X" is returned by the simulator for conflicts between "1" and "0," "Z" describes tristate outputs in hardware structures.

MIMOLA's approach to typing requires special care for integer literals and array indexes. All integer literals have to be converted to their bit vector representation. This is feasible only if the length of the bit vector is known. Context information is used to calculate the length of bit vectors representing literals.

MIMOLA's approach to typing also means that operator overloading is not feasible. By default, all operators are assumed to denote two's complement operators. Operators for naturals are delimited by vertical bars or exclamation marks. For example, !<! denotes an unsigned "less than" operation and !*! denotes an unsigned multiply operation (this notation was borrowed from DACAPO [32]). For addition, no distinction between unsigned and signed two's complement operations is necessary, since they return the same bit vectors. Avoiding operator overloading simplifies the matching between required system behavior and available component behavior: simple checks for the operation name and the bit widths are sufficient.

3.2.3 Program Definition

The behavior to be implemented in a module is captured in the program definition part of a MIMOLA description. This part forms the specification for synthesis.

Which language features should be allowed in a specification language for synthesis? The design of MIMOLA is based on the assumption that fast implementations of algorithms are to be generated. Algorithms are typically described in a standard programming language. A computer running a compiled version of a program written in a standard programming language can be seen as a legal synthesis result. This means that we should not exclude any language elements of such a language for synthesis.

In order to make the learning of MIMOLA as easy as possible, the program part of MIMOLA was designed to be similar to a standard programming language. When MIMOLA was designed, PASCAL was popular. Hence, we tried to keep the set of differences between PASCAL and MIMOLA behavioral descriptions as small as reasonably possible. In particular, we did not want to exclude language elements like nested or even recursive procedures. Of course, this means that a run-time stack may be required to implement a certain behavior.

Only a few changes are required to translate a PASCAL program into a MIMOLA program.

Example 3.1

For the sake of simplicity, we will be using the following simple multiplication program with no input and output as a running example for an algorithm at the application level.[1] A more realistic example could read in parameters via some input port.

```
program Mult;
var a, b, c: integer;
begin
    a := 5; b := 7; c := 0;
    repeat
        c := c + a; b := b - 1;
    until b = 0;
end.
```

Designers usually have some clever ideas about essential elements of the design. It would be silly not to take advantage of the designer's knowledge. We have demonstrated the effect of such knowledge on the efficiency of the resulting design [33]. MIMOLA contains several language features which facilitate capturing the designer's knowledge. The following hints can be added to the behavioral description:

- **Manual operation to operator binding**
  Designers are frequently able to provide valuable hints about which hardware component instance should be used to perform certain operations. Such hints can be included in MIMOLA behavioral descriptions. Instance identifiers can be appended to operators.

Example 3.2

Let alu be the name of an instance of some arithmetic/logic unit type. Then, the notations +_alu or +'alu can be used to bind the add operator to this unit.

- **Manual variable to storage binding**
  MIMOLA provides an extension to PASCAL to indicate such a binding.

---

[1]At the time of writing this chapter, working installations of MSS2 ceased to exist. Therefore, the examples are based on the language reference manual and other existing documentation.

Example 3.3
Assume Mem is a memory. A variable called `zero` can be bound to location 0 of this memory by the declaration:

```
var zero : word at Mem[0];
```

■ Manual identification of parallelism

Automatic program parallelization was one of the weaker areas of MSS2. An automated tool did not work as well as expected, and it was therefore usually necessary to manually identify possible parallel execution. Toward this end, MIMOLA provides parallel blocks. Such blocks are delimited by the keywords **parbegin** and **parend**. All statements contained in such a block are assumed to be executed in parallel. Parallel blocks involving the same storage locations on the left-hand side of the assignments and in the expressions are assumed to be executed deterministically by first evaluating all expressions and then performing all storage operations. All MSS2 synthesis tools check if parallel execution is feasible. If not, the block is broken up into a set of smaller blocks.

Example 3.4
The following is an instruction set interpreter:

```
program Interpreter;
   type word = (15:0);      (* bit vector (15:0)*)
   var   pc:=0;             (* program counter *)
         ri,               (* instruction register *)
         ac,               (* accumulator *)
         ix : word;        (* index register *)
         M : Array [0..#FFFF] OF word; (* '#' means hex *)
   begin
     pc:=0;
     repeat
       parbegin ri:=M[pc]; pc:= pc+1; parend;
       case ri.(15:8) of (* (15:8)=OpCode *)
         #00 : parbegin ac:=M[ix]; ix:=ix+1 parend;
                 (* load word with post-increment *)
         #01 : parbegin M[ix-1]:=ac; ix:=ix-1 parend;
                 (* store word with pre-decrement *)

           ..
       end;
     until False;
   end.
```

Parallel blocks also allow us to denote generated scheduled programs in the same language as the unscheduled ones.

Programs can be written either at the application level (like in the case of the multiplication program) or represent an instruction interpreter for a given instruction set. In the first case, there will be no typical instruction set in the final design. The final design will directly implement user applications. In the second case, synthesis will generate a microcoded interpreter for the instruction set.

While the program part of MIMOLA and PASCAL are quite similar, some differences remain. The first notable difference is that for MIMOLA, data types have to be defined by using bit vectors as the basic data type. The second difference is that unsigned operations have to be delimited by vertical bars or exclamation marks. Values returned from a function must be identified by a return statement. A number of PASCAL features is not available in MIMOLA. These include real numbers, files, variant records, enumeration types, numeric labels, predefined operations on complex data types, and impure functions. Mechanisms for implementing subprogram calls and passing parameters must be defined by transformation rules. Furthermore, synthesis hints are not available in PASCAL.

3.2.4 Structure Definition

The structure definition part is introduced by the keyword **structure**. It describes the local components a module is composed of, and how they are interconnected. The latter is done in a **connections** definition, which contains sources and sinks for the nets.

Available component types and their exported operations

We distinguish between local component types and their instances. Component type descriptions start with the keyword **module**. For any module type, MIMOLA allows the description of operations that this module potentially provides to its environment. To make things easier for the user, the syntax of exported operations resembles that of PASCAL procedures. Exported operations are specified as (signal) assignments to output ports or to internal variables. Operations can be selected either conditionally or unconditionally. Unconditionally selected operations are denoted as simple assignments involving at most one of the operation symbols. Conditionally selected operations are written in the form of a case expression, for which a control input provides the selecting value. In MIMOLA 3.45, special input modes **adr**, **fct**, and **clk** denote address, control, and clock inputs, respectively. Multiple concurrently exported operations can be listed in concurrent blocks. Concurrent blocks are delimited by the keywords **conbegin** and **conend**. The list of exported operations cannot contain anything but a set of concurrent signal assignments or assignments to local variables. Only a single assignment is allowed for each of the outputs and local variables.

Example 3.5
The following example contains a list of available arithmetic/logic unit types.

```
module BAlu1(in a,b:word; fct s:(0); out f:word);
  begin
    f <- case s of (* signal assignment *)
      0: a + b;
      1: a - b;
    end;
  end;
module BAlu2(in a,b:word;fct s:(1:0);out f:word;out ovf:(0));
  conbegin
    f <- case s of (* signal assignment *)
      0: a; (* feed through *)
      1: b;
      2: a + b;
      3: a - b;
    end;
    ovf <- ovf_add(a,b) (* predefined overflow operation *)
  conend;
```

The example contains the list of exported operations of modules BAlu1 and BAlu2. Each case line describes an operation mode of the component. Case labels denote control codes that are required to select an operation mode. The approach is general enough to specify arguments and control codes involved in a component operation. Approaches just listing the exported operations and not including the bit width, port, and control code information exist, but are definitely not sufficient. On the other hand, more general approaches (e.g., arbitrary behavioral descriptions of local components) could not be used by any of our three main tools.

The following modules describe registers and memories. MIMOLA 3.45 requires certain module-type identifiers to start with certain characters.

```
module Regi(in a:word; fct s:(0); clk clock:(0); out f:word);
  var state:word
  conbegin
    at clock do case s of
      0: state:=a;(* state assignment *)
      1: ; (* empty *)
    end;
    f<-state;
  conend;
```

```
module Srom(adr addr:word; out f:word);
    var code:array [0..#FFFF] of word;
    begin
        f<-code[addr] after 10;
    end;
```

Delays can be attached to all operations (see module Srom). However, MIMOLA does not support units like milli- or nanoseconds.

The approach allows a very elegant description of multi-port memories: they are described by concurrent read and write operations for the different ports.

The ability to describe library components and programs in the same language has always been one of the strong points of our approach and has been exploited by our tools.

Parts and net lists

Instances of local components are declared in the **parts** section of a MIMOLA description. The **connections** section specifies the interconnection of module instances. The information in this section corresponds to that of VHDL net lists.

Example 3.6

A partial hardware structure containing some RAM named Mem, an instruction ROM I and a program counter pc can be described as follows:

```
module simple_hardware; (* module head *)
    structure
        type word = (15:0);
        module BAlu1 ... (* module types *)
        module BAlu2 ...
        module Regi...
        module Srom ...
        module Sram ...
        module Rflipflop ...
        parts
            I        : Srom; (* module instances *)
            pc,reg   : Regi;
            Mem      : Sram;
            cc       : Rflipflop;
        connections
            pc.f -> I.addr; ...
    end_structure;
```

The description starts with the head of the current design object, introducing a name for it (simple_hardware). The structure definition starts with the keyword

structure. Available parts are listed after the keyword **parts**. Their interconnections are listed after the keyword **connections**. Note the connection of output pc.f to the address input l.addr of the ROM.

3.2.5 Linking Behavior and Structure

Synthesis is usually not a fully automatic process, but relies on user guidance. Such guidance is frequently provided through the help of control files or pragmas (pseudo comments). Since MIMOLA is a synthesis-oriented language, language elements for user guidance can be built into the language. As a result, there is a reduced risk of inconsistencies.

Program transformation rules

It is frequently necessary to map certain parts of the behavioral description to other behavioral descriptions or to hardware components. Instead of using fixed mappings built into the synthesis system, MIMOLA allows an explicit definition of such rules.

Example 3.7

> **replace** &a = &b **with** (&a - &b)=0 **end**;

The ampersand sign denotes rule parameters.[2] The rule causes all occurrences of equal operations to be replaced by subtracts and a check for a zero result.

Rule sets are also used to implement some of the standard high-level language elements.

Example 3.8

```
replace
   &lab: repeat &seq until &cond
with
   &lab:begin
      &seq; &lab_loop: if &cond then incr(pc) else goto &lab
   end;
end;
```

Similar rules for procedure calls enable a very flexible implementation of calling mechanisms. Calling conventions are neither built into the language nor into the tool set. This allows the user to have full control over parameter passing and return location storage. Theoretically, full-blown run-time systems for PASCAL-like languages

---

[2] This notation was borrowed from job control languages of mainframes.

can be described. In practice, available standard rules implement a FORTRAN-like calling mechanism with fixed return locations per procedure.

The rules just mentioned do not directly reference hardware components (other than the pc), but they map descriptions to a level which is "closer" to hardware and therefore they prepare the final mapping for hardware.

According to our knowledge, MIMOLA is the only hardware description language that has the capability of describing such rules.

Identification of special purpose storage locations

Major emphasis of the MSS2 is on the support of programmable instruction set architectures (e.g., application-specific instruction set processors (ASIPs) and core processors). For these, some of the registers and memories serve a special purpose: for example, a register is used as a program counter, a certain memory is used to hold instructions, and so on. Tools usually cannot figure out which of the registers and memories serve a certain purpose. Therefore, we add hints to the descriptions. These are introduced by the keyword **locations**.

Example 3.9

locations_for_variables	Mem[1024..65535];
locations_for_temporaries	cc, reg, Mem[512..1023];
locations_for_programcounter	pc;
locations_for_instructions	I[0..#FFFF];

Initialization

Several of our tools generate requirements for the initialization of memory locations. For example, our retargetable code generator basically just generates such requirements, called binary code. It is desirable to store these requirements independently of structural descriptions. Therefore, MIMOLA comprises a special language element for initializations.

Example 3.10

```
init
  Mem    :=0;
  pc     :=100;
  I[100] :=#FE7B;
  I[101] :=#37F5;
  I[102] :=#7653;
end;
```

In contrast to initializations in the program part, these initializations are not considered to be a part of the specification of the required behavior. This language element allows for a clean separation of requirements for synthesis and initializations for actual runs of the generated hardware (a kind of special stimuli).

3.2.6 Putting Things Together

The following description shows how the different parts of a MIMOLA description introduced in Fig. 3.2 are used in a source file:

```
module simple_hardware;
  structure;
    module ... ;
    module ... ;
    parts ... ;
    connections ... ;
  <program transformation rules>
  <Identification of special purpose storage locations>
  <initialization>
  <program>
```

3.3 TOOLS AND RESULTS

3.3.1 Design Flow

The MSS2 is a set of tools for the design of digital programmable hardware structures. The MSS2 contains tools for high-level synthesis [7, 34], for retargetable code generation [9, 10], for test generation [8, 13, 15, 16], and for simulation and schematics generation [35]. The tools can be used for software/hardware codesign.

Let us consider a flow starting with architectural synthesis and using the other tools for refinements and design space exploration. As an example, we demonstrate a potential design flow for the simple_hardware shown earlier.

3.3.2 The Front-end and Internal Design Representations

MSSF is the first tool to be employed in such a flow. MSSF is the front-end of the MSS2. MSSF reads in MIMOLA descriptions. It is based on recursive descent parsing and contains the usual checking for errors in the MIMOLA descriptions. MSSF generates a LISP-like internal representation, called TREEMOLA [36]. The purpose of TREEMOLA is similar to that of XML: TREEMOLA provides persistent storage of all design information, parsers for reading, and dump programs for writing this information. All other tools of the MSS2 are reading in TREEMOLA and many are also generating TREEMOLA.

TREEMOLA is general enough to represent the input to tools and the additional design information (binary code, net lists) generated by these tools. This additional information can also be represented in MIMOLA. Hence, it makes sense to translate TREEMOLA back into MIMOLA. Generation of MIMOLA from a TREEMOLA representation is possible with MSSM.

3.3.3 Mapping to Register Transfers

The next tool in the chain is MSSR. MSSR maps all high-level language elements to the register transfer level. This is done by an application of program transformations. A set of default transformations is provided by MSS2, but the designer is free to choose his own set. Program transformations are iteratively applied until no further match is found. The sequence of applications is well defined.

MSSR also performs if-conversion: three possible implementations of if-statements are considered: conditional branches, conditional expressions, and conditional (guarded) assignments [37]. Applicable implementations are stored in the TREEMOLA-file as *versions* of the source code. The following tools are able to select the best version.

In addition, MSSR performs optimizations like constant folding.

For the running multiplication example, MSSR would replace the repeat-loop by branches.

```
program Mult;
var  a, b, c: integer;
begin
    L_0001: Mem[1024] := 5; Mem[1025] := 7; Mem[1026] := 0;
    L_0002: Mem[1026] := Mem[1026]+Mem[1024];   Mem[1025] := Mem[1025]-1;
    L_0003: pc:= if (Mem[1025] = 0) then incr(pc) else L_0002;
    L_0004: stop
end.
```

Note that at this step, all variables have been replaced by references to memory locations. Some of the information available at the RT-level is not explicitly shown: bit width information and names of variables. Names of variables are stored as tags of addresses and are used in memory disambiguation. Automatically generated labels reflect source code line numbers. incr(pc) and the label of the next block can be used interchangeably. **stop** statements are automatically added to conveniently terminate simulations. They are ignored in synthesis.

3.3.4 Simulation

The semantics of TREEMOLA are well defined [38] and can be simulated, even though MIMOLA is mainly targeting synthesis. We can either simulate just the behavior of the program or simulate the entire structure, taking initializations into account.

The first type of simulation is easy to perform: it is based on translating TREEMOLA to PASCAL with a tool called MSSS and compiling the resulting program. The only major problem is to achieve bit-true simulation in PASCAL. Calls to an appropriate abstract data type are generated during the translation. This simulation is untimed. For the running example, the product of a and b would be found in the memory location Mem[1026] allocated for c.

Simulation of MIMOLA structures is performed in a discrete event simulator called MSSU. MSSU considers the precise timing of hardware components. The simulator loads generated binary code stored in the initialization section into the appropriate memories. The purpose of detailed simulation is to predict the resulting performance and achieve some additional confidence about the correctness of the design since the MSS itself is not formally verified.

3.3.5 Architectural Synthesis

Architectural synthesis is one of the three key components of the MSS2. The program definition is the main input to synthesis. The list of available hardware components and their exported operations forms the second input to synthesis. This approach adds some bottom-up information to the top-down design process: our synthesis algorithms assume that only predesigned modules should be used in the final design. This restriction guarantees that only efficient modules will be used. This restriction is especially useful if a library of complex cells (ALUs, memories) is available.

Architectural synthesis generates a completed structural net list and the necessary memory initializations (binary code).

Three synthesis algorithms have been implemented in the MSS environment: two experimental tools ([39, 40] and [41]) and TODOS (= TOp DOwn Synthesis) [7,34] implemented in MSSH, a stable tool used for a number of internal and external designs. The following presentation refers to TODOS. Like most other high-level synthesis tools, TODOS consists of a number of steps: scheduling, resource allocation, resource binding, and final net list generation.

Note that the description of simple_hardware already includes the memories of the final design. This approach is based on the assumption that the designer will typically be well aware of the limited number of options that exist for the memory design. Synthesis focuses on the design of combinatorial logic and interconnect. However, synthesis also adjusts the bit width of the instruction memory as needed to implement the control of the data path.

Scheduling

As mentioned earlier, automatic program parallelization in MSS2 was not very effective. Hence, the user of MSS2 has the option of either accepting a small level of parallelism or a manual identification of parallel blocks. In the first case, each statement in the program definition is considered being a parallel block by its own. In the second case, several statements may be contained in one parallel block.

In any case, TODOS performs scheduling by considering parallel blocks. TODOS checks if resource constraints allow parallel blocks to be actually executed in parallel. If there are not enough resources to execute input blocks in parallel, these blocks are broken up into smaller parallel blocks meeting design constraints. The resulting smaller blocks are scheduled using a simple list scheduler.

We consider constraints concerning a limited number of memory references already in this early phase. The number of memory ports is typically the most serious constraint. The reason for this is that the MSS2 assumes that most of the variables are allocated to memory and therefore the number of parallel accesses to variables is limited by the number of memory ports. This approach avoids the large number of registers generated by approaches, assuming that each variable is allocated to a register. It also avoids complicated interconnect structures for registers. However, the proper allocation of temporaries is the main difficulty here, especially if temporaries are again allocated to memories with a limited number of ports.

Example 3.11

The multiplication program can be transformed into the sequence of control steps shown here:

```
L_0001_00:  parbegin Mem[1024] := 5; pc:=L_0001_01 parend;
L_0001_01:  parbegin Mem[1025] := 7; pc:=L_0001_02 parend;
L_0001_02:  parbegin Mem[1026] := 0; pc:=L_0002_00 parend;
L_0002_00:  parbegin reg:=Mem[1024]; pc:=L_0002_01 parend;
L_0002_01:  parbegin
                cc := "=0" (Mem[1025]-1) ; Mem[1025] := Mem[1025] - 1;
                pc := incr(pc);
            parend;
L_0002_02:  parbegin
                Mem[1026] := Mem[1026] + Reg;
                pc := if cc then incr(pc) else L_0002_00
            parend;
L_0004_00:  stop
```

Limited access to the memory is the key reason for the size of the parallel blocks, in particular for splitting the loop into control steps L_0002_00, L_0002_01, and L_0002_02. cc and reg are registers required for storing intermediate values. "=0" is a monadic function in MIMOLA 3.45. Note that an optimization merged the assignments for the loop body and the loop test (L_0003 disappeared). A straightforward implementation would have required three control steps for the loop body and two for the test.

Resource allocation

Resource allocation in TODOS is based on finding matchings between the operations in the program and the operations exported by available components. A number of details (like the bit widths, commutativity, neutral elements of operations, etc.) are considered for such matchings. Matchings are reflected in a relation "operation o in the program can be performed on module type m." This relation is used to find the number of instances of module types such that the total cost of module instances is minimized and such that a sufficient number of instances exists for each of the parallel blocks. This minimization is based on integer programming.

Let M be the set of component types and let $m \in M$ be any component type. Let c_m be the cost and let x_m be the number of instances of type m. The objective then is to minimize

$$c = \sum_{m \in M} (x_m * c_m) \tag{3.1}$$

We now have to compute the set of constraints for x_m. Let o be an operation type in the program. Define a relation *matches* such that o *matches* $m \iff m$ provides operation o. Let $f_{i,j}$ be the number of operations of type j used in control step i. Let $F_i = \{ j \mid f_{i,j} > 0 \}$ be the set of operations used in control step i. Let F_i^* be the powerset of F_i, that is, the set of all subsets of F_i. Let $a_{g,m}$ be 1 if some operation in g can be performed in module-type m and 0 otherwise. Then, a condition for a sufficient number of copies is that

$$\forall i, \forall g \in F_i^* : \sum_{m \in M} (a_{g,m} * x_m) \geq \sum_{j \in g} f_{i,j} \tag{3.2}$$

This means: for all control steps and for each subset of operations, the number of instances of components that are able to perform any of the operations in the subset must at least be equal to the operation frequency in the control step.

Generating constraints for each control step is not really necessary, since the constraints of different control steps can be combined. Let g be a set of operations types and let b_g be the maximum number of occurrences of this set of operations in any control step:

$$b_g = \max_i \left(\sum_{j \in g} f_{i,j} \right) \tag{3.3}$$

Let F^* be the set of operation-type combinations that are used in control steps:

$$F^* = \bigcup_i F_i^* \tag{3.4}$$

Then, from (3.2) it follows that

$$\forall g \in F^* : \sum_m (a_{g,m} * x_m) \geq b_g \tag{3.5}$$

is also a sufficient set of constraints. Constraints (3.5) can be interpreted as follows: for every combination of operations that is present in some control step, the total number of these operations cannot exceed the number of instances of component types which provide at least one of these operations.

Resource allocation for (combinatorial) components therefore reduces to minimizing (3.1) subject to the set (3.5) of constraints. This is a classical integer programming problem. More details are provided in a paper [42].

For the simple running example, it is sufficient to allocate a single adder/subtracter alu of type BAlu1.

Resource assignment

Resource assignment is responsible for binding operations to resources. The two previous steps guarantee that a legal binding exists. The key issue to be considered in finding a good binding is to minimize interconnect. TODOS uses a branch-and-bound-type method to find a binding with small interconnect costs. For complexity reasons, it is not feasible to consider all blocks at once during this optimization. They have to be considered one at a time. We sort blocks by their complexity (number of operations) and allocate resources in order of decreasing complexity.

Generation of the final net list

Generation of the final net list includes the insertion of required multiplexers and connecting clock inputs to a single master clock. Control inputs are connected to instruction fields at the output of the instruction memory.

This step also involves the minimization of the width of the instruction memory based on instruction fields that are redundant in some of the control steps. The instruction field for a combinatorial part is redundant in a particular control step if the part is unused in that control step. Instruction fields for memories and registers are never redundant, since unintentional writing has to be prevented.

For the running example, Fig. 3.3 is a graphical representation of the resulting net list.

The data path includes data memory Mem, register reg, the arithmetic/logic unit, a zero detector, a condition code register, and two multiplexers. Mem.A and Mem.B are memory outputs and inputs, respectively. The alu is able to add and to subtract.

The controller consists of the pc, the multiplexer at the input of pc, the incrementer, and control memory I. The multiplexer at the input of pc has two modes: a transparent mode for input c (used for incrementing pc) and an **if**-mode: c is selected if a is true, otherwise b is selected (this mode is used for **else**-jumps). Note that neither **then**-jumps nor unconditional jumps can be implemented. The former would require an **if**-mode with c and b reversed, the latter would require a transparent mode for b. Dashed lines denote signals I.xxx coming from the controller.

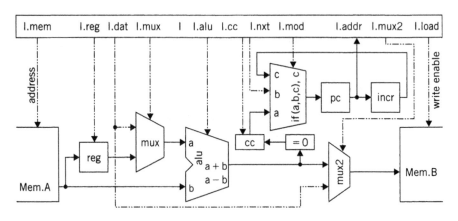

FIGURE 3.3

Solution to the synthesis problem (clock inputs not shown).

Synthesis algorithms integrating scheduling, resource allocation, and assignment have been proposed as well, the most recent one using VHDL [43].

3.3.6 Test Program Generation

Frequently, the designs resulting from synthesis are well testable by self-test programs. However, this is not guaranteed. We have designed tools automatically generating test programs.

For the running example, we would generate a self-test sequence testing reg for stuck-at errors:

```
(* Self test program for register reg using default patterns *)
Mem[0]:=#5555;                  (* binary 0101.. pattern *)
reg:=Mem[0];                    (* load into register *)
cc:="=0" (Mem[0] - reg);        (* check if ok *)
pc:=if cc
      then incr(pc)
      else ErrorExit;           (* jump to error report *)
Mem[0]:=#AAAA;                  (* binary 1010.. pattern *)
reg:=Mem[0];
cc:="=0" (Mem[0] - reg);
pc:=if cc
      then incr(pc)
      else ErrorExit;
```

A pattern of alternating ones and zeros is stored in the memory. Then, we try to copy this pattern into the register. Next we compare memory and register content. If they are the same, copying has succeeded and we proceed to the next test. In the case of a failing copy operation, the generated code would cause a jump to an error report routine. This test is repeated for swapped ones and zeros. The default patterns #5555 and #AAAA have been selected to test all bits for stuck-at-one and stuck-at-zero errors and to also have a limited test for shorts between adjacent bits.

We assume that there is a sufficient number of methods for the computation of test vectors for each of the RT-level components. These patterns can be stored in a test-pattern library. If such a library exists, the test patterns stored in the library will replace the default constants #5555 and #AAAA. The fault coverage depends on the set of patterns stored in the library.

In the case of the condition code register cc, we would generate a test for stuck-at-zero errors:

```
cc:= "=0" (0);          (* try to generate 1 at cc *)
pc:=if cc
    then incr(pc)
    else ErrorExit;     (* jump to error report *)
```

The constant 0 at the output of alu would be generated by subtracting some constant from itself, like in the test program for register reg.

Unfortunately, our current hardware structure cannot be tested for a stuck-at-one at cc:

```
cc:= "=0" (#5555);      (* try to generate 0 at cc *)
pc:=if cc
    then ErrorExit      (*jump to error report*)
    else incr(pc);
```

Such a test would need a **then**-jump, which cannot be implemented by the multiplexer at the input of pc. The user could decide to solve this problem by extending the jump hardware such that unconditional jumps are possible. Unconditional jumps together with **else**-jumps can be used to emulate **then**-jumps:

```
cc:= "=0" (#5555);      (* try to generate 0 *)
pc := if cc
    then incr(pc)       (* stuck-at-one *)
    else Cont;          (* ok *)
pc := ErrorExit;        (* jump to error report *)
Cont: <next test>;
```

The first tool for test program generation within the MSS2 (called MSST) was designed by Krüger [8, 15]. The approach suffered from difficult memory allocation techniques in PASCAL. A more recent tool with a similar scope was designed by Bieker [13, 16]. It is implemented in PROLOG. The initial effort was larger than that for PASCAL, but the resulting tool was much more flexible.

3.3.7 Code Generation

The user could also decide to make additional modifications to the generated hardware. These manual post-synthesis design changes are frequently required, for example, in order to conform to some company standards or to overcome some limitations of the synthesis algorithm. Verification of this step will be described in this section.

In the case of our sample hardware, the user could decide to reroute immediate data to be stored in memory Mem. He could omit multiplexer mux2 at the memory input and select an instance of ALU-type BAlu2 which has transparent modes for both inputs. The modified hardware structure is shown in Fig. 3.4.

Rerouting immediate data possibly influences the resulting performance: parallel assignments to Mem and to cc would now both store the result computed by alu. The precise effect of this change on the performance depends upon the original program. MSS2 allows the user to compute the resulting performance by retargetable compilation (see the following paragraphs).

Hardware structures generated by correct synthesis systems are automatically correct ("correctness by construction"). In this context, "correct" means: the structure together with the binary code implement the behavior.

Manually modified structures are potentially incorrect. MSS2 allows verifying the design by attempting compilation for predefined structures. The compiler in MSS2 tries to translate programs into the binary control code of a given hardware structure. If the binary code can be generated by the compiler, the structure together with the binary code implement the behavior. If no code can be generated by the compiler, then either the structure is incorrect or the compiler does not have enough semantic knowledge. Several methods are used to convey semantic knowledge to the compiler. The most important feature of the compiler is the fact that

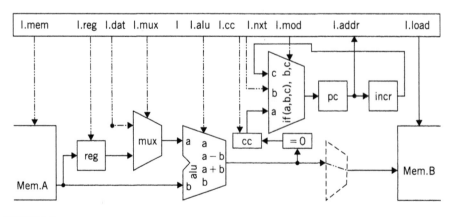

FIGURE 3.4

Modified hardware structure.

it is retargetable. This means the code for a different machine can be generated by changing the structure being used as input.

Three retargetable compilers have been designed:

1. MSSV [6, 11] was the first retargetable compiler within the MSS2. MSSV is based on tree matching.

2. MSSQ [9, 10], based on graph matching, is much faster than its predecessor. The different phases of code generation are more integrated and code alternatives are better supported in MSSQ. Pattern matching is performed at the RT-level.

   There are examples for which MSSQ is two orders of magnitudes faster than MSSV. VLIW machines with many orthogonal instruction fields are the main target for the MSSQ code generator. Code was generated for various processors including AMD-2900-based designs. MSSQ was the only compiler that was able to handle the benchmark at MICRO-20. For our running example, MSSQ would generate the binary code for the modified architecture.

3. For the third compiler, RECORD, Leupers [14] proposed to use instruction set extraction. The key idea is to derive descriptions at the instruction set architecture (ISA) level automatically from the descriptions at the RT-level. Once an ISA description is derived, standard compiler generation tools such as dynamic programming-based instruction selectors like iburg [44] can be employed. For RISC- and CISC-instruction sets, this approach is faster than MSSQ.

Instruction set extraction starts at the input of memories and traverses the data path until memory outputs are found. Every time some data transforming component is found, the corresponding data transformation is added to the pattern describing the instruction. If some control input requires a particular value, methods for generating this value are also analyzed. Methods typically include certain encodings of the instruction.

Example 3.12

Example: for our modified hardware structure, we would find the instructions listed in Table 3.2.

All instructions can be combined with available branch modes.

3.3.8 Overall View of the Dependence among MSS Tools

MSS2 comprises some utility tools also processing TREEMOLA. MSSB is a tool for dumping binary code in a readable format. Also, experimental programs for schematics generation from TREEMOLA structural net lists have been designed [35].

An overall view of the dependence among MSS2 tools is provided in Fig. 3.5. For the sake of simplicity, conversions between different representations of

Table 3.2 Instructions extracted for the running example.

Name	Semantics	Fields of the instruction involved in the encoding
store immediate	Mem[I.mem]:=I.dat	I.mem,I.alu,I.mux,I.dat,I.load
add register	Mem[I.mem]:=Mem[I.mem]+reg	I.mem,I.alu,I.mux,I.load
add immediate	Mem[I.mem]:=Mem[I.mem]+I.dat	I.mem,I.alu,I.mux,I.dat,I.load
...	...	...

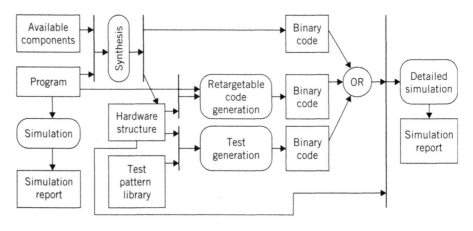

FIGURE 3.5

Activity diagram for designing with MSS2.

design information, for example, by MSSF, MSSM, and MSSB, are not included in the figure.

Any execution of tools that is consistent with these dependences can be performed.

3.3.9 Designs Using MSS2

Designs using the MSS2 include:

- The design of the asynchronous, microprogrammed SAMP by Nowak [45].

- The design of a CISC-processor at the University of Hamburg: according to Rauscher et al. [46], the synthesized chip is only 7% larger than the manually generated design. Rauscher et al. argue that the size of the synthesized design could be reduced by about 5%. One reason for the small overhead is that the TODOS design uses register files instead of separate registers.

- The design of the PRIPS PROLOG-machine [47] at the University of Dortmund.

3.4 CONCLUSIONS

3.4.1 Evolution of Ideas and Directions

Where did the work on the MIMOLA project lead to? How have research directions that were considered during the MIMOLA project been explored in later work?

First of all, textual hardware description languages have become the standard technique for describing designs. The first MIMOLA language reference was published in 1977. VHDL was standardized by IEEE in 1987. It seems like work on Verilog was started around 1984 [48]. It became an IEEE standard in 1995. SystemC became an IEEE standard in 2005 [49]. During the design of all three languages, synthesis was neglected and had to be considered later.

Second, key ideas of MIMOLA-based synthesis approaches were extended in the OSCAR high-level synthesis system [43]. Like MIMOLA-based approaches, OSCAR places emphasis on the modeling of both the behavior to be achieved and the available components. Also, extending the rule-based approach of MIMOLA, OSCAR incorporates expression transformations. The integer programming-based resource allocation of the MSS2 is generalized into integer programming-based scheduling, resource allocation, and binding. Successful commercial products did not become available until after decades of waiting. The Catapult C synthesis tool [50] implements many of the dreams of the MIMOLA project. We aimed at having high-level synthesis from a standard programming language and Catapult achieved this goal by using C/C++ as the specification language.

Third, retargetable compilation has become commercially available. Target Compiler Inc. has extended data structures with some influence by MSSQ [51]. Others have followed a similar route [52]. Tensilica offers compilers with some retargetability [53]. The Trimaran framework [54] is a framework for retargetable compilation for a parameterized processor architecture. Retargetable compilation is also available with tools centered around the LISA language [55], which is also described in this book.

Fourth, automatic test program generation became a new direction and is now used more frequently [56–59].

VLIW processors also have become a major direction in computer architecture. They still cannot replace processors in systems running legacy code. However, processors such as the NXP TriMedia included in the Nexperia processor [60] and the TMS320C64x [61] are successful in the embedded domain.

3.4.2 What Went Wrong and What Went Right

Zimmermann's proposal to start high-level synthesis—something completely new—was certainly very smart. It gave the involved researchers a good amount of lead time before synthesis became mainstream. Fundamental research with applicable results is certainly what research institutions should head for (and not for incremental improvements of products available on the market). Entering this new territory was also a lot of fun and a source of enthusiasm for the involved researchers. Cooperation between the people involved was really great. There was a low administration workload, and this stimulated research.

There were also some things that went wrong. First of all, the overall effort for making all the new ideas useful in practice was largely underestimated. Second, most of us were taught to publish only close to perfect results. Results of work in progress and papers on particular aspects of the MSS2 should have been published to make the ideas more widely known. Third, the lack of adequate computing resources and adequate programming languages and tools (using a mainframe with an old-fashioned OS) for the first ten years of the project was certainly a handicap. Finally, automatic parallelization techniques should have received more attention.

3.4.3 Summary

The MIMOLA project, involving the design of the language and the design of corresponding tools, was a long-term project which provided milestones in a number of new research directions. MIMOLA is one of the first languages specifically designed for synthesis and not just for simulation. This approach avoided time-consuming considerations of synthesizable subsets and differences between synthesis and simulation semantics. MSS1 and MSS2 high-level syntheses are among the first approaches to high-level synthesis. The idea of starting with a behavioral description similar to that of a programming language, usage of formal mathematical models for high-level synthesis, the incorporation of program transformation rules, and the emphasis on the library of available hardware components did become commercially available only decades after the first ideas within the MIMOLA project. MSS2 retargetable compilation is one of the first approaches toward compilation for flexible architectures. It is the key enabling technology for design space exploration of processor architectures and currently in use for more or less flexible architectures. The MSS2 test program generator was the first automatic test program generator for self-test programs starting from the RT-level.

REFERENCES

[1] G. Zimmermann. A method for designing digital computers using the programming language MIMOLA (in German). *Springer Informatik Fachberichte*, 6:465–478, 1976.

[2] G. Zimmermann. Report on the computer architecture design language MIMOLA. Technical Report 4/77, Institut für Informatik & P.M., University of Kiel, 1977.

[3] G. Zimmermann. The MIMOLA design system: A computer aided digital processor design method. *16th Design Automation Conference*, pages 53–58, 1979.

[4] P. Marwedel. The MIMOLA design system: Detailed description of the software system. *Proc. of the 16th Design Automation Conference*, pages 59–63, 1979.

[5] P. Marwedel. A retargetable microcode generation system for a high-level microprogramming language. *ACM Sigmicro Newsletter*, 12:115–123, 1981.

[6] P. Marwedel. A retargetable compiler for a high-level microprogramming language. *ACM Sigmicro Newsletter*, 15:267-274, 1984.

[7] P. Marwedel. A new synthesis algorithm for the MIMOLA software system. *23rd Design Automation Conference*, pages 271-277, 1986.

[8] G. Krüger. Automatic generation of self-test programs: A new feature of the MIMOLA design system. *23rd Design Automation Conference*, pages 378-384, 1986.

[9] L. Nowak. Graph based retargetable microcode compilation in the MIMOLA design system. *20th Annual Workshop on Microprogramming (MICRO-20)*, pages 126-132, 1987.

[10] P. Marwedel and L. Nowak. Verification of hardware descriptions by retargetable code generation. *26th Design Automation Conference*, pages 441-447, 1989.

[11] P. Marwedel. Tree-based mapping of algorithms to predefined structures. *Int. Conf. on Computer-Aided Design (ICCAD)*, pages 586-593, 1993.

[12] P. Marwedel and G. Goossens, editors. *Code Generation for Embedded Processors*. Kluwer Academic Publishers, 1995.

[13] U. Bieker and P. Marwedel. Retargetable self-test program generation using constraint logic programming. *32nd Design Automation Conference*, pages 605-611, 1995.

[14] R. Leupers. Retargetable generator of code selectors from HDL processor models. *European Design and Test Conference (ED & TC)*, 1997.

[15] G. Krüger. A tool for hierarchical test generation. *IEEE Trans. on CAD*, 10:519-524, 1991.

[16] U. Bieker, M. Kaibel, P. Marwedel, and W. Geisselhardt. STAR-DUST: Hierarchical test of embedded processors by self-test programs. Technical report, University of Dortmund, CS Department, Report No. 700, 1998.

[17] P. Marwedel and W. Schenk. Cooperation of synthesis, retargetable code generation and test generation in the MSS. *EDAC-EUROASIC'93*, pages 63-69, 1993.

[18] J. A. Darringer and Jr. W. H. Joyner. A new look at logic synthesis. In *17th Design Automation Conference*, pages 543-549, NY, USA, ACM Press, 1980.

[19] M. Barbacci. Instruction set processor specifications for simulation, evaluation, and synthesis. *16th Design Automation Conference*, pages 64-72, 1979.

[20] A. Parker, D. Thomas, D. Siewiorek, M. Barbacci, L. Hafer, G. Leive, and J. Kim. The CMU design automation system: An example of automated data path design. *16th Design Automation Conference*, pages 73-80, 1979.

[21] R. Piloty. CONLAN report. *Report RO 83/1*, Institut für Datentechnik, Technische Hochschule Darmstadt, 1983.

[22] J. Mermet, P. Marwedel, F. J. Rammig, C. Newton, D. Borrione, and C. Lefaou. Three decades of hardware description languages in Europe. *Journal of Electrical Engineering and Information Science*, 3:699-723, 1998.

[23] J. Hennessy, N. Jouppi, S. Przybylski, C. Rowen, T. Gross, F. Baskett, and J. Gill. MIPS: A microprocessor architecture. *SIGMICRO Newsletter*, 13(4):17-22, 1982.

[24] P. Treleaven. Future computers: Logic, ..., data flow, control flow. *IEEE Computer*, pages 47-55, 1984.

[25] T. Baba and H. Hagiwara. The MPG system: A machine independent microprogram generator. *IEEE Trans. on Computers*, 30:373-395, 1981.

[26] R. G. G. Cattell. Formalization and automatic derivation of code generators. Technical report, PhD thesis, Carnegie-Mellon University, Pittsburgh, 1978.

[27] R. S. Glanville and S. L. Graham. A new method for compiler code generation. In *POPL '78: Proc. of the 5th ACM SIGACT-SIGPLAN symposium on Principles of programming languages*, pages 231-254, NY, USA, ACM Press, 1978.

[28] M. Ganapathi, C. N. Fisher, and J. L. Hennessy. Retargetable compiler code generation. *ACM Computing Surveys*, 14:573-593, 1982.

[29] S. M. Thatte and J. A. Abraham. Test generation for microprocessors. *IEEE Trans. on Computers*, pages 429-441, 1980.

[30] D. Brahme and J. A. Abraham. Functional testing of microprocessors. *IEEE Trans. on Computers*, pages 475-485, 1984.

[31] R. Jöhnk and P. Marwedel. MIMOLA reference manual—version 3.45. Technical Report 470, Computer Science Dpt., University of Dortmund, 1993.

[32] Fa. DOSIS. DACAPO II, User Manual, Version 3.0. *DOSIS GmbH*, Dortmund, 1987.

[33] P. Marwedel and W. Schenk. Improving the performance of high-level synthesis. *Microprogramming and Microprocessing*, 27:381-388, 1989.

[34] P. Marwedel. An algorithm for the synthesis of processor structures from behavioural specifications. *Microprogramming and Microprocessing*, pages 251-261, 1986.

[35] K. Kelle, G. Krüger, P. Marwedel, L. Nowak, L. Terasa, and F. Wosnitza. Tools of the MIMOLA hardware design system (in German). Report 8707, University of Kiel, Dept. of Computer Science, 1987.

[36] R. Beckmann, D. Pusch, W. Schenk, and R. Jöhnk. The TREEMOLA language reference manual—version 4.0—Technical Report 391, Computer Science Dpt., University of Dortmund, 1991.

[37] P. Marwedel and W. Schenk. Implementation of IF-statements in the TODOS-microarchitecture synthesis system. In: G. Saucier, J. Trilhe (ed.): *Synthesis for Control Dominated Circuits*, North-Holland, pages 249-262, 1993.

[38] U. Bieker. On the semantics of the TREEMOLA-language version 4.0. Technical Report 435, Computer Science Dpt., University of Dortmund, 1992.

[39] O. Broß, P. Marwedel, and W. Schenk. Incremental synthesis and support for manual binding in the MIMOLA Hardware Design System. *4th International Workshop on High-Level Synthesis, Kennebunkport*, 1989.

[40] W. Schenk. A high-level synthesis algorithm based on area oriented design transformations. *IFIP Working Conference On Logic and Architecture Synthesis, Paris*, 1990.

[41] M. Balakrishnan and P. Marwedel. Integrated scheduling and binding: A synthesis approach for design-space exploration. *Proc. of the 26th Design Automation Conference*, pages 68-74, 1989.

[42] P. Marwedel. Matching system and component behaviour in MIMOLA synthesis tools. *Proc. 1st EDAC*, pages 146-156, 1990.

[43] B. Landwehr and P. Marwedel. A new optimization technique for improving resource exploitation and critical path minimization. *10th International Symposium on System Synthesis (ISSS)*, pages 65-72, 1997.

[44] C. W. Fraser, D. R. Hanson, and T. A. Proebsting. Engineering a simple, efficient code-generator generator. *ACM Letters on Programming Languages and Systems*, 1(3):213-226, September 1992.

[45] L. Nowak. SAMP: A general purpose processor based on a self-timed VLIW-structure. *ACM Computer Architecture News*, 15:32-39, 1987.

[46] N. Hendrich, J. Lohse, and R. Rauscher. Prototyping of microprogrammed VLSI-circuits with MIMOLA and SOLO-1400. *EUROMICRO*, 1992.

[47] C. Albrecht, S. Bashford, P. Marwedel, A. Neumann, and W. Schenk. The design of the PRIPS microprocessor. *4th EUROCHIP-Workshop on VLSI Training*, 1993.

[48] D. K. Tala. History of Verilog. *http://www.asic-world.com/verilog/history.html*, February 2008.

[49] IEEE. IEEE Standard SystemC Language Reference Manual. *http://standards.ieee.org/getieee/1666/download/1666-2005.pdf*, 2005.

[50] Mentor Graphics. Catapult synthesis. *http://www.mentor.com/products/esl/high_level_synthesis/catapult_synthesis/index.cfm*, 2007.

[51] J. V. Praet, D. Lanneer, W. Geurts, and G. Goossens. Method for processor modeling in code generation and instruction set simulation. US patent 5918035, *http://www.freepatentsonline.com/5918035.html*, June 1999.

[52] S. A. Gupta. Programmatic synthesis of a machine description for retargeting a compiler. US patent 6629312, *http://www.freepatentsonline.com/6629312.html*, September 2003.

[53] Tensilica. XPRES C-to-RTL Compiler. *http://www.tensilica.com/products/lits_whitepapers.htm*.

[54] L. N. Chakrapani, J. Gyllenhaal, Wen-mei W. Hwu, Scott A. Mahlke, K. V. Palem, and R. M. Rabbah. *Trimaran: An Infrastructure for Research in Instruction-Level Parallelism*. Springer Lecture Notes in Computer Science, Vol. 3602, 2005.

[55] M. Hohenauer, H. Scharwaechter, K. Karuri, O. Wahlen, T. Kogel, R. Leupers, G. Ascheid, H. Meyr, G. Braun, and H. van Someren. A methodology and tool suite for C compiler generation from ADL processor models. *Design, Automation and Test in Europe*, pages 1276–1281, 2004.

[56] A. Krstic and S. Dey. Embedded software-based self-test for programmable core-based designs. *IEEE Design & Test*, pages 18–27, 2002.

[57] N. Kranitis, A. Paschalis, D. Gizopoulos, and Y. Zorian. Instruction-based self-testing of processor cores. *Journal of Electronic Testing*, 19:103–112, 2003.

[58] N. Kranitis, A. Paschalis, D. Gizopoulos, and G. Xenoulis. Software-based self-testing of embedded processors. *IEEE Trans. on Computers*, pages 461–475, 2005.

[59] P. Bernardi, Rebaudengo, and S. M. Reorda. Using infrastructure IPs to support SW-based self-test of processor cores. *Workshop on Fibres and Optical Passive Components*, pages 22–27, 2005.

[60] NXP. Nexperia PNX 1500. *http://www.nxp.com/acrobat_download/literature/9397/75010486.pdf*, February 2008.

[61] Texas Instruments Inc. TMS320C64x/C64x+ DSP CPU and instruction set reference guide (Rev. F). *http://focus.ti.com/lit/ug/spru732f/spru732f.pdf*, February 2008.

nML: A Structural Processor Modeling Language for Retargetable Compilation and ASIP Design

4

Johan Van Praet, Dirk Lanneer, Werner Geurts, and Gert Goossens

4.1 INTRODUCTION

nML is a hierarchical and highly structured architecture description language (ADL), at the abstraction level of a programmer's manual. It models a processor in a concise way for a retargetable processor design and software development tool suite. nML has been carefully designed to contain the right amount of hardware knowledge as required by these tools for high-quality results.

nML is the processor modeling language used by IP Designer (Chess/Checkers), an industry-proven retargetable tool suite that supports all aspects of application specific processor (ASIP) design: architectural exploration and profiling; hardware generation and verification; and—last but not least—software development based on highly optimizing C compilation, instruction-set simulation, and debugging technology.

Being commercially available products from Target Compiler Technologies, nML and IP Designer have been used to design and program ASIPs for a wide array of applications: for example, for portable audio and hearing aid instruments, video coding, wireline and wireless modems, and network processing.

Sections 4.2, 4.3, and 4.4 in this chapter explain the nML language, as it is used now in the IP Designer tool suite. Section 4.2 gives an overview of nML, Section 4.3 is about the structural skeleton part of nML, and Section 4.4 explains how to model the instruction set and the processor behavior in nML. Section 4.5 describes how to specify pipeline hazards in nML. In Section 4.6 the historical evolution of nML is described, from its original conception in academia to its latest commercial version.

Section 4.7 gives a short overview of Target's processor design and software development tools that are using nML: a C compiler, a simulator generator, a **65** synthesizable (register-transfer-level) hardware description language generator, and a test-program generator. Further, Section 4.8 describes some typical design applications that have been implemented using nML and the IP Designer tool suite. Finally, Section 4.9 concludes this chapter.

4.2 THE nML PROCESSOR DESCRIPTION FORMALISM

In nML, a structural skeleton of the target processor is first specified, which declares, among others, all the processor's storage elements. This structural skeleton is the topic of Section 4.3.

The second part of an nML processor description contains a grammar that defines the machine language of the processor, and thus models its instruction set. Every sentence in the language corresponds to one instruction in the instruction set. *AND-rules* in the grammar describe the composition of orthogonal instruction parts; *OR-rules* list alternatives for an instruction part.

The behavior of the instructions—that is, the semantics of the sentences that can be derived from the grammar—is held in the attributes of the instruction-set grammar. The *action* attribute describes which actions are performed by an instruction part, and gives a register-transfer-level view on the processor's data path. It is complemented by the *value* attribute to efficiently describe memory and register addressing modes. The *syntax* attribute specifies the assembler syntax for the corresponding instruction part and the *image* attribute defines its binary encoding. The instruction-set description part of nML is described in Section 4.4.

If desired, the structural skeleton and the instruction-set grammar can be interleaved in an nML description, to group grammar rules of a processor part together with the declaration of storage elements they are connected to. Depending on the complexity of the processor, this can be a good style. However, conceptually we always keep these two parts of nML separated.

In the following sections, we will draw examples from an nML description for a simple example processor, called *tctcore*. The architecture of this processor is shown in Fig. 4.1 and its instruction set is summarized in Table 4.1. Both will be explained in parallel with their translation into nML. The complete nML description for *tctcore* is available from our web site [1].

4.3 A STRUCTURAL SKELETON OF THE PROCESSOR

All the storage elements of a processor are declared *globally* in the first part of the nML description. They form a *structural skeleton* that defines the connection points for the primitive operations that are defined on the processor. In fact, the values in the storage elements determine *the state of the processor*, and the instructions execute primitive operations to change this state.

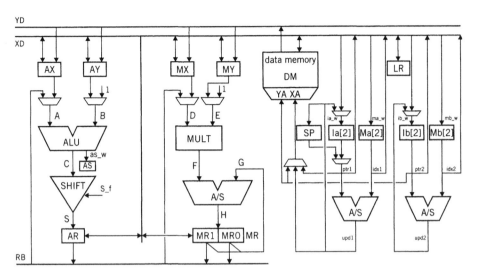

FIGURE 4.1

Architecture of the example *tctcore* processor.

4.3.1 Memories and Registers

Each storage element is typed with a data type for the values it contains and, if applicable, with an address type for its addresses. The valid addresses for a storage element must also be specified, from which its capacity can be derived.

Memories are declared with the keyword mem, and registers with the keyword reg, as shown below for the example processor of Fig. 4.1:

```
mem PM[1024]<pm_type, addr>;   // program memory
mem DM[0..1023,1]<num, addr>;  // data memory
reg XY[4]<num,b2u>;            // input registers
reg RR[3]<num,b2u>;            // result registers
```

The program memory PM contains 1024 values of type pm_type and is addressed by values of type addr. The data memory DM holds values of type num and is also addressed by values of type addr, in the range 0..1023, with address step 1. If the range is given by a single value n, it is equal to specifying a range from 0 to $n - 1$. When not specified, the address step defaults to 1. The register files RR and XY (not shown in Fig. 4.1) carry the same data type as the data memory, and are addressed by a 2-bits data-type b2u.

None of the data types, pm_type, addr, num, and b2u, are built-in in the nML language, but they must be declared as C++ classes in the so-called processor header file. This is done using standard C++ syntax, but preprocessor macros are available to save typing work. In Target's IP Designer tool suite, example class libraries are also available.

The accumulator register of the *tctcore* processor combines two registers in a record:

```
reg MR0<num>;
reg MR1<num>;
reg MR<acc>{          // record register
    MR0;
    MR1;
};
```

The width of the record register must be equal to the sum of the widths of its components. The width of a storage is defined by the bits specifier of its data type, in the C++ processor header file. All storage elements, including memories and register files, can be combined in records.

4.3.2 Storage Aliases

Some storage declarations provide just another view for already declared storage. For example, in the *tctcore* description, the X/Y input registers of the ALU and multiplier have been grouped in a register file called XY, and the individual register fields have been defined as range aliases:

```
reg AX<num> alias XY[0];    // ALU regs
reg AY<num> alias XY[1];
reg AR<num> alias RR[0];
reg MX<num> alias XY[2];    // MACC regs
reg MY<num> alias XY[3];
reg MR0<num> alias RR[1];
reg MR1<num> alias RR[2];
```

After the alias keyword the start address of the alias in the parent register file is indicated. The alias and parent register must have the same data type. The alias register can also contain multiple fields, for example,

```
reg XY_alu[2] alias XY[0];  // alias of XY[0..1]
```

A record alias can provide a record view for already declared storage. For example, to specify concisely how to load or store data of the acc type in the data memory DM, the alias DMacc can be specified, specifying an address step of 2:

```
mem DMacc[1024,2]<acc, addr> alias DM; // data memory
```

4.3.3 Transitory Storage

Memories and controllable registers are called *static* storage elements, because they store each of their values until they are explicitly overwritten. The same value can thus be read several times from the same static storage element.

Connections such as buses and nets (wires) fall in the class of *transitory* storage elements, which are defined in nML to pass a value from input to output without delay. *Transitories* are declared with the keyword trn. Pipeline registers, declared with the keyword pipe, are a special kind of transitory storage having a delay of one cycle. These are used to describe multi-stage actions. Some examples from *tctcore*:

```
trn A<num>;   // ALU input
trn B<num>;   // ALU input
trn C<num>;   // ALU output
trn XD<num>;  // X data bus
pipe F<acc>;  // pipeline between multiply and accumulate
```

It is not needed to declare *all* nets and buses of the processor as transitories. Transitories must only be declared as they are needed to connect primitive operations or to model resources on which conflicts may occur. Indeed, in the C compiler of the IP Designer tool suite, transitories are used to model resources that may lead to *hardware conflicts* in code generation: the compiler allows at the most one operation at a time to write to a transitory.

It can be desirable to restrict the access to a given memory or register to a list of so-called *access transitories*. For example, the *tctcore* example contains the following declaration:

```
mem DM[dm_size]<num,addr> read(XD YD) write(XD);
```

The memory DM can only be read through transitories XD and YD, and can only be written through XD. Here the access transitories are buses in the processor (see Fig. 4.1), but the designer can choose to specify a new set of access transitories representing the physical ports of memory DM:

```
mem DM[dm_size]<num,addr> read(DM_rw DM_r) write (DM_rw);
```

4.3.4 Immediate Constants and Enumeration Types

Immediate constants that are encoded in an instruction are declared using the keyword cst with the corresponding processor data type, e.g.:

```
cst c_3 <fact>;    //  3-bit immediate
cst c_8 <b8s>;     //  8-bit immediate
```

An enumeration type contains a set of named binary constants, which by default have consecutive values, starting from zero. This default can be overridden.

```
enum alu {add " + ", sub " - ", and " & ", or " | "};
```

All the constants in an enumeration type have the same number of bits, namely, the minimum number required to represent the largest constant. So the values for the constants in enum alu are, respectively, 00, 01, 10, and 11. The symbols between double quotes in an enumeration type are mnemonics for the constants, to be used in assembly syntax attributes—see Section 4.4.6.

One enumeration type with the predefined name stage_names is used to define the names of the pipeline stages of the processor:

```
enum stage_names { IF, ID, EX1, EX2 };
```

4.3.5 Functional Units

Functional units can be declared in nML, using the keyword fu, to group primitive operations that are physically executed on the same hardware unit. For example:

```
fu alu;          // arithmetic logic unit (ALU)
fu mult;         // multiplier
```

Since all hardware conflicts are modeled as access conflicts on transitories, functional units are optional. If present, they are used to assign operations to modules for hardware generation. The syntax to do this is explained in Section 4.4.4.

4.4 INSTRUCTION-SET GRAMMAR

The second part of an nML description is a specification of the instruction set and the execution behavior of the processor. Complex architectures may allow hundreds or thousands of legal combinations of operations and addressing modes to compose the instruction set. The size of the description is reduced dramatically by introducing instruction-set hierarchy and by sharing similarities among a variety of instructions. In an nML description, an attribute grammar is used for that purpose. The production rules in the grammar introduce the structure of the instruction set into the processor description. The topology (or connectivity) of the data path is contained in the grammar attributes, together with the instruction's assembly syntax and binary encoding.

To illustrate the concepts of this part of an nML description, some examples will be taken from the instruction set of the *tctcore* example processor of Fig. 4.1. This instruction set is depicted in Table 4.1 in a compact way: each bit string that can be formed from left to right forms a valid instruction. For example, by concatenating the top entry of all columns, one obtains the instruction "0 00 00 0 00 00 0 000 00 00," which adds the values in the registers AX and AY, and in parallel loads a value from the memory field addressed by la0 into AX.

Further examination of the structure of the instruction-set table shown in Table 4.1 reveals that it falls apart in a number of subtables, each having a title printed in bold. For example, the first subtable contains arithmetic operations and parallel load/store(s) with indirect addressing. Instructions in this format start with "0."

In Table 4.1 each subtable representing a format in its turn contains some subtables, and so on. This kind of hierarchical structure can conveniently be represented in a grammar.

Table 4.1   Instruction set of the example *tctcore* processor.

Arithmetic operations in parallel with 1 or 2 indirectly addressed loads/stores

0	00	00	00:+ 01:− 10:and 11:or	0:AY 1:"1"	00:AX 01:AR 10:MR0 11:MR1	00: single	0:load 1:store	000:AX 001:AY 010:MX 011:MY 100:LR 101:AR 110:MR0 111:MR1	00:Ia0 01:Ia1 10:Ib0 11:Ib1	00: +0 01: +1 10: +Ma/bi 11: −Ma/bi

X-bus / Y-bus move section:

01:load/load 10:store/load	X-bus move			Y-bus move		
	00:AX 01:AY 10:MX 11:MY	0:Ia0 1:Ia1	0: +1 1: +Mai	00:AX 01:AY 10:MX 11:MY	0:Ib0 1:Ib1	0: +1 1: +Mbi

01	0:+ 1:−	0:>>1 1:<<1		
10	nnn <<n (−4..3)			
11	00:*,+ 01:*,− 10:*	0:MY 1:"1"	00:MX 01:AR 10:MR0 11:MR1
	11:rnd/nop	0:rnd 1:nop	xx	11: nop xxxxxxxx

SP indexed immediate load/store

10	00	0: load 1:store	000:AX 100:LR 001:AY 101:AR 010:MX 110:MR0 011:MY 111:MR1	nnnnnnnnnn: SP offset

Directly addressed load/store

10	01	0: load 1:store	000:AX 100:LR 001:AY 101:AR 010:MX 110:MR0 011:MY 111:MR1	aaaaaaaaaa: memory address

Generate 10 bits constants in AGU registers

10	10	0	000:Ia[0] 100:Ma[0] 001:Ia[1] 101:Ma[1] 010:Ib[0] 110:Mb[0] 011:Ib[1] 111:Mb[1]	nnnnnnnnnn: constant

Direct branch instructions

10	10	1	000:EQ 100:LT 001:NE 101:LE 010:GT 110:OV 011:GE 111:true	aaaaaaaaaa: branch target

Hardware Do-loop

10	11	00	00: AX 10: MX 01: AY 11: MY	aaaaaaaaaa: loop end address

Generate 8/16 bits constants in data registers

10	11	01	00: AX 10: MX 01: AY 11: MY	0x	nnnnnnnn: 8-bit sign-extended constant
				1x	nnnnnnnn: high order part of constant

Pointer initialization relative to SP

10	11	100	0: Ia0 1: Ia1	nnnnnnnnnn: SP offset value

SP modify

10	11	101	x	nnnnnnnnnn: SP modify value

Jump to subroutine

10	11	110	x	aaaaaaaaaa: branch target

Register move

10	11	1110	xx	source				destination
				0000:AX 0100:LR 1000:Ia[0] 1100:Ma[0] 0001:AY 0101:AR 1001:Ia[1] 1101:Ma[1] 0010:MX 0110:MR0 1010:Ib[0] 1110:Mb[0] 0011:MY 0111:MR1 1011:Ib[1] 1111:Mb[1]				cf. source

Return from subroutine

10	11	1111	xxxxxxxxxx

x : don't care bits in instruction word   nn...n, aaa...a : immediate bits in instruction word

4.4.1 Breaking Down the Instruction Set: AND Rules and OR Rules

An nML description is best written using a top-down approach. After defining the structural skeleton of the data path, the instruction set must be analyzed, and its structure is captured by writing down production rules for the underlying grammar. Each possible derivation from these rules (i.e., each possible sentence in the grammar) represents one *legal* instruction. There are two kinds of rules:

- *OR rules* list all *alternatives* for an instruction part. These alternatives are *mutually exclusive*, meaning that only one alternative can be executed at a time.

- *AND rules* describe the *composition* of instruction parts. In this case, the composing instruction parts are *orthogonal*, meaning that the concatenation of any legal derivation for every instruction part forms a legal derivation for the AND rule itself.

The fact that Table 4.1 consists of a number of subtables is expressed in nML by means of an OR rule:

```
opn tctcore (
    arith_mem_ind_instr |   // Parallel arithmetic + indirect load/store
    mem_sp_idx_imm_instr |  // Stack pointer indexed immediate load/store
    mem_dir_instr |         // Direct load/store
    imm_load_addr_instr |   // Immediate loads to address registers
    branch_instr |          // (Conditional) branch instruction
    doloop_instr |          // Hardware loop instruction
    imm_load_data_instr |   // Immediate loads to data registers
    ptr_init_sp_instr |     // Pointer initialise relative to SP
    sp_mod_instr |          // Stack pointer modify
    jsr_instr |             // Jump to subroutine
    rmove_instr |           // Register moves
    return_instr);          // Return from subroutine
```

The rules that are referenced as alternatives in the OR rule, and are separated by a vertical bar, are again either AND rules or OR rules. For example, the rule mem_dir_instr is again an OR rule:

```
opn mem_dir_instr (load_direct | store_direct);
```

The tctcore rule is the start rule for each valid instruction derivation, which is specified in the first part of the nML description by declaring the *start-symbol* of the grammar:

```
start tctcore;
```

The first format is detailed by the AND rule named arith_mem_ind_instr:

```
opn arith_mem_ind_instr (ar : arith_instr, mi : mem_ind_instr)
```

As indicated in its *parameter list*, arith_mem_ind_instr is composed of two orthogonal instruction parts, namely arith_instr and mem_ind_instr, which respectively correspond to the left and the right halves of the first subtable of Table 4.1. The declaration of a *parameter* consists of an *instantiation name* and a *reference* to some other rule or to a constant. The instantiation name will be used in the attributes of the AND rule (see in the following paragraphs). Again, the rules that are referenced are either AND rules or OR rules.

This process of analyzing the instruction set can recursively be applied until all entries of Table 4.1 are covered by grammar rules. It is important to note that although the grammar structures the instruction set and introduces hierarchy in the description, it does not contain any other information about the target processor. The execution behavior of the processor, that is, its primitive operations, their connectivity, and the actual instructions, is described in the grammar attributes.

4.4.2 The Grammar Attributes

The specification and combination of grammar attributes are the basic mechanism used to specify the execution behavior and the instruction set of a target processor in nML.

For an AND rule three kinds of attributes are defined:

- The *action attribute* (Section 4.4.4) describes which register-transfer actions are performed by an instruction or instruction part. Each AND rule must have *one* action attribute.

  There is a special kind of nML AND rule, called *mode* AND rule, to describe register and memory addressing modes for which also a *value attribute* (Section 4.4.7) is defined that specifies a storage location.

- The *syntax attribute* (Section 4.4.6) specifies the assembler syntax (mnemonics) for the corresponding instruction (part). It must only be present if one intends to derive an assembler or disassembler tool from the nML description. An AND rule may have multiple syntax attributes if needed.

- The *image attribute* (Section 4.4.5) defines the binary encoding for the corresponding instruction (part). In some cases, multiple image attributes may be needed. An instruction encoding can also be derived automatically by the nML front-end tool, in which case no image attributes must be specified.

An OR rule does not need any explicitly defined attributes, but implicitly passes attributes between its left- and right-hand sides. However, it is allowed to define image attributes for an OR rule—normally in terms of the image attributes of rules it references as alternatives—but *not* to define action or syntax attributes.

4.4.3 Synthesized Attributes

While the instruction-set structure is represented in nML using the top-down approach of Section 4.4.1, the grammar attributes are built in a bottom-up fashion.

Attributes of lower-level rules may be "instantiated" in the attributes of another rule higher in the nML hierarchy. The attributes of the latter rule are then said to be *synthesized* from the attributes of the lower-level rules. The contents of attributes for leaf AND rules in the nML hierarchy will only be explained in the next sections, but the concept of synthesized attributes can already be seen in the following example:

```
opn arith_mem_ind_instr (ar : arith_instr, mi : mem_ind_instr)
{
    action {
        ar.action;
        mi.action;
    }
    syntax : ar.syntax ", " mi.syntax ;
    image : ar.image::mi.image;
}
```

The action attribute of arith_mem_ind_instr consists of the actions of both rules that are referenced in its parameter list by the instantiation names ar and mi. The syntax attribute specifies that the assembly syntax of arith_mem_ind_instr consists of the assembly syntax of both rules arith_instr and mem_ind_instr, separated by a comma. The image attribute shows that this instruction part is encoded by concatenating the encodings (image attributes) of both referenced rules. The :: operator concatenates bit strings. It is uniquely determined which attributes can be used to synthesize other attributes, so the "action," "syntax," and "image" keywords are normally not written to refer to an attribute.

By computing the attributes of the start symbol, for a grammar derivation corresponding to a particular instruction, one obtains the complete execution behavior of the instruction, together with its complete binary encoding and assembly syntax.

4.4.4 Action Attribute

The action attribute describes the *register-transfer* behavior of an instruction part. It is defined in terms of assignments between elements of the structural skeleton, for example, reading from a register or memory and assigning the value to a transitory, assigning a value from a transitory to a transitory, or writing a value from a transitory to a register or memory.

An action attribute may also instantiate primitive operations of the processor that are declared in the processor header file. In the action attribute, the connections of a primitive operation are described using the C++ syntax for function calls, but with names of transitories instead of variables.

To explain the nML grammar attributes, we use the nML rule describing the ALU of the *tctcore* processor (see Table 4.1, top left):

```
opn alu_instr (op : alu, al : alu_left_opd, ar : alu_right_opd)
{
    action {
      stage EX1:
```

```
        A = al.value;
        B = ar.value;
        switch (op) {
          case add : C = add(A, B, AS) @alu;
          case sub : C = sub(A, B, AS) @alu;
          case and : C = A & B @alu;
          case or  : C = A | B @alu;
        }
        S = C @sh;
        AR = S;
      }
    syntax : AR " = " al op ar;
    image  : op::ar::al;
  }
```

The syntax and image attributes are explained in Sections 4.4.5 and 4.4.6.

For the action attribute, a switch statement is used to select the primitive operation to be executed. The selector of the switch statement is declared by the "op:alu" parameter, which refers to the enumeration type alu explained in Section 4.3.4. Each entry in a switch statement contains a case label and one or more actions, which may be of any kind—either primitive operations, or references to actions of other rules, or even other switch statements. The case labels must *not* be mutually exclusive, and it is also possible to use logic expressions to compute them. However, unlike in C, there is *no* "fall-through" mechanism.

The switch statement in the action attribute of the alu_instr rule contains the actual ALU operations. For example, when the selector op has the value sub (= 01), the primitive operation C = sub(A,B,AS) is executed. It is assigned to the functional unit alu (see Section 4.3.5) by the annotation @alu. The result C is connected to register AR in this rule.

The connection of the operands A and B is specified by value attributes of a mode rule, which is further explained in Section 4.4.7. As explained in Section 4.4.3, an action attribute may also instantiate other action attributes, for example, to specify the operands for A and B instead of instantiating those value attributes.

The order of statements in an action attribute is unimportant. They are *concurrent*, and it is the connectivity to the structural skeleton that matters.

Note that the data types of the storage elements in the instantiation of a primitive operation must exactly match those of the declaration in the processor header file. If the types do not match, intermediate transitories of the right type are needed, as implicit type conversions are supported for assignments between transitories.

Guard statements

Next to switch statements, if/else statements are also supported. These are all dependent on enumeration-type values that are encoded in the instructions. To describe data-dependent, conditional behavior in nML, so-called control primitives can be declared and used, or one can use guard statements. A guard statement tests the value of a transitory, instead of testing an immediate enum parameter. It can be used high in the nML hierarchy, to impose an execution condition on a complete instruction format, or locally, for example, to implement a conditional operation.

Guard statements are also typically used in a special-purpose always rule, which contains actions to be executed every clock cycle, independently of the instructions being executed. These actions can, for example, model connections to the outside world, or even small peripherals for the processor, with guard statements controlling them.

Pipeline stages

In the *tctcore* example, the primitive operations on the alu all are executed in stage EX1, so it is sufficient to specify this stage at the start of the action attribute of rule alu_instr.

The multiply and accumulate operations are also single-stage operations, but they are executed in different stages, as shown in rule macc_instr here, with F being declared as a pipeline register (D, E, G, and H are transitories):

```
opn macc_instr (op : macc, ml : mult_left_opd, mr : mult_right_opd)

{
    action {
      stage EX1:
        D = ml;
        E = mr;
        F = mult(D,E) @mult;
      stage EX2:
        switch (op) {
        case multAdd : H = add(G = MR, F) @as;
        case multSub : H = sub(G = MR, F) @as;
        case mult    : H = F @as;
        }
        MR = H;
    }
    syntax : MR op ml " * " mr;
    image  : op::mr::ml;
}
```

When using primitive operations that span several pipeline stages, it is possible to specify a range in a stage definition. To specify cycle accurate timing, the pipeline stages in which the storage elements are accessed have to be annotated explicitly, between *backward quotes*, in a so-called *access stage* annotation. The access stage must be contained in the range of the stage definition. Consider, for example, a multi-stage macc operation that incorporates the pipeline register F:

```
opn mult_acc_prim()
{
    action {
      stage EX1..EX2:
        G `EX2` = MR `EX2`; // one stage later than mult. operands
        H `EX2` = macc(D `EX1`, E `EX1`, G `EX2`) rsrc(F `EX1`);
    }
}
```

This example also shows that it is possible to annotate an extra (abstract) resource to a statement in the action attribute, for example, to model a hardware conflict on *internal* resources of a primitive operation. If the macc operation would use resource F in both stage EX1 and stage EX2, this can be annotated as rsrc(F 'EX1,EX2'). From this, the tools can then derive that a new macc operation can only be started every two cycles

Section 4.5.2 explains how to solve pipeline conflicts for an example where the operand fetch and result write-back actions are happening in different stages.

4.4.5 Image Attribute

The image attribute specifies the binary encoding of the instruction parts. All instantiation names in the parameter list of the alu_instr rule, shown in Section 4.4.4, occur in its image attribute. This is indeed synthesized by concatenating the (binary) value of the selected constant in the alu enumeration type and the image attribute of the alu_left_opd and alu_right_opd rules.

It is also possible to explicitly define image attributes for an OR rule, similar to the definition of image attributes for an AND rule. For example:

```
opn arith_instr (alu_instr | as_sh_instr | shift_instr
                 | macc_rnd_nop_instr)
{
    image : "00"::alu_instr  |
            "01"::as_sh_instr  |
            "10"::shift_instr  |
            "11"::macc_rnd_nop_instr;
}
```

There is an important difference with the image attribute of an AND rule: in an AND rule the instance name—for example, op and al—is used to refer to another rule, while in an OR rule the name of the referenced rule itself is used in the image attribute—for example, alu_instr.

The bit-length of the image attribute for the different alternatives of an OR rule does not have to be the same. This allows to describe variable-length instruction sets, where not all instructions have the same number of bits.

It is possible to leave it up to the nML front-end tool to derive format bits for specific OR rules, or more generally, to derive the image attributes of *all* rules. This means that it is allowed to write nML code without defining any image attribute. It is also possible to *mix* automatically generated image attributes with hand-coded image attributes, both for AND rules and OR rules.

4.4.6 Syntax Attribute

The syntax attribute shows the assembly syntax of the instruction parts. Everything between double quotes must literally appear in an assembly language statement for the corresponding rule. Identifiers that appear in the parameter list (like al and ar in the alu_instr rule of Section 4.4.4) are expanded according to the syntax attribute they refer to. For an enumeration type, the symbols between double quotes in an enumeration type are mnemonics for the constants; if no quoted mnemonics are present, the constant name is used as a mnemonic. Other identifiers that are not between double quotes, like AR, must be storage element names, and must also appear literally in the assembly language. White space in the syntax attribute will *not* literally appear in the assembly syntax, *unless* it is contained within quotes. According to its syntax attribute, an example assembly syntax for alu_instr is AR = AX - AY.

A rule can have multiple syntax attributes (see the example at the end of Section 4.4.7), and a syntax attribute can specify alternative syntaxes with optional matching conditions. There is also support for assembly languages where the instruction fields (e.g., instructions slots in a VLIW architecture) may be specified in arbitrary order. In short, virtually any assembly language syntax can be described in nML.

4.4.7 Mode Rules and Value Attributes

For convenience of notation, nML supports a special type of rule called a *mode* rule. In the example alu_instr rule of Section 4.4.4, both parameters alu_left_opd and alu_right_opd are mode AND rules. The definition of mode rule alu_left_opd follows next:

```
mode alu_left_opd (sel : alu_left)
{
    value {
        switch (sel) {
        case AX:  AX;
        case AR:  RR[0];
        case MR0: RR[1];
        case MR1: RR[2];
        }
    }
    syntax : sel;
    image : sel;
}
```

Compared to a normal AND rule, a mode AND rule has an additional *value attribute* to specify how (a location in) a storage element is addressed, either to fetch or to store a value. The value attribute of a mode rule can refer to a single storage element, or to several storage elements using a switch construct, such as in alu_left_opd. Possible side effects of addressing a storage location, such as post-modification of the address, are described in the action attribute of the mode rule.

A mode rule is instantiated by assigning its value attribute to a transitory (see A = al.value; in rule alu_instr of Section 4.4.4), or by assigning a transitory to its value attribute. Indeed, the main advantages of using a mode rule are that it can be used in both directions, and that it can be reused as an operand or result location selector for different transitories. This way, the recurring uses of the same storages as a source or destination in the actions of many rules can be *factored* out into a single mode rule, which greatly reduces the complexity of the processor model.

Mode AND rules can also be combined into mode OR rules, as shown here for the OR rule ind_mem_acc describing single indirect memory access for *tctcore*. It selects either mode ind_mem_acc1 with address calculation unit acu1 or mode ind_mem_acc2 (not shown) with acu2:

```
mode ind_mem_acc (ind_mem_acc1 | ind_mem_acc2);

enum sgl_addr_mode { indirect "0", incr "1", add "M", sub "-M" };

mode ind_mem_acc1(i : c_1, m : sgl_addr_mode)
{
    value : stage EX1..EX2: DM[XA`EX1`]`EX2`;
    action {
      stage EX1:
        XA = ptr1 = Ia[i];
        switch (m) {
        case indirect:
        case incr: Ia[i] = upd1 = add(ptr1, idx1 = 1) @acu1;
        case add : Ia[i] = upd1 = add(ptr1, idx1 = Ma[i]) @acu1;
        case sub : Ia[i] = upd1 = sub(ptr1, idx1 = Ma[i]) @acu1;
        }
    }
    syntax : "DM[Ia" i "]" ;
    syntax : "Ia" i " += " m ;
    image  : "0"::i::m;
}

opn load_ind_instr (m: ind_mem_acc, r : reg_sel)
{
    action : stage EX1..EX2: r`EX2` = XD`EX2` = m`EX2`;
    syntax : r " = " m ", " m.1 ;
    image  : "0"::r::m;
}
```

```
opn store_ind_instr (m : ind_mem_acc, r : reg_sel)
{
   action : stage EX1..EX2: m`EX2` = XD`EX1` = r`EX1`;
   syntax : m " = " r ", " m.1;
   image  : "1"::r::m;
}
```

The OR mode ind_mem_acc passes along the attributes of its alternatives, and both the value and action attributes are instantiated by the parameter m in the action attribute of the load rule load_ind_instr and store rule store_ind_instr.

Note the timing specification for the memory access in the value attribute of mode ind_mem_acc1. *Tctcore* is assumed to have synchronous data memories with timing specification as follows. The address is applied in stage EX1, while the memory core access happens in stage EX2. The timing of the memory read port is stage EX2, as specified by XD'EX2' in AND rule load_ind_instr. For the memory write port, it is stage EX1, because of XD'EX1' in rule store_ind_instr. As explained in Section 4.3.3, instead of directly using bus XD, one may add intermediate transitories to access the memories, for a better modeling of memory port conflicts.

Note also the two syntax attributes for the mode rule ind_mem_acc1, and their use in the load and store instructions. m or m.0 refers to the first syntax attribute, m.1 refers to the second one. One can verify that the assembly syntax for an indirect load with binary encoding all-zeroes is: AX = DM[Ia0], Ia0 += 0.

4.4.8 Inherited Attributes

In contrast to the *synthesized attributes* described in Section 4.4.3, which can be evaluated during a bottom-up traversal of the nML hierarchy, so-called *inherited attributes* propagate information top-down in that hierarchy.

The latter attributes are used in an nML description to provide extra parameterizability to the nML rules, for example, to instantiate the same subtree of nML rules for a different part of the structural skeleton, for the concise description of parallel structures. This is not further elaborated in this chapter.

4.5 PIPELINE HAZARDS: STALLS AND BYPASSES

Pipeline hazards are artifacts of the pipelined execution of instructions. Following hazards are commonly identified in literature [2]:

- *Structural hazards* occur when a hardware resource (a transitory in nML) is used in different pipeline stages.

- *Data hazards* occur when a register is accessed in different stages by different instructions. A typical example is the read-after-write (RAW) hazard that occurs when a register is written in a late stage while a next instruction reads that register in an early stage.

■ *Control hazards* occur when jump instructions change the program counter.

A processor pipeline consists of an *instruction fetch pipeline*, that is, the instruction fetch stage (IF) optionally preceded by one or more prefetch stages, and an *execution pipeline*, starting with the instruction decode stage (ID) and one or more execute stages, which may include memory access and write-back stages.

In nML, the execution pipeline is explicitly modeled in the action attributes, while the instruction fetch pipeline is implicit for the compiler, with some nML properties identifying the program memory access stages. The detailed behavior of the instruction fetch pipeline is implemented in the processor controller module for instruction-set simulation and hardware generation.

4.5.1 Control Hazards

Control hazards mainly relate to the instruction fetch pipeline, and are solved by specifying abstract instruction properties. Consider, for example, the conditional jump instruction on the *tctcore*:

```
opn cond_branch (t : c_10, c : cond)
{
    action {
      stage EX1:
        tT = t;          // prepare direct absolute address
        switch (c){
        case EQ :
          tC = eq(AS);   // compute condition
          cjump(tC, tT);// decide actual jump
          ...
        }
    }
    syntax : "IF " c " JUMP TO " t;
    image  : c::t, cycles(3|2);
}
```

The Boolean condition is derived from status register AS in stage EX1, so the branch target can be computed by primitive operation cjump. Indeed, control instructions are modeled just like arithmetic instructions, with the only difference being that their action attributes contain primitive operations that modify the processor's control flow, like jumps, calls, and returns.

Fig. 4.2 shows the pipeline diagram for conditional jumps in different cases. In each of these cases, the branch target is computed in the EX1 stage of the jump instruction.

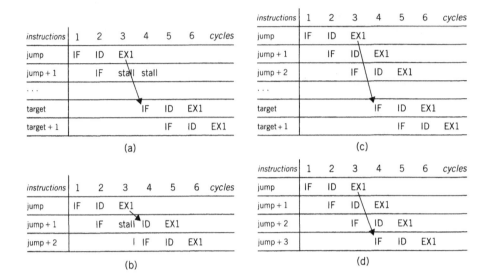

FIGURE 4.2

Pipeline diagram for conditional jumps: (a) taken; (b) not taken; (c) delayed, taken; (d) delayed, not taken.

If the jump is taken, see Fig. 4.2(a), the target instruction can only be fetched the next cycle and the pipeline must be stalled for two cycles. If not taken, see Fig. 4.2(b), the already fetched instruction in the incremental path can be issued to the decode stage, so the pipeline must be stalled for only one cycle. The total number of cycles for an instruction (1 + number of stalls) can be annotated to the image attribute with the cycles keyword. If needed, the taken and not-taken cases are separated by |, as in cycles(3|2).

For a so-called delayed jump [2] the next two instructions in the incremental path are always issued without stalling, both for the taken and the not-taken cases—see Fig. 4.2(c) and (d). The annotation to the image attribute would then be cycles(1), delay_slots(2).[1] Note that in some cases the compiler will have to fill in NOPs in the delay slots, which is not better than stalling.

4.5.2 Structural and Data Hazards

Structural and data hazards relate to the execution pipeline. They are analyzed from the pipeline stage timing in the action attribute and reported by the nML front-end tool.

The designer can explicitly specify in nML how to solve each of these hazards, using either a software stall (NOP insertion in the code), a hardware stall

---

[1]The default cycles(1) must not be annotated.

(interlocking), or a bypass (feed forward path) [2]. Hazards for which no solution is specified in nML are solved by software stalls, if a stall is needed.

The compiler always tries to solve hazards without cost, by reordering instructions or by using bypasses if these are present. It will only use stalls as a last resort.

The simulator and hardware generation tools will automatically implement the necessary hardware for bypasses and hardware stalls, based on their specification in nML.

Stalls

Most arithmetic instructions on *tctcore* write their result to register RR during stage EX1, so the next instruction can read it one cycle later during its own stage EX1. However, the (delayed) load and the multiply-accumulate instructions write to register file RR during stage EX2, so there is a data hazard for the following instruction during its stage EX1 when it wants to read their result. The instruction must be stalled for one cycle.

In nML, this stall condition is specified for a software stall by an sw_stall rule:

```
sw_stall 1 cycles () {
    stage EX2 : RR[#] = ...;
}
-> {
    stage EX1 : ... = RR[#];
}
```

The hazard rule applies to any two instructions matching the two instruction patterns specified between the braces. With ... being a wildcard, the hazard rule applies to any instruction writing to register file RR in EX2 followed by any instruction reading from the same address in RR in stage EX1. The *n* cycles annotation in the hazard rule header means it applies when there is an offset of *n* cycles between the start point of the two instructions, so $n = 1$ means that the two instructions directly follow each other. The "#" symbolizes the fact that both instructions must use the same address for the stall to trigger.

One may write very detailed instruction patterns, specifying the exact access transitory and add more register-transfer statements, so the rule matches only one or a few instructions, or write very broad patterns as shown in the example earlier. A pattern may even just indicate that a certain resource is occupied, for example, to model a structural hazard on a bus XD:

```
sw_stall 1 cycles () {
    rsrc(XD `EX2`);  // any instruction using XD in stage EX2
}
-> {
```

```
              rsrc(XD `EX1`);   // any instruction using XD in stage EX1
      }
```

To solve the hazard with a hardware stall, only the keyword sw_stall must be changed in hw_stall.

Bypasses

Bypassing or forwarding only applies to read-after-write (RAW) data hazards. Different from hardware or software stalls, bypassing or forwarding completely avoids the RAW data hazard, in this way avoiding any pipeline stalls.

Bypassing is typically applied in a processor where the operands are read and the result is written in a separate stage. Consider the following nML example, where pA, pB, pC are pipeline registers, tC a transitory, R a register file, and c_4u refers to a 4-bits constant:

```
opn add(dst : c_4u, src1 : c_4u, src2 : c_4u)
{
    action {
      stage ID:
        pA = R[src1];
        pB = R[src2];
      stage EX;
        pC = tC = pA + pB;
      stage WB:
        R[dst] = pC;
    }
    syntax : "r"dst "= r"src1 " + r"src2;
    image : dst::src1::src2;
}
```

Without any bypassing, a 3-cycle offset is needed between two data-dependent additions. Fig. 4.3(a) illustrates this. The first instruction updates r3 in stage WB in cycle 4. Only the last instruction (started after 3 cycles) reads the updated r3 value in stage ID in cycle 5. By adding bypass paths, the two next instructions can already read the addition result to be written in r3. These bypasses in the data path structure are illustrated in Fig. 4.3(b): two paths for the left operand and two paths for the right operand. The left multiplexer selects either the tC transitory (1-cycle offset between dependent instructions), the output of the pC pipe (2-cycle offset between dependent instructions), or the output of the register file R[] (minimal offset of 3 cycles). In nML, the two bypass paths for the left operand are specified by means of the following bypass rules:

```
bypass 1 cycles () {
    stage EX: pC = #tC;      // starts at tC
    stage WB: R[#] = pC;
```

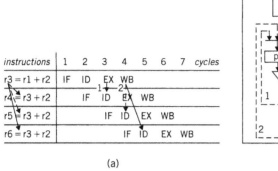
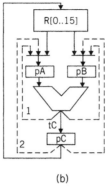

(a)   (b)

FIGURE 4.3

(a) Bypasses avoiding read-after-write data hazards and (b) the bypasses in the data path structure.

```
} -> {
    stage ID: #pA = R[#];   // ends at input of pA
}

bypass 2 cycles () {
    stage WB: R[#] = #pC;   // starts at output of pC
} -> {
    stage ID: #pA = R[#];   // ends at input of pA
}
```

A bypass rule has the same structure as a stall rule. The first example rule applies for 1-cycle offset between the two instructions, the second for a 2-cycle offset. A bypass rule only applies when the two instructions access the same register field, as expressed by the "#" symbol between the register file square brackets. In addition, in a bypass rule the "#" symbol is used to specify the start- and end-point (transitory or pipe) of the bypassed path, specified in the two instruction patterns.

4.6 THE EVOLUTION OF nML

The first version of the nML description formalism was developed at the Technical University of Berlin, in the early 1990s [3]. At first it was intended to specify processors for a simulator generator [4]. Later, an attempt was made to use it in the context of a compiler [5]. Independently from this project, that version of nML was recognized in Reference [6] to be well suited as a processor specification for retargetable instruction-set simulators and assemblers, but its use for retargetable compilation was questioned, mainly because a usable nML description for simulation typically contained too much detail for compilation.

Several other research projects in the 1990s were inspired by nML, mostly focussing on instruction-set simulation.

In the mid 1990s, while doing research on retargetable compilation at Imec, we redesigned nML to specify precisely the information needed in the instruction-set graph (ISG) processor model [7], used for compilation, simulation, and hardware generation.

From 1997 onward, the development of nML was continued by Target Compiler Technologies, mainly focussing on broadening the architectural scope supported by the language and adding more support for C compilation and hardware generation.

The main differences between the very first version of nML and the current version described in this chapter, are as follows:

- In the original version of nML, no distinction was made between storage elements; all were called memory. To make nML suitable for retargetable compilation and hardware generation, the concept of a *structural skeleton* with *memories*, *registers*, and *transitories* has therefore been introduced, on which hardware conflicts can be checked [8]. This skeleton ties an nML description to the data path of the target processor where needed.

- The concept of *mode rules* had to be rethought to incorporate read and write ports, and to make the use of modes consistent with the modeling of hardware conflicts.

- In the original nML language a number of C-like primitive operations and data types were predefined. There was an escape route to describe processor-specific primitive operations by function calls, but exactly the same function call then had to appear in the application program, which is often inconvenient.

  In the current version of nML, the specification of primitive operations—which can directly be performed on the processor—has therefore been separated from the nML description. They are specified in C++ syntax in the so-called processor header file. The primitive operations are abstract labels for the compiler, and the behavioral information needed for simulation is encapsulated in their behavioral models.

  In our approach, a so-called compiler header file contains the mapping of C/C++ built-in data types and operators to the processor data types and primitive operations.

- The original version of nML had no support for pipelined processors. We added pipeline stage annotation to the action attributes, and developed the concept of hazard stall and bypass rules, to specify how pipeline hazards must be solved in hardware or in software.

- A number of other refinements, both on the syntactic and on the semantic level, are less important, but were nevertheless needed to make an nML description really usable as a processor model for a broad range of architectures to be used for compilation, simulation, and hardware generation.

The main idea behind all enhancements is to guide the designer to specify the right amount of hardware details that is relevant for the tools to generate high-quality results.

4.7 A RETARGETABLE TOOL SUITE FOR ASIPs

This section introduces IP Designer (Chess/Checkers), an industry proven retargetable tool suite, enabling the design of application-specific processors (ASIPs) in multi-core Systems on a Chip [9]. It supports all aspects of ASIP-based design: architectural exploration and profiling; hardware generation and verification; and— last but not least—software development kit (SDK) generation based on highly optimizing C compilation, instruction-set simulation, and debugging technology. Using nML, an architecture designer can quickly define the instruction-set architecture of a processor. After reading the nML description, the different IP Designer tools are automatically targeted to the specified architecture. Fig. 4.4 gives an overview of the tool-set.

4.7.1 Chess: A Retargetable C Compiler

Chess is a retargetable C compiler that translates C source code into machine code for the target processor. Different from conventional compilers such as GCC [10],

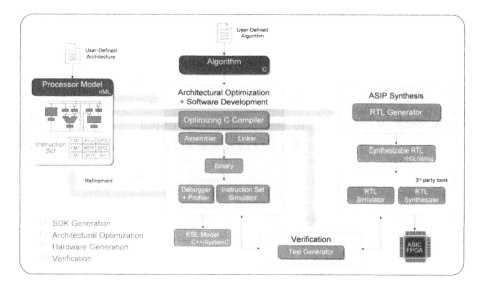

FIGURE 4.4

IP Designer (Chess/Checkers) retargetable tool suite.

Chess uses graph-based modeling and optimization techniques [7] to deliver highly optimized code for specialized architectures exhibiting peculiarities such as complex instruction pipelines, heterogeneous register structures, specialized functional units, and instruction-level parallelism. The compiler comes with a retargetable assembler and disassembler, and a retargetable linker.

4.7.2 Checkers: A Retargetable Instruction-set Simulator Generator

Checkers is a retargetable instruction-set simulator (ISS) generator that creates a cycle and bit accurate ISS for the target processor. The ISS can be run in a standalone mode or be embedded in a cosimulation environment through an application programming interface (API), optionally including SystemC wrappers. Checkers includes a graphical debugger that can connect both to the ISS, and to the available processor hardware for on-chip debugging. The connection is made via a JTAG interface. Source-level debugging and code profiling are supported.

4.7.3 Go: A Hardware Description Language Generator

Go is a hardware description language (HDL) generator that produces a synthesizable register-transfer level (RTL) model of the target processor core in VHDL or Verilog. The high-level view on the data path, which is present in the action attributes in nML, is translated into a net list of RTL blocks that can easily be linked to their origin in nML. If needed, the user can thus optimize the RTL model by modifying the structural information in the action attributes and in the structural skeleton.

Through an extensive list of configuration options, Go is very flexible in generating the RTL code according to the designer's requirements. For example, users can plug in their own implementations of functional units and of the memory architecture. Go can also optimize for low power, for example, by applying clock gating to selected registers, and operand isolation to functional units [11]. A special debug infrastructure for on-chip debugging via JTAG can be generated.

4.7.4 Risk: A Retargetable Test-program Generator

Risk is a retargetable test-program generator that can generate assembly-level test programs for the target processor with a high fault coverage, suited for simultaneous execution in the ISS and in the RTL HDL model.

4.7.5 Broad Architectural Scope

The nML front end of the IP Designer tool suite translates the processor description into an intermediate representation called instruction-set graph (ISG) [7] that captures all essential structural and timing information. All optimizations in the retargetable C compiler, instruction-set simulator, and hardware generator directly operate on this detailed hardware representation. As a result, IP Designer can support a very broad range of processor architectures. The tools enable true architectural exploration, beyond many of the limitations of configurable templates offered by intellectual property vendors. Reference [12] illustrates the architectural exploration process with IP Designer (Chess/Checkers), based on profiling information generated by the tools.

The following list of architectural options illustrates the broad architectural scope of the IP Designer tool suite:

- Data types: Using the power of C++, any user-defined data type can be supported, including, for example, integer, fractional, floating-point, Boolean, complex, and vector (SIMD) data types.

- Arithmetic functions: In addition to the built-in functions and operators of the C language, user-defined (primitive) functions are supported. The compiler supports function overloading, as well as calling user-defined functions as intrinsics in the C source code.

- Registers and interconnect: Special purpose register sets and irregular interconnection schemes can be specified in nML, and are well supported by the compiler thanks to innovative register allocation techniques.

- Memory and I/O: Both Von Neumann and Harvard architectures are supported. There is no limitation to the number of data memories or memory ports that can be specified. External memory and communication models with data-dependent latencies can be included via a memory interface API. In the ISS, this API is based on SystemC. The compiler automatically exploits a wide variety of addressing modes.

- Instruction word: Both orthogonal (VLIW) and encoded instruction sets can be specified and are supported by all tools. Also, variable-length instructions are supported. Powerful data flow analysis, scheduling, and software pipelining techniques in the compiler aim at exploiting the available instruction-level parallelism.

- Instruction pipeline: The instruction pipeline is accurately described in nML, including pipeline hazards and register bypasses. There is no limitation on the pipeline depth. The compiler ensures that the generated code is hazard free. Alternatively, the tools support hardware stalling and bypasses to avoid pipeline hazards. In this case the required interlocking and bypass hardware is automatically generated in the ISS and synthesizable RTL code.

- Program control: Supported program control features include: subroutines, interrupts, hardware do-loops, residual control (i.e., mode-dependent behavior), and predication (i.e., conditional execution of instructions). Subroutines can be implemented even in the absence of a software stack. The compiler supports optimized context switching, including inter-procedural code optimizations.

4.8 DESIGN EXAMPLES

The IP Designer tool suite has been applied to design and program processors for a variety of applications. This section presents a selection of designs made with IP Designer, illustrating typical design trade-offs made in each application domain.

4.8.1 Portable Audio and Hearing Instruments

NXP Semiconductors' CoolFlux DSP [13] is a representative example from this application domain.

While typical data rates are low enough to permit the use of general-purpose DSPs, a major challenge comes from the ultra-low power requirement of the targeted battery-powered systems, combined with the flexibility requirement imposed by multiple coding standards and a growing variety of advanced audio algorithms.

The best compromise is an ASIP that has a general-purpose instruction set, augmented with domain-specific instructions with a high degree of instruction-level parallelism. CoolFlux DSP supports up to eight parallel operations in its most parallel instructions. As a result, most audio codecs can be executed in significantly fewer instruction cycles than on a general-purpose DSP, which provides the necessary headroom for reducing the processor's clock frequency and supply voltage, thus reducing dynamic power consumption. Additionally, a compact architecture with a small register set and an encoded instruction word, as well as an optimized RTL HDL design, contribute to CoolFlux DSP's ultra-low power consumption of 60 μW/MHz at 1.2 V (130 nm CMOS). CoolFlux DSP is available with a rich software library of audio and telecom applications, developed by NXP using the Chess compiler.

4.8.2 Wireline Modems

The discrete multi-tone (DMT) engine in STMicroelectronics' Astra+ chip set for ADSL2+ customer-premises equipment modems [14] is an example from this application domain.

This application uses advanced equalization algorithms, requiring specialized computations on complex numbers, that must sustain the system's high data rate of 24 Mbps. Because of this high throughput and the nonconventional data types, such an application, would traditionally be implemented as a fixed function ASIC core, in which the required algorithmic data flow patterns are directly mapped into hardwired data paths controlled by a finite-state machine (FSM). However, by adding a few registers and multiplexers, and by replacing the FSM by a microcoded engine, a programmable ASIP was derived that offered more flexibility at almost the same cost as a fixed-function core. Such an ASIP does not offer general-purpose programmability, but it can support restricted algorithmic variations within the same application domain. The DMT engine ASIP was designed in nML and programmed using the Chess compiler. The RTL HDL implementation was automatically generated using the Go tool. The first generation of the DMT engine ASIP was only intended for the ADSL standard. A second generation, additionally supporting the ADSL2 and ADSL2+ standards, was manufactured only six months later. This short turnaround time can be attributed to the software programmability offered by the underlying ASIP architecture [14].

4.8.3 Wireless Modems

The vector-processing ASIP for an HSDPA receiver presented by Nokia [15] is an example from this application domain.

Baseband systems of next-generation wireless modems are good candidates for an ASIP implementation. The desire to build multi-standard systems necessitates the use of programmable solutions. Software-defined radio is a term that well describes this trend. However, standard processors cannot be used, due to the stringent data rate requirements (e.g., 14.4 Mbps for 3GPP HSDPA, up to 100 Mbps for IEEE 802.11n), requiring specialized computations on complex numbers. Parallel processing is the key for these applications. The OFDM-based systems provide inherent data-level parallelism because of the requirement to process multiple subcarriers. The MIMO-based systems provide options for parallelization in matrix arithmetic.

The HSDPA receiver ASIP presented in [15] is a vector (SIMD) processor that directly takes advantage of the algorithm's data-level parallelism. Each processing element implements general-purpose as well as application-specific functions, using complex number arithmetic. Vector data types are directly specified in the source code for the Chess compiler, and function overloading is applied to these types. Note that Chess does not automatically translate scalar C code into vectorized code. This is not considered a problem by application engineers, who are well aware of the parallel nature of their algorithms. The customization of the HSDPA receiver ASIP resulted in significant area and power savings compared to a more general-purpose vector processor.

4.8.4 Video Coding

Texas Instruments' video coding accelerator [16] is representative of this application domain.

The existence of multiple standards also suggests programmability, while the evolution to higher definitions imposes increasing computational speed requirements. A vector ASIP was designed that takes advantage of the inherent data-level parallelism of pixel processing algorithms.

4.8.5 Network Processing

Reference [17] describes the design of an ASIP for packet processing in a 10 Gbps Ethernet-based DSL access multiplexer. Interestingly enough, due to the economics of this market, this design targets an FPGA rather than an ASIC implementation.

Yet, the use of ASIPs implemented as soft cores in the FPGA logic proves to be an effective solution to obtain an efficient level of resource sharing in the FPGA.

Flexible packet processing requires a specialized data memory architecture with dual address generation units supporting 8-, 16-, and 32-bit access, and powerful low-latency branch instructions. The ASIP has a VLIW architecture, featuring a high degree of instruction-level parallelism to accelerate packet processing tasks. To achieve a 10 Gbps data rate, task-level parallelism is additionally employed: the FPGA contains ten identical instances of the same ASIP, for parallel processing of independent packets.

4.9 CONCLUSIONS

We presented the nML architecture description language that models a processor in a concise way for a retargetable processor design and software development tool suite. This tool suite, named IP Designer, offers fast architectural exploration, hardware synthesis, software compilation, instruction-set simulation, and verification.

To yield concise descriptions, the instruction set is described hierarchically in nML by means of a grammar, with OR rules to specify alternative instruction parts or formats and AND rules to specify orthogonal instruction parts.

nML captures the instructions' behavior in a compact register-transfer model, exposing the exact resource utilization and pipeline of the processor. This formalism allows for an accurate description of structural and timing irregularities that are typical of many application-specific processors, and is at the basis of patented compilation simulation and hardware generation techniques used in the IP Designer (Chess/Checkers) tool suite.

nML hazard rules provide efficient solutions for pipeline conflicts, either by stalls or bypasses. Their compact notation gives the designer full control on how to handle pipeline hazards, to play with the hardware–software trade-off, while being relieved from the detailed hardware implementation of interlocking and feed forward paths.

The nML language has been shown to contain the right amount of hardware knowledge as required by each of these tools for high-quality results.

This was supported by presenting designs from five different application domains, illustrating the capabilities of nML and the IP Designer tool suite for a broad range of architectures.

REFERENCES

[1] nML description of the tctcore example processor. *http://www.retarget.com/nml*, March 2008.

[2] J. L. Hennessy and D. A. Patterson. *Computer Architecture: A Quantitative Approach (3rd ed.)*. Morgan Kaufmann Publishers, San Mateo, California, 2003.

[3] M. Freericks. The nML machine description formalism. Technical Report 1991/15, Computer Science department, T.U. Berlin, Berlin (Germany), 1991.

[4] F. Löhr, A. Fauth, and M. Freericks. Sigh/Sim—an environment for retargetable instruction set simulation. Technical Report 1993/43, Computer Science department, T.U. Berlin, Berlin (Germany), 1993.

[5] A. Fauth and A. Knoll. Automated generation of DSP program development tools using a machine description formalism. In *Proc. of the IEEE International Conference on Acoustics, Speech and Signal Processing (ICASSP)*, pages 457–460, April 1993.

[6] M. R. Hartoog, J. A. Rowson, P. D. Reddy, S. Desai, D. D. Dunlop, E. A. Harcourt, and N. Khullar. Generation of software tools from processor descriptions for hardware/software codesign. In *Proc. of the Design Automation Conference (DAC-97)*, pages 303–306, Anaheim, California, 1997.

[7] J. Van Praet, D. Lanneer, W. Geurts, and G. Goossens. Processor modeling and code selection for retargetable compilation. *ACM Transactions on Design Automation of Electronic Systems*, 6(3):277–307, July 2001.

[8] A. Fauth, J. Van Praet, and M. Freericks. Describing instruction set processors using nML. In *Proc. of the European Design and Test Conference (EDTC)*, pages 503–507, 1995.

[9] Chess/Checkers: a retargetable tool-suite for embedded processors—technical white paper. *http://www.retarget.com*, 2003.

[10] R. Stallman. Gnu compiler collection internals. *http://gcc.gnu.org*.

[11] G. Goossens, J. Van Praet, D. Lanneer, and W. Geurts. Ultra-low power? Think multi-ASIP SOC! In *Proc. of the IP07 Conference*, pages 549–554, Grenoble, December 2007.

[12] W. Geurts, G. Goossens, D. Lanneer, and J. Van Praet. Design of application-specific instruction-set processors for multi-media, using a retargetable compilation flow. In *Proc. of the International Signal Processing Conference (GSPx)*, Santa Clara, October 2005.

[13] H. Roeven, J. Coninx, and M. Adé. Coolflux DSP—the embedded ultra low power C-programmable DSP core. In *Proc. of the International Signal Processing Conference (GSPx)*, September 2004.

[14] L. Dawance. To a more powerful DSP based on TCT technology for ADSL+. In *DSP Valley Annual Research and Technology Symposium*, October 2003.

[15] K. Rounioja and K. Puusaari. Implementation of an HSDPA receiver with a customized vector processor. In *Proc. of the International Symposium on System-on-Chip*, pages 20–23, Tampere, Finland, November 2006.

[16] H. Maréchal. ASIP design methodology with Target's Chess/Checkers retargetable tools. In *Proc. of the International Signal Processing Conference (GSPx)*, October–November 2006.

[17] K. Van Renterghem, P. Demuytere, D. Verhulst, P. Vandewege, and X.-Z. Qiu. Development of an ASIP, enabling flows in ethernet access using a retargetable compilation flow. In *Proc. of the DATE Conference*, pages 1418–1423, Nice, France, April 2007.

LISA: A Uniform ADL for Embedded Processor Modeling, Implementation, and Software Toolsuite Generation

Anupam Chattopadhyay, Heinrich Meyr, and Rainer Leupers

To reduce the complexity of processor design, researchers proposed an abstract modeling of the modern embedded processor via Architecture Description Languages (ADLs). The abstract processor model, in effect, serves as an executable specification for early design space exploration. The Language for Instruction Set Architecture (LISA), an uniform ADL, is elaborated in this chapter. LISA supports automatic generation of software toolsuite (C Compiler, assembler, linker, simulator, profiler) and optimized RTL description within a short time, accelerating the design space exploration. LISA is based on the concept of designing from scratch, allowing complete freedom to the designer. Various aspects of LISA specification and the tools built around it are discussed in this chapter. The chapter begins with the LISA specification (Section 5.1). This is followed by a study of automatic generation of retargetable software tools from LISA (Section 5.2) as well as an integrated flow with application-driven processor synthesis tools. The processor synthesis tool-flow consists of an architecture-independent application profiling tool and automatic custom instruction-set identification capability. The RTL implementation of the processor from ADL specification is finding increasing usage in accurate performance estimation, verification (via equivalence checking), and final tape-out. The methods for utilizing the high-level structural information from ADL to derive an optimized RTL implementation (Section 5.3) is elaborated in this chapter. A generic intermediate representation during RTL implementation from LISA allows to plug in architecture-dependent as well as architecture-independent optimizations. Verification of the processor is a growing challenge, in view of the rising Non-Recurring Engineering (NRE) cost and shortening time-in-market window. Section 5.4 in this chapter outlines the LISA-driven processor verification aspect. In today's systems,

95

it is commonplace to find multiple heterogeneous processors together with other components. Early estimation of the system performance using these processors is extremely useful. This chapter also covers this facet of LISA in perspective of system-level simulation and exploration (Section 5.5). Processor design, in general, is a fast evolving area. To allow the designer to play around with the processor—enabling modeling of novel and complex processor structures—is an important facility. The recent developments of LISA as a language are discussed in Section 5.6. Finally, several case studies are presented to show the principles of modern processor design using LISA, in Section 5.7.

5.1 LANGUAGE-BASED ASIP MODELING

The interdependent design points of ASIP make it nearly impossible to explore parts of the design space without considering the whole. Additionally, there are numerous design decisions which are difficult to parameterize. These contribute to processor design complexity, thereby paving way for several ADLs in the recent past to gain strong acceptance. This section provides a description of the language LISA (Version 2.0) [1]. Followed by this, the methodology of LISA-driven design space exploration is elaborated.

5.1.1 Intuitive Modeling Idea

Before starting with the language semantics, it is useful to understand the key ideas of the ASIP modeling. In LISA, the complete ASIP is first viewed as an Instruction Set Architecture (ISA). The ISA captures the syntax, encoding, and the behavior of the instructions. Such a model is referred as purely a behavioral model. On top of this description, the structural information is added. Alternatively, the ISA is mapped to a structural description. The structural information includes a timing model for each instruction, clocked behavior of registers, and many more such features. This process of ASIP modeling is shown in Fig. 5.1. This stepwise formation of the ASIP also reflects the increasing accuracy of the description. Starting from an instruction-accurate ISA-only description, the designer proceeds toward cycle-accurate and then phase-accurate processor description.

5.1.2 ISA Modeling

The components of LISA language used for ISA modeling are described in the following paragraphs.

LISA operation DAG

The ISA comprises of syntax, encoding, and behavior of processor instructions. LISA uses a modular modeling style for capturing the ISA. In LISA, each instruction is distributed over several operations. *LISA operations* act as basic components, which

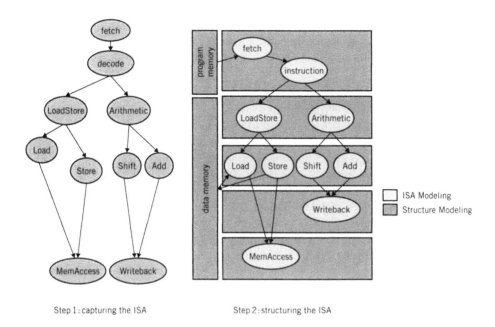

Step 1: capturing the ISA   Step 2: structuring the ISA

FIGURE 5.1

Development of ASIP model.

are connected together by *activation*. The operation contains behavior, syntax, and encoding. One common operation can be activated by multiple parent operations. Again, one operation can activate multiple children operations.

The complete structure is a Directed Acyclic Graph (DAG) $\mathcal{D} = \langle V, E \rangle$. V represents the set of LISA operations, E the graph edges as set of child–parent relations. These relations represent *activations*. The entire DAG can have a single or multiple LISA operations as root(s). A group $\mathcal{G}$ of LISA operations is defined as $\mathcal{G} := \{P|P \in V\}$ such that the elements of P are mutually exclusive to each other.

Behavior description

LISA operations' *behavior section* contains the behavior description. The behavior description of a LISA operation is based on the C programming language. By the behavior description, the combinatorial logic of the processor is implemented. In the behavior section, local variables can be declared and manipulated. The processor resources, declared in a global *resource* section, can be accessed from the behavior section as well. Similar to the C language, function calls can be made from LISA behavior section. The function can be an external C function, or an internal LISA operation. In case of a function call to LISA operation, it is referred as *behavior call*. The behavior call can be made to either a single *instance* of LISA operation or a *group* of LISA operations. Though referred as a group, the behavior call is eventually made to only one member of the group, the one being currently decoded. The grouping reflects the exclusiveness of the member LISA operations.

An exemplary LISA operation DAG with different root operations is shown in Fig. 5.2.

Instruction encoding description

LISA operations' *coding section* is used to describe the instructions' encoding. The encoding of a LISA operation is described as a sequence of several *coding* elements. Each coding element is either a terminal bit sequence with "0,""1,""don't care" bits, or a nonterminal. The nonterminal coding element can point to either an *instance* of LISA operation or a *group* of LISA operations. The behavior of a LISA operation is executed only if all terminal coding bit patterns match, all nonterminal instances match, and at least one member of each group matches. The root LISA operation containing a coding section is referred as the *coding root*. Special care has to be taken for the description of the coding root(s). A LISA model may have more than one coding root, for instance, for the ability to use program code with different

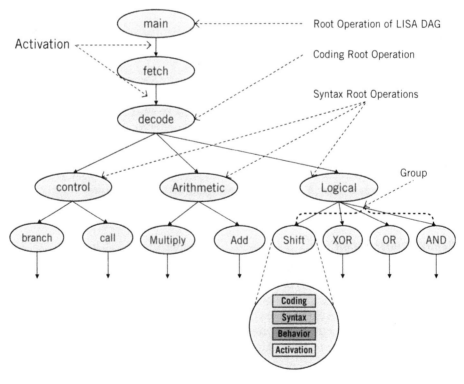

FIGURE 5.2

LISA operation DAG.

instruction word sizes. This set of coding roots $\mathcal{R}_c$ contains coding roots that are mutually exclusive to each other. For RISC architectures with a fixed instruction word, size $\mathcal{R}_c$ contains only a single element. For VLIW architectures, each coding root $r \in \mathcal{R}_c$ decodes one from a set of parallel instructions.

Instruction syntax description

The *syntax section* of LISA operation is used to describe the instructions' assembly syntax. It is described as a sequence of several *syntax* elements. A syntax element is either a terminal character sequence with "ADD," "SUB," or a nonterminal. The nonterminal syntax element can point to either an *instance* of LISA operation or a *group* of LISA operations. The root LISA operation containing a syntax section is referred as the *syntax root*.

5.1.3 Structure Modeling: Base Processor

LISA resources

LISA resources consist of general hardware resources for storage and structuring such as memory, registers, internal signals, and external pins. Memory and registers provide storage capabilities. Signals and pins are internal and external resources without storage capabilities. The LISA resources can be parameterized in terms of sign, bit-width, and dimension. LISA memories can be more extensively parameterized. There the size, accessible block size, access pattern, access latency, and endian-ness can be specified. The LISA resources are globally accessible from any operation. Resources can be designated as clocked indicating that it is updated at the beginning of each clock cycle. Alternatively, a latch-like behavior is inferred.

The LISA language allows direct access to the resources multiple times in every operation. From inside the behavior section, the resources can be read and written like normal variables. Memories are accessed via a predefined set of interface functions. These interface functions comprise of blocking and nonblocking memory access possibilities.

LISA pipeline

The processor pipeline in LISA can be described using the keyword PIPELINE. The stages of the pipeline are defined from left to right in the actual execution order. More than one processor pipelines can be defined. The storage elements in between two pipeline stages can be defined as the elements of PIPELINE_REGISTER. With the pipeline definition, all the LISA operations need to be assigned in a particular pipeline stage. An exemplary LISA pipeline definition is shown in Fig. 5.3.

Activations

Activations schedule the execution of LISA operations. In an instruction-accurate LISA model, the activations are simply triggered along the increasing depth of the LISA operation DAG. For a cycle-accurate LISA model, the activations are triggered

```
RESOURCE {
  PIPELINE pipe = {FE; DC; EX; WB };
  PIPELINE_REGISTER IN pipe
  {
      unsigned int address1;
      unsigned int address2;
      unsigned int operand1;
      unsigned int operand2;
      unsigned int result;
  }
}
```

```
OPERATION fetch in pipe.FE
{
    BEHAVIOR { .. }
    ACTIVATION { decode }
}

OPERATION decode in pipe.FE
{
    CODING { .. }
    BEHAVIOR { .. }
    ACTIVATION { .. }
}
```

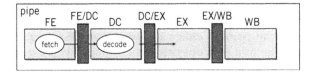

FIGURE 5.3

Pipeline definition in LISA.

according to the pipeline stage assignment of an operation. A LISA operation can activate operations in the same or any later pipeline stage. Activations for later pipeline stages are delayed until the originating instruction reaches the stage.

An activation section contains a list A_L of activation elements A_e. All elements $A_e \in A_L$ are triggered concurrently. A_e may be a group, a LISA operation, or a conditional statement. For groups, only the correctly decoded LISA operation is activated. Conditional activations $A_c = \langle A_{if}, A_{else} \rangle$ are IF/ELSE statements that again consist of the activation elements A_{if} and A_{else}. Therefore, any level of nested conditional activation is possible. Within the conditional statement of an activation, the current state of LISA resources can be checked, giving a high degree of flexibility to the processor designer.

5.1.4 Levels of Abstraction in LISA

The design space exploration of a processor often begins by identifying some key processor instructions. Subsequently, the processor resources are designed and several custom processor features are integrated. This method of iterative improvement on a basic processor is strongly supported by LISA. LISA enables a designer to develop the processor at various levels of abstraction. While a high level of abstraction allows the designer to perform an early exploration, decreasing abstraction levels permits making more accurate measurements and proceeding toward the implementation. In LISA, abstraction can take place in two orthogonal domains, namely, *spatial* abstraction and *temporal* abstraction.

Fig. 5.4 captures different levels of abstraction in the two aforementioned domains during processor development. For spatial domain of abstraction, a designer may begin with pseudo resources and can improve it stepwise to pipeline definition, clocked register modeling, and interrupt modeling. For temporal domain

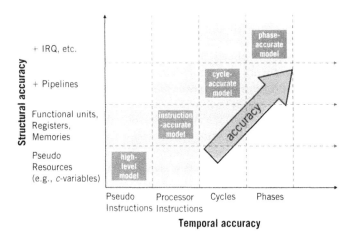

FIGURE 5.4

Levels of abstraction: spatial and temporal.

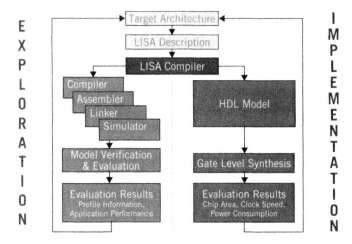

FIGURE 5.5

LISA-based processor design exploration and implementation.

of abstraction, the instructions can begin with a wholesome behavior description; improving toward pipelined model and finally attaining phase-based execution.

5.1.5 **LISA-based Processor Design**

With the aid of LISA, the designer can attempt the complex task of processor modeling, exploration, and implementation in the way suggested in Fig. 5.5. The processor design is viewed as the collection of two iterative tasks, exploration, and implementation. The role of exploration is to indicate the performance of the processor while executing a given target application. The performance is measured by cycle count during high-level instruction-set simulation. The role of implementation is to provide a processor description (e.g., synthesizable RTL) so that it can be processed with commercial physical synthesis flow. The RTL implementation of the processor can also be used to derive the performance estimation in terms of power, area, and clock frequency. Therefore, the implementation task can be viewed as an extended exploration with increased estimation accuracy.

The LISA exploration platform aids the processor designer by generating the following tools/descriptions during processor exploration and implementation phase.

- **C compiler** C is one of the most common High Level Language (HLL) used for programming embedded applications. A C compiler is required to translate the C description to assembly language (of the processor's ISA). LISA allows semiautomatic generation of C compiler from the LISA description [2, 3].

- **Assembler** The role of an assembler is to transform the human readable assembly language to the processor-specific binary object code.

- **Linker** The linker combines binary object code, which is emitted by the assembler, with processor's data memory contents. This results into the executable for the processor.

- **Instruction-set simulator** An instruction-set simulator (ISS) serves a dual purpose of debugging and profiling during processor exploration phase. The ISS, automatically generated from a LISA description [4], possess extensive profiling capabilities like loop count, instruction coverage, and resource utilization statistics usable for both the application and the architecture. The debugging features such as breakpoint and LISA-level line stepping provide a way to quickly fix the errors in the application or architecture modeling.

- **Synthesizable RTL** Apart from the software toolsuite, LISA supports generation of synthesizable, optimized RTL description in popular HDL formats (VHDL, Verilog) along with corresponding simulation and synthesis scripts [5–7].

5.2 AUTOMATIC SOFTWARE TOOLSUITE GENERATION

In this section, the two single-most important tools for processor design exploration, C compiler and instruction-set simulator, are elaborated. This discussion is followed by some peripheral tools, which are deployed together with LISA first, for the derivation of custom instructions from a high-level application, and second, to determine the instruction opcodes automatically during ISA design.

5.2.1 Instruction-set Simulator

Depending on runtime and memory restrictions, four different types of instruction-set simulation are supported by LISA. These are as described here.

Interpretive simulation

An interpretive simulator can be considered as a virtual processor. The simulator completely mimics the functionality of a processor, as it is implemented in the most simple hardware form. Every instruction is fetched from the memory, decoded in the runtime, and executed. The simulation steps are performed at runtime, providing maximum flexibility. With this flexibility comes the performance overhead. Clearly, the control flow modeled in software consumes significant runtime rendering the interpretive simulation slow in comparison to other simulation techniques. The advantage is that it consumes much less memory due to its runtime execution.

Compiled simulation

To speedup the simulation performance, compiled simulation technique can be adopted for LISA-based ISS. For compiled simulation, the fetching and decoding of an instruction is done prior to the simulation run. The results of decoding are stored beforehand and utilized during simulation. This technique improves the simulation speed by omitting the recurring fetch and decode operations. In a sense, this is exactly opposite to what is provided by interpretive simulation. This limits flexibility by not allowing dynamic memory updates. Furthermore, in order to store the complete application's decoded results, the compiled simulator's memory requirement on the host machine gets huge.

Just-In-Time Cache Compiled Simulation (JIT-CCS)

The JIT-CCS is introduced with the goal of combining the advantages of both the aforementioned simulation techniques and possibly to determine the fine balance between simulation speed and host memory requirements. For JIT-CCS, the processor instructions are fetched and decoded prior to a simulation step. The decoding results are stored in a software cache. In every subsequent simulation step, the existing decoding results in the cache are searched. Depending on the degree of locality in an application, the decoding result can be reused—thereby saving valuable simulation time. More important, the size of the software cache in JIT-CCS can be varied from 1 to 32,768 lines, where each line corresponds to a decoded instruction. The maximum cache size corresponds to 16 MB on the host machine. Even then, the memory requirement of JIT-CCS is much smaller compared to that in a compiled simulation.

The effect of JIT-CCS in bridging the compiled and interpretive simulation is reflected in Fig. 5.6. In this figure, the performance of JIT-CCS for an ARM7 core is captured. The host machine used for this simulation is a 1200 MHz Athlon PC, 768 MB RAM running the Microsoft Windows 2000 operating system. In the figure, a cache size of one line basically means that the JIT-CCS is effectively performing like an interpretive simulator. Every instruction is fetched, decoded, and executed, as

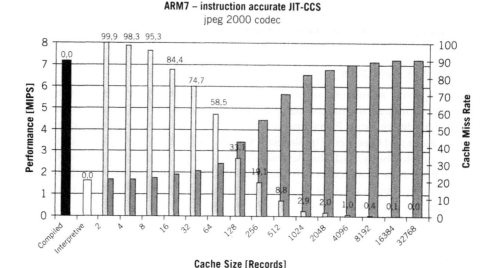

FIGURE 5.6

Performance of the Just-In-Time Cache Compiled Simulation [4].

the results for more than one instruction cannot be stored in such a small software cache. With increasing cache size, the JIT-CCS performance (depicted using wide bars) draws closer to that of compiled simulation and reaches a point of saturation. Another interesting point to note is that with increasing simulation speed the miss rate for the software cache is decreasing (shown with narrow bars). This conclusively presents a correlation between the simulation speed and the cache size, which can be fine-tuned using JIT-CCS technique.

Hybrid simulation

Normally, a software developer does not need high simulation accuracy all the time. Some parts of an application just need to be functionally executed to reach a zone of interest. For those parts, a fast but inaccurate simulation is completely sufficient. For the region on interest, however, a detailed and accurate simulation might be required to understand the behavior of the implementation of a particular function. The Hybrid Simulation (HySim) [8] concept—a new simulation technique—addresses this problem. It allows the user to switch between a detailed simulation, using previously referred ISS techniques, and direct execution of the application code on the host processor. This gives the designer the possibility to trade simulation speed against accuracy. The tricky part in hybrid simulation is to keep the application memory synchronous between both simulators. By limiting the switching to function borders, this problem becomes easier to handle. At a function border only global variables and function parameters need to be passed from one simulation to the other. By knowing the calling convention of both C compilers employed (i.e., host and target compilers) this can be easily implemented. Local variables are confined to functions, thus, they do not need to be synchronized between both simulators. In order to identify global variable accesses, the application source code is statically instrumented, and this guarantees a synchronous memory state. From the user's perspective, HySim behaves just like a standard processor simulator. Besides the normal simulation functionality it provides an additional feature: the *fast forward break-point*. By setting this breakpoint the designer indicates the point(s) of interest in the entire application to the simulator. Thus, HySim will execute as many functions as possible directly on the host to reach the desired state quickly. After reaching the fast forward breakpoint, the user can continue to debug his program using the ISS in order to obtain detailed information for debugging or performance evaluation. For evaluating the number of simulation cycles during host-based execution, a cost annotation is done [9].

5.2.2 The Compiler Designer

The large pool of software developers for embedded systems design their software using a high-level language, typically C. It is immensely beneficial to have the C compiler ready for the ASIP for several reasons. First, the task of software development becomes easier with C. Otherwise, the developers have to painstakingly learn the machine-specific assembly language. For large applications, the complexity of assembly-level programming is prohibitive. Second, the finer traits of the processor architecture must be effectively exploited by the assembly programmer to get maximum benefit out of it. The compiler can contain both target-independent and processor-specific optimizations to help the processor designer accurately estimate the performance of the application. To keep the compiler in the architecture-exploration loop requires it to be easily retargetable to the changing processor configurations. The compiler, derived from LISA processor description, depends on CoSy compiler development system from ACE [10]. The modular design style of CoSy allows various configuration possibilities to generate production quality compiler for a wide range of processors. A LISA description is subjected to a semiautomatic GUI-based tool (referred as Compiler Designer) for generating Code Generator Description (CGD) files. CoSy generates the compiler from these CGD files. Within the CGD, following things are embedded:

- Target processor resources, e.g., registers, functional units.
- A scheduler table describing instruction latencies and resource usage.
- A description of mapping rules, to be used in the code selection phase of compiler.
- Calling conventions, C data-type sizes, and their alignment.

The generation of the aforementioned information from a LISA description is nontrivial. Some of these information, for example, the target processor resources, can be readily obtained from the description. On the contrary, the information about

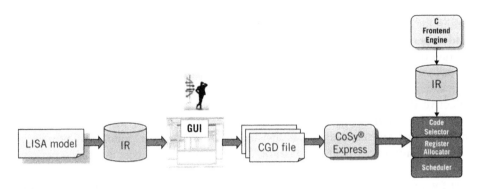

FIGURE 5.7

Compiler generation from LISA.

the C data-type or the calling convention is not present in the LISA description at all. In between these two extremes, the information like latency of individual instruction can be extracted from the LISA description by applying advanced algorithms. The method adopted for overall compiler generation is tuned to obtain an optimized compiler without restricting the freedom of design in the LISA description. The Compiler Designer reads in the LISA model, extracts the relevant information and presents it in form of an easy-to-use GUI. This information can be further altered/enhanced by the designer. Once done, the designer can get the CGD files from the GUI automatically for processing with CoSy (refer Fig. 5.7). An overview of the approaches taken to extract different compiler-related information from LISA description is presented here. The first information, about the target processor resources, is omitted as it is straightforward to obtain.

Instruction scheduler

The task of instruction scheduler is to determine the sequence of instructions to be executed on the target processor. The key information required in the instruction scheduler is the latency of each instruction under various dependency constraints and resource usage. This information is captured in the scheduler table, which is automatically generated from the LISA description [2]. To ensure a functionally correct scheduler, sometimes a conservative estimation of the latency value is done. This can be manually modified in the Compiler Designer GUI. Using this scheduler table, a list scheduler is generated.

Code selector

The purpose of the code selection is to perform a mapping of the compiler's Intermediate Representation (IR) operators to the processor's assembly instructions. These mapping rules, once available via the CGD files, are used by CoSy to deliver a code selector based on tree pattern matching. The mapping rules can be directly obtained from the semantics of the instructions described in LISA [3, 11]. Though

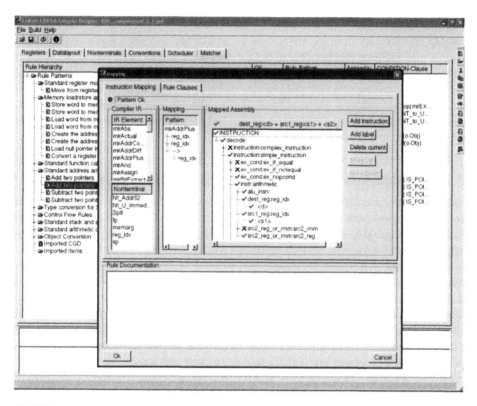

FIGURE 5.8

Compiler Designer GUI.

these direct extraction is sufficient to get a working prototype compiler, these mapping rules can be further improved by manual editing. The information extracted from the LISA description is presented in the Compiler Designer GUI (shown in Fig. 5.8). A typical mapping dialog is displayed at the forefront. On the left part of the mapping dialog, the IR operators are listed. By drag-and-drop mechanism, mapping rules can be composed from the IR operators (as shown in the center window). For a particular pattern, a processor instruction needs to be selected. This can be done by linking the arguments of the mapping rule to the operands of machine instructions (right window). For each of these mapping rules, created automatically or manually, the compiler generator automatically creates the code emitter to be used in the back-end of the compiler.

Data layout, register allocator, and calling conventions

Once the available processor resources are known, the Compiler Designer GUI provides a convenient mechanism to specify various configurations. First, the set of allocatable registers need to be selected from the available processor registers.

Dedicated registers for stack and frame pointers are to be specified. Registers, which cannot be saved to memory, are also to be noted. Second, the calling conventions, that is, the state of the processor before and after a function execution require to be properly described for a working compiler. This includes determination of registers for passing function parameters and return values. All these specifications can be done via different panes of the Compiler Designer GUI. The parameters like C data-type details and type alignment of the target processor can be entered in the GUI table.

5.2.3 Custom Instruction Synthesis for LISA-based Processor Design

Although there are embedded processors being designed completely from scratch to meet stringent performance constraints, there is also a trend toward partially pre-defined, configurable embedded processors, which can be quickly tuned to given applications by means of custom instruction and/or custom feature synthesis. The custom instruction synthesis tool needs to have a frontend, which can identify the custom instructions from an input application under various constraints and a flexible back-end, which can retarget the processor tools and generate the hardware implementation of the custom instruction quickly. With LISA as the processor designer back-end, a powerful custom instruction synthesis tool-flow is proposed [12]. The tool-flow is described briefly here. The discussion is divided into two phases, Custom Instruction (CI) identification and software tool retargeting. The hardware implementation of the CIs are achieved by generating the CIs in form of LISA description.

CI identification, in general, can be conceived as a problem of optimally partitioning the DFG of an application between different CIs (and possibly between CIs and core processor instructions) that maximizes application speedup. However, any arbitrary cluster of nodes in a DFG cannot qualify as a CI in general. A CI is architecturally feasible only when the constituent DFG fragment follows a certain set of constraints. Such constraints can be broadly classified into the following three categories.

- *Data-flow constraints*: A set of DFG operations grouped into the same CI must satisfy a convexity constraint [9] that prevents cyclic dependencies between one CI and other DFG operations. Similarly, schedulability constraints have to be met that prevent deadlocks between sets of CIs during the compiler's instruction scheduling phase.

- *Latency and area constraints*: As a rule of thumb, the speedup due to CIs is proportional to their complexity. Therefore, CI synthesis generally tends to put as many DFG operations as possible into a single CI. However, there are limitations on permissible CI size due to area, power consumption, and/or latency constraints. In particular, the CI combinational critical path must not exceed the core processor's clock period. Pipelining of CIs can soften this constraint. Since most embedded ASIPs need to meet very tight area budgets, a constraint can also be imposed by the designer on the maximum silicon area of a CI.

■ *Architectural constraints*: Assuming the CI is synthesized on a dedicated Custom Functional Unit (CFU), communication frequently takes place via the core's General Purpose Register (GPR) file. Due to the limited instruction word length, generally, only a few GPRs are available for this purpose (e.g., 2 input GPRs and 1 output GPR per CI). Furthermore, a CI may or may not access the data memory, and other constraints may hold for immediate constants, etc. Although these constraints impose tight limitations, the CFU may provide internal resources or clustered register access [13] to support more complex CIs.

In the proposed flow, the aforementioned constraints are modeled as an Integer Liner Programming (ILP) problem and ILP solver is used to obtain CIs from a given application. A GUI framework aids the designer to select the application fragment and enter the constraints. The custom instructions are generated in form of LISA description, so that the software tools are easily retargeted.

Once the CIs are identified, the application must be subjected to a modified processor toolsuite to exploit those. The LISA description of the CI allows quick retargeting of the simulator, assembler, and linker. To retarget the C compiler, a compiler intrinsic named *inline assembly* is utilized. The inline assembly feature of CoSy compiler development system allows the assembly instructions to be directly plugged inside the C code with several additional settings. These settings allow the designer to specify the resource usage, latency, and syntax of the inline assembly. During compilation, the compiler generates assembly instructions for the complete application and simply replaces the inline assembly call with the given syntax. The latency and resource usage of the inline assembly is respected during register allocation and scheduling phase. With this feature, the retargeting of C compiler can be done in a straightforward manner. The CIs are embedded in the target application using inline assembly. The retargeted software tools can be used for cycle-true core/CFU simulation to back-annotate the exact speedup in terms of cycle count achieved by the CIs. The implementation of the CIs is achieved by running HDL generation on the LISA implementation of those CIs.

5.2.4 Instruction Opcode Synthesis

Most of the existing ASIP design frameworks require tedious manual specification of binary instruction opcode. To have a working version of the ASIP tools, this is required even at very early design stages. With LISA, an instruction opcode synthesis tool is integrated [14] to perform this automatically.

The central algorithm adopted for automatic opcode synthesis starts with a tabular representation of all the processor instructions. The table is organized with instructions having similar bit-width of operands. The instructions with highest operand width are first allocated the opcodes. These opcodes are reused for the instructions with lower operand widths. The reusage of opcodes requires only one additional distinguishing bit. This process is repeated to obtain variable length opcodes for all the instructions.

In the context of LISA, the instructions are arranged hierarchically in the form of LISA operations. The opcode synthesis algorithm is applied to different hierarchies to obtain the local group's opcode. However, it is observed that the local encoding is suboptimal in comparison with the global context. On the other hand, applying the synthesis algorithm in a global context can produce better results, but is confronted with the issue of high runtime complexity. This issue is dealt with by identifying (manually or heuristically) a set of key operations, which unambiguously represents a single instruction of the processor. After identifying all key operations in the whole architecture description, the operand lengths are determined by accumulating all operand lengths of the associated nonterminal operations on the path to the key operation. Subsequently, the set of key operations are encoded.

5.3 AUTOMATIC OPTIMIZED PROCESSOR IMPLEMENTATION

A processor description in LISA can be subjected to an HDL generator, which is capable of producing optimized RTL implementation of the target processor. Unlike traditional processor modeling, where the software toolsuite and the RTL implementation were developed separately, this approach provides important advantages. First, the single LISA specification eliminates the problem of discrepancy in different processor tools and suboptimal processor performance because of that. Second, the quick generation of RTL description allows a more accurate performance estimation in the early design space exploration phase. Third, for manually written RTL descriptions, this automatically generated RTL description can serve as a reference description for verification. Finally, the RTL description can be further subjected to commercial physical synthesis flow to manufacture the processor. To qualify the automatically generated RTL for all these, an important prerequisite is that it should be highly optimized in terms of area-timing and power. The HDL generator tool of LISA utilizes high-level structural information, embedded in the LISA description, to make the quality of RTL description equivalent or even better than a manually described RTL model. In the following subsections, the central principle of LISA-based HDL generation and its inherent optimization techniques are discussed.

5.3.1 Automatic Generation of RTL Processor Description

The HDL generation process from LISA [5, 15] is graphically depicted in Fig. 5.9. The LISA description is initially parsed and represented internally in a representation termed as Unified Description Layer (UDL). In the UDL, the structural as well as behavioral details of the processor are stored in an easy-to-use format. At

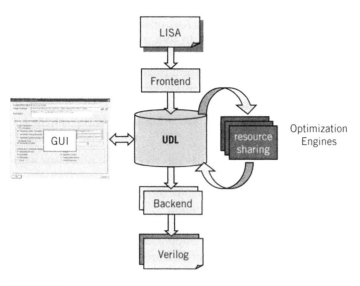

FIGURE 5.9

Automatic RTL description generation from LISA.

this level, a GUI-based user interface allows the designer to navigate the synthesis flow. The designer may opt for various optimization engines, as well as set up definite optimization constraints. The user may also generate gate-level synthesis and RTL simulation scripts tuned to commercial tools. Depending on the user options, the optimization engines are triggered. From the optimized UDL representation, language-specific back-ends are called to generate the RTL description in synthesizable form of VHDL, Verilog, or SystemC. The entire HDL generation process is modular and extensible. For example, new optimization engines or new language back-ends can be conveniently plugged in.

5.3.2 Area and Timing-driven Optimizations

In order to generate an RTL description for the purpose of implementation, it is imperative to match the quality of it to the manually optimized implementation. A definite step for that is to plug in optimizations driven by area and timing. Though the commercial gate-level synthesis tools perform such optimizations, several key structural information available in the high-level LISA description (and preserved in the UDL) are difficult to determine from the RTL description. Furthermore, given the complexity of a modern embedded processor, manually performing optimizations all over the design is a tedious job. The area and timing-driven optimizations performed during LISA-based HDL generations are proposed to exactly solve this problem. Surely, an experienced designer can reap further benefits of this model by doing manual RTL modifications.

A key information preserved in the UDL format for optimizations is the exclusiveness information available in the LISA description. In LISA, the processor instructions are hierarchically arranged in groups of *LISA operations*, among which a detailed exclusiveness relation exists. These exclusiveness information is stored in the UDL. The datapaths of the basic behavioral blocks (corresponding to each LISA operation) are also stored in the UDL in Control Data Flow Graph (CDFG) and Data Flow Graph (DFG) format. Based on these information, the following major optimizations are applied:

- *Structural simplifications*: LISA designer usually models the processor without delving into structural fine-tuning. This leaves a scope of several structural simplifications [16]. One such example is *decision minimization*. This is essentially a variant of condition-independent code motion. By decision minimization, condition-independent hardware operations are moved outside the conditional blocks. Another structural simplification is called *signal scope localization*. In this optimization, the signals declared all over the LISA model are examined and if a signal is accessed within the scope of a single simulation cycle without the necessity of storing its value, it is converted to a local variable. The same optimization can also be performed for registers, where instead of grouping all the register files together in a dedicated hardware block, locally used register resources are moved close to the corresponding functional units.

- *Path sharing*: To allow ease of programming at the high-level abstraction of LISA, register resources can be read and written multiple times from a single LISA operation. This produces multiple register access paths, which in turn results in huge multiplexers in the register file. By analyzing the exclusiveness of all the basic blocks in the processor, the read and write paths to the same register resources are shared.

- *Resource sharing*: In a LISA model, costly hardware operators like multiplication and addition can be extensively used all over. This grants an ease-of-programming in high-level abstraction at the expense of heavy area consumption. The designer can turn on *resource sharing* optimization to share the resources among exclusive behavioral blocks. The sharing is performed iteratively. At each step of sharing, a set of shareable resources is formed and the effect of sharing on the neighboring logic is estimated. The estimation is done on the basis of an abstract cost model. Depending on the estimation and the designer-specified area, and also timing constraints, the resource sharing is performed with little effect on the critical path.

5.3.3 Energy-driven Optimizations

Due to the stagnant battery capacity, the power budget of modern embedded processors is tight. This makes the energy-driven optimization an important part of processor implementation flow. In the HDL generation from LISA, several energy-driven optimizations are available, as discussed here.

■ *Clock gating*: To reduce the dynamic power of a processor, it is important to gate the storage elements when no new data is being read. Due to its overwhelming effect on overall power consumption, commercial gate-level synthesis tools are currently supporting automatic clock gating. However, to determine the gating signal, that is, the signal to indicate when a new data is not available, is an important research problem. In the perspective of LISA, it is identified by tracing the register write operations to LISA operations [17]. The clock gating for registers as well as pipeline registers is performed by inserting dedicated gating blocks during HDL generation.

■ *Operand isolation*: In operand isolation, the redundant switching activity in the functional blocks of a circuit is reduced by blocking the inputs during idle state. Similar to the clock gating, this requires an understanding of the processor's data flow and determination of the idleness condition on the basis of *Observability Don't Care* of the functional blocks' outputs. Given the access to the structural organization of the LISA processor's functional blocks, this is obtained in a straightforward manner [7].

Apart from these energy-driven optimizations during HDL generation, energy can be saved by minimizing the consumption of power in the instruction and data buses in a processor. In the context of LISA, this is also performed by altering the instruction opcode and modifying the register allocation [18].

5.3.4 Automatic Generation of JTAG Interface and Debug Mechanism

The LISA-based HDL generation framework allows automatic generation of processor features such as hardware debug mechanism [19]. This feature is important for the designers willing to access and modify the processor state via an additional interface, commonly referred to as JTAG interface. Such an interface and debug mechanism enables the designer to debug the processor after fabrication. Implementing this feature manually in RTL is clearly a prohibitive task with the trade-off between debugging capability and performance overhead, for example, area and timing, caused by this. In LISA, the JTAG interface is debug mechanism settings can be performed via the GUI during HDL generation. The area impact of various choices is obtained immediately. With this, the trade-off between debugging flexibility and performance can be done early in the design phase. Depending on the configuration, the complete interface and resources necessary for debug mechanism are automatically instantiated in the generated RTL.

5.4 PROCESSOR VERIFICATION

Validation and verification are the two essential phases of a processor design. While validation ensures that the design-intent is correctly reflected in the design specification, the task of verification is to check that the design matches with the specification. In the context of application-specific processor design, both these include the check if the target application(s) are correctly executed. In the LISA-based flow, this can be done across multiple abstraction levels down to RTL via simulation.

5.4.1 Equivalence Check via Simulation

Simulation of target application(s) and corresponding equivalence check across various levels of LISA abstraction is graphically displayed in Fig. 5.10. Using an instruction-accurate LISA model, a designer can easily verify the correct application execution using the LISA simulation capabilities. The simulator is aided with wide-ranging debugging features, for example, memory tracing, LISA-level behavior stepping, and setting breakpoints. With the high-level LISA description acting as a golden model, the processor states (registers, memory, status flags) can be dumped at the end of each cycle in a widely used Value Change Dump (VCD) format. The VCD file can be regenerated by running the application with a changed LISA model. Obviously, while executing the application for a cycle-accurate LISA model, adjustments in the application are necessary compared to the one running for an instruction-accurate LISA model. The VCD file from the original golden model and the new LISA model can be directly compared via perl script to locate mismatches. Further down the implementation track, the RTL implementation of the processor can be subjected to the same application. The VCD file generated from RTL simulation can again be compared with the VCD file created during LISA simulation.

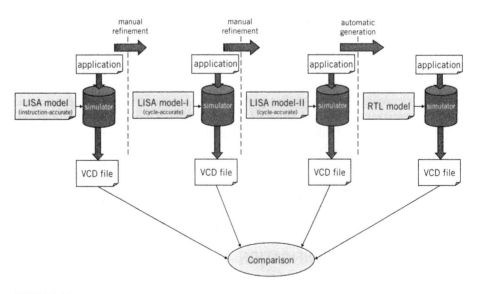

FIGURE 5.10

LISA-based verification via simulation.

5.4.2 Generation of Test Vectors from LISA Descriptions for Instruction-set Verification

Considering a broader perspective, the designed processor needs to be tested for applications yet unknown. This can be guaranteed by formally verifying that each individual instruction performs as expected for all possible processor states. Given the enormity of this verification space, current formal verification tools fall short of this task. Consequently, major ADLs focus on uncovering hidden bugs in the design by verifying the design with various test vectors. The test vectors are designed with two definite goals: first, to improve the coverage of the design, that is, to test as many likely scenarios as possible. Second, to feed the design with corner-case test vectors. In both the cases, the expertize of designer improves the test vector generation by reducing the redundancy, thereby saving costly verification time. Therefore, the LISA-based test vector generation focuses on a semiautomatic approach, where the designer/verification engineer can guide the process of test vector generation from LISA. In the following paragraphs, two different test vector generation approaches based on LISA are presented. While the first one is more focussed on creating corner-case test suite, the second approach is motivated toward achieving full coverage test vectors.

Integration with Genesys

The test generation environment from IBM, termed Genesys, is integrated with LISA to derive intelligent test cases. Genesys has a wide range of functionalities to target the verification of cache protocols, multi-processor configurations, etc. Though it was originally conceived to deliver test cases for large-scale processors [20], Genesys has been successfully applied to the verification of DSPs and ASIPs [21].

The building blocks of Genesys test generation environment are shown in Fig. 5.11. The architecture-specific details, for example, the testing knowledge, simulator, and instruction-set are fed into the test generator. The simulator is automatically generated from a LISA description. Besides the regular simulation

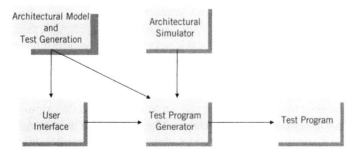

FIGURE 5.11

Genesys test generation environment.

capabilities, the simulator also needs to include special Genesys API, which is automatically generated during simulator generation. The instruction set is captured in the form of instruction trees. The general structure of each instruction is presented via the instruction tree. During test generation, generic instructions are created from the instruction tree. These generic instructions are stored in the knowledge base (testing knowledge), which are used to specialize the instructions based on user directions. For example, a generic instruction for multiplication can be specialized to test for a resulting overflow. The architecture simulator is used in this context to predict the results of instruction execution and generate complex test cases. Given the direct influence of the verification engineer via the testing knowledge, interesting corner cases can be generated. In particular, the test generation can be controlled in the following ways:

- A set of instructions can be selected to be used within generated test suite.
- The probability of exceptions can be controlled.
- Instruction sequences, which reuse the same processor resources, can be generated.
- User-specified functions can be used to generate operands of an instruction.
- Specific test patterns to drive the processor in a definite state can be specified.
- Test scenarios for generating various kinds of loops can be specified.

Genesys facilitates a design-driven pseudo-random test pattern generation. With the testing knowledge, expert test generation is possible. By integrating LISA model with Genesys test generation framework, high-quality test generation for LISA models is possible.

Full-coverage test vector generation

To simply ensure that the complete processor description is covered during verification (for a given initial processor state), it suffices to enumerate all legal instructions. The immediate operands of the instructions can be filled up at random. By further analyzing the behavior of each instruction, it is possible to predict the value of processor resources, under which the instruction gets fully covered. In the context of instruction-accurate LISA models, this is accomplished by designing a set of equations, the solution of which leads to the test cases guaranteeing full code coverage of the model [22].

For generating full-coverage test vectors for cycle-accurate LISA models, a more rigorous approach is undertaken [23,24]. The processor instructions are represented in Backus-Naur Form (BNF) of grammar. Table 5.1 shows parts of an *instruction grammar*. For this example, the instruction word width is 32 bits and there are 16 available registers indexed by src_reg and dst_reg. From a given LISA description, the instruction grammar is automatically generated. The instruction grammar, in turn, can be fed to a Test Generation Engine (TGE) with appropriate constraints for generating test vectors.

In order to attain full statement coverage for the processor, the coverage of the conditional blocks in it is naturally targeted. These conditional blocks are

Table 5.1	Exemplary instruction grammar.
insn	: add dst_reg src_reg src_reg
	‖ sub dst_reg src_reg src_reg ‖ nop
add	: 0000 0001
sub	: 0000 0010
src_reg	: 0000 xxxx
dst_reg	: 0000 xxxx
nop	: 0000 0000 0000 0000 0000 0000 0000 0000

internally represented as multiplexers after parsing the LISA description. The values under which alternative multiplexing conditions are satisfied are first ascertained. Following that, the corresponding values are traced back to definite processor resources. This backtracking of values executes across processor pipeline stages rendering specific values for processor resources for example, memory, pipeline registers, registers, and immediate fields. The memory and registers values are set by triggering dedicated load-store instructions. Immediate fields are set during generation of corresponding instructions. Pipeline registers are set by triggering the instructions in previous pipeline stage, without conflicting with the current instruction. By this method of backtracking, a set of instructions running in parallel at different pipeline stages are conceived. The instructions are generated by feeding appropriate constraints to the test generation engine.

5.5 SYSTEM-LEVEL INTEGRATION

Modern System-on-Chips (SoCs) contain an increasing number of heterogeneous components of which, programmable processor is an important part. To balance the conflicting demands of flexibility and performance, the system components must be coexplored early in the design phase. This requires the ADL to have the capacity of being effortlessly plugged in a system-level modeling and exploration framework. Like the standalone processor modeling with LISA, a successive refinement flow of the system-level model with LISA processor(s) is supported. The design framework allows other components described in SystemC Transaction Level Model. The successive refinement flow includes development of both processing components and the communication architecture [25].

The integration of LISA processor models in a system-level exploration framework is presented in Fig. 5.12. The enabling technology of this approach is retargetable integration of arbitrary LISA-based processor models with SystemC-based communication platform models. As shown in the figure, existing library of SystemC bus architectures are deployed for exploring the communication architecture. The designer can integrate a custom architecture for communication exploration as well.

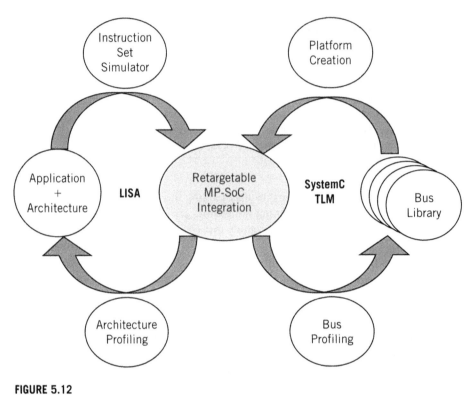

FIGURE 5.12

System-level exploration with LISA.

5.5.1 Retargetable Processor Integration

For performing system-level exploration, the automatically generated processor simulators need to be integrated with the SoC via buses and memories. These components are accessed within LISA model through generic Application Programming Interfaces (APIs). In order to map the generic LISA API to corresponding Transaction-Level Model (TLM) API, a bus interface class library is implemented. For the various abstraction levels, the LISA memory API functions are implemented to map the memory accesses onto the TLM bus API. The functional LISA interface has to make sure that all SoC modules stay synchronized to each other, and also that the LISA behavioral model can still rely on successful memory accesses, even in case of memory resources external to the respective processor module. Three different abstractions of LISA API are conceived, as described in the following list:

- *Ideal LISA interface*: The ideal (or debug) LISA interface is needed for debugging purposes. The ideal interface performs the memory access without delays. This kind of interface can be directly mapped to the debug interface of the bus model, to which the processor is connected.

- *Blocking LISA interface*: A blocking interface function call advances system simulation time as far as necessary to process with the function call. During this time, the processor simulator is suspended. In terms of a SystemC system simulation, the Interface Method Call (IMC) implementation executes *wait()* until the memory access is completed. This access time includes time for bus arbitration, memory access request, and memory device reply. In case of cache misses, the blocking LISA interface waits till the correct memory access. This interface is mapped to the transaction TLM API of the connected bus.

- *Cycle-accurate LISA interface*: The cycle-accurate LISA interface is mapped onto the transfer TLM API of the respective bus. It may be called nonblocking interface, since none of these interface calls can *block* the processor, which means advance system simulation time (in any SoC module) before returning the control back to the processor. For the cycle-accurate API mapping, in many cases a bus interface state machine is implemented. This gives the processor designer a powerful tool which detects and, if necessary, also corrects timing violations. New data coming too early from the bus or the processor side can be buffered, and if data or a command is not arriving in time, the bus or processor pipeline is stalled. Using such an adapter, a bus cycle-accurate processor simulator can still be connected to a bus model it was not originally intended for. Additionally, if a transaction TLM API is not available in a bus model, then such a bus interface state machine also allows mapping the functional LISA API to the transfer TLM API of the bus model.

Since the generic LISA API is to be mapped to a bus-specific TLM API, such a coupling has to be implemented for every TLM bus model that the LISA processor simulators should be embedded into. Currently, several bus interfaces are already implemented in such a way including AMBA AHB, AMBA AHBLite, and simple bus. Using these APIs, the processor, the coupling, and the bus model can be coexplored as well as refined in a stepwise fashion. After refining one processor module to a lower abstraction level, the entire system can be simulated without changing the communication architecture.

5.5.2 Multiprocessor Simulation

To debug and profile the software in a complex multi-processor scenario, it is important to have the simulation view of the entire system. In LISA, this is supported by a multiprocessor GUI-based debugging facility. The multiprocessor debugging GUI allows full control of different components it comprises of during runtime. For example, to check the software's effect on a particular processor, the processor needs to be dynamically connected to the multiprocessor debugger GUI. The rest of the components, which are not currently in view, keep on simulating at a maximum speed without the GUI connection. The software developer can further access resources, which are external to the current processor though are mapped to the processor's address space. These resources can be viewed and modified like

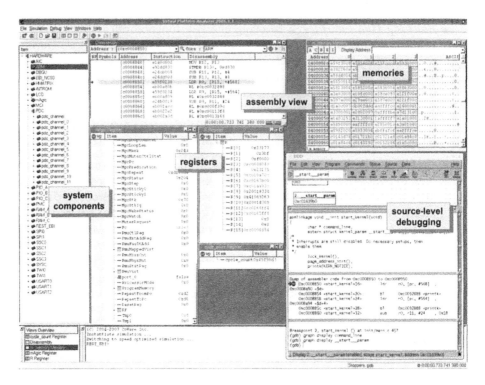

FIGURE 5.13

Multi-processor system-on-chip debugging.

peripheral registers and external memories (shown in Fig. 5.13). The designer can dynamically connect to particular processors, set breakpoints, disconnect from the simulation, and automatically reconnect when the breakpoint is hit. The level of observability and controllability, achieved in this multiprocessor debugging facility, is as fine as offered by a singular processor simulation.

5.6 COMPACT MODELING OF ADVANCED ARCHITECTURES

To support the increasing order of SIMD execution, more and more parallel slots are used in modern multimedia processors. Modeling such an architecture at a high level is an important boost to the design productivity, given their modeling complexity. The LISA language allows a concise and modular VLIW description formalism. Another interesting processor architecture emerging in the embedded community is partially reconfigurable one. For such an architecture, it is necessary to define arbitrary design partitions at high-level abstraction. The LISA's latest extensions also support such facility. In this section, these two key language extensions are described briefly.

5.6.1 **VLIW Architecture Modeling**

To conveniently model the various possible instruction combinations without introducing redundancy in the modeling, LISA supports the description of instruction *bundles*, as shown in Fig. 5.14. In the example code, it is shown that using different LISA operations existing below the coding root (decode), different combinations of instructions can be prepared. Each composition, referred as a bundle, can be decoded, based on the incoming instruction. Additionally, the bundles can accommodate immediate values in it. For a new instruction composition, the decode operation simply needs to be enhanced with a bundle and a new case-condition be placed.

In addition to the bundle composition, it is also useful to describe the same operation occurring in multiple slots in a modular fashion. That is facilitated in LISA by having a template-based operation description style (refer Fig. 5.15). With this, an operation is described with an identifier. The operations further activated/called from this template operation may or may not bear an identifier. In case it contains an identifier, it is tagged to the chain of the current template operation. The top-level template operations are instantiated with a definite identifier value, which is usually related to the slot number, as exemplified in the Fig. 5.14. Accordingly, during simulator/RTL generation, multiple instantiations of the entire operation tree for each slot are created. Note that an operation with a template identifier can trigger

```
OPERATION decode {
 DECLARE {
   ENUM type = { bundle_1, bundle_2 };
   GROUP slot_1 = { control_op|| alu_op<1> || ldst_op || mul_op<1>};
   GROUP slot_2 = { alu_op<2> };
   GROUP slot_3 = { alu_op<3> || mul_op<3> };
   GROUP slot_4 = { logic_op};
 }

 SWITCH (type) {
   // bundle with only one instruction
   CASE bundle_1 : {
     CODING { 0b0 slot_1 }
     SYNTAX { slot_1 ";" }
     ACTIVATION { slot_1 }
   }
   // bundle with four instruction
   CASE bundle_2 : {
     CODING { 0b1 slot_1 || slot_2 || slot_3 || slot_4 }
     SYNTAX { slot_1 "|" slot_2 "|" slot_3 "|" slot_4 ";" }
     ACTIVATION { slot_1, slot_2, slot_3, slot_4 }
   }
 }
}
```

FIGURE 5.14

Bundle composition for VLIW.

```
OPERATION alu_op<id> IN  pipe.DC
{
   DECLARE
     {
         GROUP opcode= { add<id> || sub<id> || and<id> };
         GROUP src1 = { reg || dreg };
         INSTANCE bypass_src1<id>, bypass_src2<id>, alu_op_ex<id>;
     }
   CODING { 0b0 opcode 0b00000 src1 src2 dst }
   SYNTAX { opcode ~" " dst"," src1 "," src2 }

   BEHAVIOR
     {
         bypass_src1();
         bypass_src2();
     }
   ACTIVATION { alu_op_ex }
}
```

FIGURE 5.15

Template-based operation description for VLIW.

an operation with definite identifier value, as will be done typically for a functional unit residing into one slot being activated by different instructions from different slots.

5.6.2 Partially Reconfigurable Processor Modeling

For partitioning the processor description in high-level of abstraction, two different language extensions are done in LISA. The first one is for partitioning the ISA, and the second one for partitioning the processor structure.

Partitioning the ISA

The instructions, to be mapped to the reconfigurable block, need to be allocated a definite space in the overall ISA. This allocation is similar to any regular ISA branch. The speciality is attributed when this particular branch is decoded locally in the reconfigurable block. Under such circumstances, the parent node of the LISA coding tree (which is a resident in the base processor) needs to be aware that the following branch may incorporate new nodes even after the fabrication. This is ensured by denoting the top-most *group* of LISA operations to be mapped to FPGA, as a FULLGROUP.

Partitioning the structure

The entire processor structure developed using the keywords so far need to be partitioned into fixed part and reconfigurable part. The partitioning can be done in several levels, as listed here.

- *Set of LISA operations*: A set of LISA operations can be grouped together to form an UNIT in LISA. This unit can be termed as RECONFIGURABLE in order to specify that this belongs outside the base processor. More than one unit within one or in multiple pipeline stages can be put into the reconfigurable block in this way.

- *A pipeline*: An entire pipeline structure can termed as RECONFIGURABLE, and thereby moved outside the base processor. This allows modeling of custom instructions with local-pipeline controlling capability.

- *Set of LISA resources*: The special purpose and general purpose registers residing within the base processor can be moved to the reconfigurable block by setting an option to *localize register* in the LISA RTL synthesis step. By having this option turned on, all the registers that are accessed only by the reconfigurable operations (i.e., operations belonging to the reconfigurable pipeline or unit) are moved to the reconfigurable block. In case of units of adjacent pipeline stages being reconfigurable, the pipeline registers accessed in between the units are also mapped to the reconfigurable block.

- *Instruction decoder*: The decoding of LISA operations from an instruction word is done in a distributed manner over the complete processor pipeline. A designer may want to have the decoding for reconfigurable instructions to be performed in the reconfigurable block itself. This can be done by setting an option to *localize decoding* in the reconfigurable block. By this, the decoder of reconfigurable operations is moved to the reconfigurable block.

- *Clock domains*: The base processor and the FPGA can run under different synchronized clock domains. The integral dividing factor of the clock is denoted by the LISA keyword LATENCY. For the custom instructions mapped onto the FPGA, the root LISA operation need to have a latency for marking a slower clock domain.

5.7 CASE STUDY

Several academic and industrial processor developments are done with LISA. In this section, two latest examples of ASIP design space exploration and optimized implementation are presented. These reflect two important aspects of LISA-based processor design platform. First, the ability to explore numerous design points, which LISA provides. Second, to derive optimized, industry-quality tools/implementations from LISA so that the traditional manual processor design style can be largely bypassed.

5.7.1 ASIP Development for Retinex-like Image and Video Processing

Wide-ranging consumer electronics and scientific computing applications are currently based on different image- and video-processing algorithms. In such a scenario, generic image enhancement algorithms like Retinex [26] are finding widespread usage. Adding software programmability to the core algorithm allows it to be used for consumer electronics, video surveillance, avionics, etc. However, it is important to meet the performance criteria of real-time image processing. This is achieved using LISA-based ASIP design environment, as elaborated in the following paragraphs.

Application analysis and architectural decisions

In Retinex theory, an image is captured as the pixel-by-pixel product of the ambient illumination and the reflectance of it. The values of reflectance can possibly be determined by the ratio between the image and estimated illumination. By this, the illumination and the reflectance can be processed independently. While illumination improves the brightness of the image, the details of it are enhanced by processing reflectance. The structure of Retinex-like filters is quite similar to that shown in Fig. 5.16. The product and division are replaced by addition and subtraction, respectively, for logarithmic sensors. The filter is responsible for illumination estimation, while the Γ and β blocks process illumination and reflectance. A similar structure is also used for processing video sequences, where the video is processed frame by frame. To avoid the effect of abrupt temporal correction effects, a temporal filter is additionally used.

Memory organization

The heavy data traffic of Retinex demands an efficient memory organization with various possibilities. The most simple memory organization can be done with a single one holding the complete frame. The pipelining of operations in the frame level can be removed, but that decreases the performance. The parallelism of two different algorithmic blocks again require independent memories. To satisfy the timing requirements, two frame memories are used. To improve the timing without introducing further memory blocks, the pipelining of operations is done at pixel level. The detail architectural structure is given in Fig. 5.17. The data RAMs are

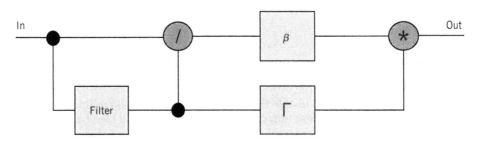

FIGURE 5.16

Algorithmic structure for Retinex-like filters.

initially loaded with the image to produce the illumination component. During this process, the intermediate results are held at Y RAM and the original image is stored at X RAM. After this, the reflectance component is determined by scanning these two memories and performing division. Then, the Γ and β transformations are performed, and finally these components are recombined to produce the final image. During the frame processing, the operations are performed on the subsequent pixels to speedup the process.

ISA decisions

To meet the throughput demands, a mandatory ISA customization is done to perform the piecewise linear approximation of the Γ and β operations quickly. By collapsing subsequent data processing operations, an instruction is composed to perform nonlinear transformations with single-cycle throughput. The instruction is made to load an operand from data memory, perform the Γ or β transformation in the piecewise linear form, and store the data back to the memory. The pipeline is designed to facilitate these operations seamlessly. Furthermore, it is observed that these operations are performed repeatedly in loops and access subsequent memory locations. To avoid the overhead in loops, special instructions to load the loop count and loop size are designed. To enhance the memory accesses for subsequent locations, automatic address increment facility is provided in several instructions. The updating of addresses for next computation, therefore, is done in parallel with the current one. A dedicated address generation unit is also modeled in LISA, which calculates the next address for the data memory by increasing the pixel pointer. Overall, 42 instructions are implemented, which can be grouped according to their functionality as following: there are 6 color conversion instructions, 9 instructions for nonlinear transformations, 11 arithmetic instructions, 9 memory access instructions, 6 instructions related to initialization, and 1 loop-control instruction.

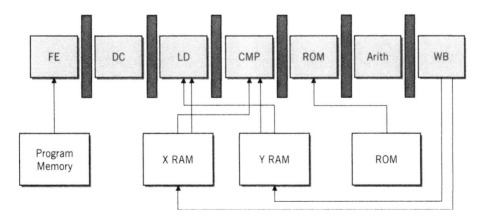

FIGURE 5.17

Processor structure.

Processor structural organization

The pixel-by-pixel processing is performed in a pipelined architecture. As per the ISA customization requirements, the operations are split in subsequent pipeline stages. To organize the corresponding memory operations, the latency of memory access is also taken into consideration. Increased number of pipeline stages also increases the area, as well as the complexity of pipeline control. After analyzing various alternatives, the best trade-off between complexity and throughput is achieved using seven pipeline stages, as shown in the Fig. 5.17. This pipeline organization leads to the difficulty of data hazards. Data hazards can be tackled via architectural solutions, for example, pipeline interlocking and/or bypassing—both of which can be conveniently modeled in LISA. In case of Retinex, it is observed that for the given allocation of instruction parts in the pipeline stages, most of the instructions cannot produce the result before the ARITH stage. Therefore, bypass mechanism is employed from the ARITH and WB stage to the relevant earlier pipeline stages i.e., LD, CMP, ROM and ARITH (from WB) stages.

A typical input and output image of the implemented Retinex ASIP is shown in Fig. 5.18.

Synthesis Results

The final LISA description is processed with the HDL generation tool to obtain fully synthesizable VHDL, Verilog descriptions. The RTL description is further synthesized using Synopsys Design Compiler in a $0.18\,\mu m$ 1.8V CMOS standard cell technology. The core processor consumed an area of 96 KGates with 18,544 bytes of ROM for implementing LUT-based operators. For various formats of input image, the data SRAM size and processing time vary, which are listed in the Table 5.2. It is observed that for supporting various Retinex-like filters an instruction SRAM size of 1 KByte is sufficient, which is implemented in this case. The maximum clock frequency achieved is 100 MHz, allowing for a maximum throughput of about

FIGURE 5.18

Demonstration of retinex ASIP.

Table 5.2 SRAM and processing time requirements.

	QCIF (176×144)	(256×256)	VGA(640×480)
Processing Time (Million Cycles)	1.3	3.4	16.3
Data SRAM Size (Bytes)	88704	229376	1075200

1.9 million pixels/s. The corresponding energy cost for the ASIP core is 50 nJ/pixel to process the 2-branch Retinex algorithm applied to color images. With the given performance figures, the ASIP can be used to process large still images (e.g., SXGA, WXGA) within 1 second and achieve up to 29 fps for real-time 256×256 videos. When compared to off-the-shelf DSPs and dedicated ASIC implementations, the current ASIP delivers another trade-off between energy and flexibility. For example, in Reference [27], a software optimized version of single-scale Retinex filter is implemented on a TI C6711 DSP. The performance achieved with that is 21 fps for a monochrome 256×256 image. With the processor running at 150 MHz, realized in 0.18μm 1.8 V CMOS technology, the energy cost is much higher (580 nJ/pixel). A dedicated ASIC implementation achieves better energy figures, but at the cost of flexibility.

5.7.2 Retargetable Compiler Optimization for SIMD Instructions

With multimedia applications acting as a major driving domain for embedded systems, the importance of quickly retargeting the C compiler with SIMD optimization is obvious. The C compiler, (semi)-automatically generated from LISA, is capable of delivering such domain-specific optimizations without compromising the convenient retargetability [28], as shown in the following case study.

For this case study the Philips Trimedia 32 processor [29] is chosen as the architecture, which is modeled in LISA. From the LISA model, the SIMD configuration options are automatically derived. The configuration options are plugged in the compiler specification of CoSy compiler development platform. The required retargeting effort for SIMD support is several man-days. As driver application, multimedia application kernels, mostly taken from the DSP-Stone benchmarks, are compiled. Few more complex DSP algorithms are also included in the target application set, as given in Table 5.3.

Architecture study and speedup results

The Trimedia is a 5-slot VLIW architecture supporting a number of SIMD instructions. However, utilizing SIMD instructions does not guarantee a speedup in all cases, due to several reasons. First, there are resource restrictions which must be followed. Due to the processor organization, one can issue five parallel ADD instructions whereas only two dual-ADD SIMD instructions can be issued. Clearly, with a smart scheduler the non-SIMD version can turn out to be more advantageous. Second, the SIMD

Table 5.3 Benchmark description.

Benchmark	Description
quantize	matrix quantization with rounding
compress	discrete cosine transformation to compress a 128 x 128 pixel image by a factor of 4:1, block size of 8 x 8
idct_8x8	IEEE-1180 compliant inverse discrete cosine transformation
viterbigsm	GSM full rate convolutional decoder

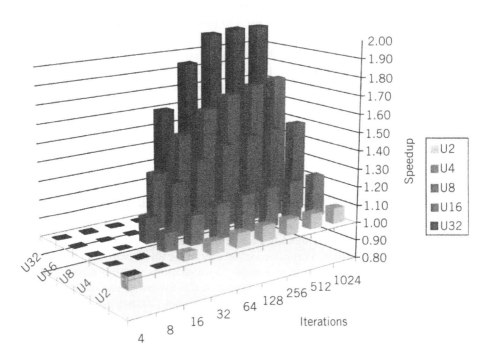

FIGURE 5.19

Speedup for dotproduct.

instructions are tagged with a higher latency compared to non-SIMD ones. Finally, there can be resource pressure as well as alignment issues affecting the potential gain by SIMD optimizations. These effects are captured together for a simple benchmark, namely, *dotproduct*, where vector elements are accessed by means of array accesses in the C code:

```
for(i=0;i<N;i++) sum += a[i] * b[i];
```

The automatically generated compiler from Trimedia LISA model, aided with SIMD optimizations, gives the speedup results, as shown in Fig. 5.19.

The speedup results are drawn together with the unroll factor (U) and the number of loop iterations. The number of loop iterations is set to a minimum value for compensating the setup overhead. The speedup, as can be observed, increases with increasing unroll factor and increasing loop iterations. This is due to the increasing number of SIMD instructions. For a higher number of unroll factors, the resource pressure increases, making the SIMD optimization more effective. The speedup approaches an asymptotic value of 2, corresponding to the theoretical maximum in this case.

The speedup results for target benchmarks are summarized in Table 5.4, described both in array and pointer-oriented style. A fixed iteration count of 1024 is used for all the benchmarks with loop unrolling enabled. In most cases a significant speedup can be noticed, even after accounting for the dynamic alignment checks in the SIMD loop versions. The speedup values for complex DSP routines are generally lower, as a small fraction of those codes can only be mapped to the SIMD instructions. In some cases, a super-linear speedup (more than 2) or even a slowdown is observed. A closer inspection reveals that these effects are caused by scheduler in the compiler back-end, which can be controlled by further manual treatments. Similar to the speedup, an effect on code size decrease is observed (not presented here in detail). On an average, the code size is reduced by a factor of 0.6 compared to the unrolled benchmarks running without the SIMD optimization.

Table 5.4   Speedup results.

Benchmark	style	U2	U4	U8	U16	U32
vector addition	array	1.55	1.71	1.83	1.91	1.96
fir	array	1.09	1.23	1.49	1.72	2.11
n_real_updates	array	1.64	1.78	1.88	1.94	1.96
n_complex_updates	array	0.89	0.87	0.88	0.87	0.88
dot product	array	1.09	1.23	1.49	1.72	2.11
matrix1	array	1.07	1.36	1.58	1.82	2.40
mat1x3	array	1.08	1.28	1.49	1.74	1.85
vector addition	ptr	1.66	1.79	1.88	1.93	1.96
fir	ptr	1.11	1.39	1.81	2.31	2.35
n_real_updates	ptr	1.75	1.85	1.92	1.96	1.96
n_complex_updates	ptr	1.06	1.08	1.08	1.08	1.08
dot product	ptr	1.11	1.39	1.81	2.31	2.35
matrix1	ptr	1.11	1.39	1.81	2.30	2.58
mat1x3	ptr	1.11	1.39	1.81	2.31	2.47
quantize (I = 64)	array	1.53	1.59	1.63	1.65	1.66
compress (I = 8)	array	1.07	1.08	1.23	n/a	n/a
idct_8x8 (I = 8)	array	1.08	1.09	1.11	n/a	n/a
viterbigsm (I = 8)	array	1.09	1.10	1.11	n/a	n/a

In some cases, a code size increase is observed (1.5 times for *matrix1x3*) due to the dynamic alignment check between SIMD and non-SIMD version of the loop in the code.

5.8 CONCLUSIONS

This chapter introduced the language LISA along with its associated toolsuite for exploring the design space of modern embedded processors. The ability of LISA to model a wide range of architectures (with continuous encompassing of upcoming architectural trends) is accompanied by a powerful framework, which delivers production-quality software tools and optimized RTL implementation among others. The versatility of the language formalism inspires two overlapping use-cases in Electronic System Level (ESL) design automation landscape. The first one is for abstract architectural simulation, where the focus is on software tools generation and system-level integration. The second use-case is that of designing application-specific processors, micro-controllers, and customized accelerators, where the hardware generation and custom instruction selection are exploited. In both the scenarios, LISA is extensively used in academia as well as in industry via the successful commercial offering of CoWare Processor Designer [30].

REFERENCES

[1] A. Hoffmann, H. Meyr, and R. Leupers. *Architecture Exploration for Embedded Processors with LISA*. Kluwer Academic Publishers, 2002.

[2] O. Wahlen, M. Hohenauer, R. Leupers, and H. Meyr. Instruction Scheduler Generation for Retargetable, Compilation. In *IEEE Design & Test of Computers*, pages 34–41, IEEE, January–February 2003.

[3] J. Ceng, M. Hohenauer, R. Leupers, G. Ascheid, H. Meyr, and G. Braun. C Compiler Retargeting Based on Instruction Semantics Models. In *Proc. of the Conference on Design, Automation & Test in Europe (DATE)*, pages 1150–1155, Munich, Germany, March 2005.

[4] A. Nohl, G. Braun, O. Schliebusch, R. Leupers, H. Meyr, and A. Hoffmann. A Universal Technique for Fast and Flexible Instruction-set Architecture Simulation. In *DAC '02: Proc. of the 39th Conference on Design Automation*, pages 22–27, NY, USA, ACM Press, 2002.

[5] O. Schliebusch, A. Chattopadhyay, D. Kammler, R. Leupers, G. Ascheid, and H. Meyr. A Framework for Automated and Optimized ASIP Implementation Supporting Multiple Hardware Description Languages. In *Proc. of the ASPDAC – Shanghai*, pages 280–285, China, 2005.

[6] E. M. Witte, A. Chattopadhyay, O. Schliebusch, D. Kammler, G. Ascheid, R. Leupers, and H. Meyr. Applying Resource Sharing Algorithms to ADL-driven Automatic ASIP Implementation. In *ICCD '05: Proc. of the 2005 International Conference on Computer Design*, pages 193–199, Washington DC, USA, 2005, IEEE Computer Society.

[7] A. Chattopadhyay, B. Geukes, D. Kammler, E. M. Witte, O. Schliebusch, H. Ishebabi, R. Leupers, G. Ascheid, and H. Meyr. Automatic ADL-based Operand Isolation for Embedded Processors.

In *DATE '06: Proc. of the Conference on Design, Automation and Test in Europe*, pages 600-605, 3001 Leuven, Belgium, 2006, European Design and Automation Association.

[8] S. Kraemer, L. Gao, J. Weinstock, R. Leupers, G. Ascheid, and H. Meyr. HySim: A Fast Simulation Framework for Embedded Software Development. In *CODES + ISSS '07: Proc. of the 5th IEEE/ACM International Conference on Hardware/Software Codesign and System Synthesis*, pages 75-80, NY, USA, ACM Press, 2007.

[9] T. Kempf, K. Karuri, S. Wallentowitz, G. Ascheid, R. Leupers, and H. Meyr. A SW Performance Estimation Framework for Early System-Level-Design Using Fine-grained Instrumentation. In *Design, Automation & Test in Europe (DATE)*, pages 468-473, Munich, Germany, March 2006.

[10] Associated Compiler Experts. *http://www.ace.nl*, 15th February, 2008.

[11] M. Hohenauer, H. Scharwaechter, K. Karuri, O. Wahlen, T. Kogel, R. Leupers, G. Ascheid, H. Meyr, G. Braun, and H. van Someren. A Methodology and Tool Suite for C Compiler Generation from ADL Processor Models. In *Proc. of the Conference on Design, Automation & Test in Europe (DATE)*, pages 1276-1281, Paris, France, February 2004.

[12] R. Leupers, K. Karuri, S. Kraemer, and M. Pandey. A Design Flow for Configurable Embedded Processors Based on Optimized Instruction Set Extension Synthesis. In *DATE '06: Proc. of the Conference on Design, Automation and Test in Europe*, pages 581-586, 3001 Leuven, Belgium, 2006, European Design and Automation Association.

[13] K. Karuri, A. Chattopadhyay, M. Hohenauer, R. Leupers, G. Ascheid, and H. Meyr. Increasing Data-Bandwidth to Instruction-Set Extensions through Register Clustering. In *IEEE/ACM International Conference on Computer-Aided Design (ICCAD)*, pages 166-171, 2007.

[14] A. Nohl, V. Greive, G. Braun, A. Hoffmann, R. Leupers, O. Schliebusch, and H. Meyr. Instruction Encoding Synthesis for Architecture Exploration Using Hierarchical Processor Models. In *Proc. of the 40th Design Automation Conference*, pages 262-267, ACM Press, 2003.

[15] O. Schliebusch, A. Hoffmann, A. Nohl, G. Braun, and H. Meyr. Architecture Implementation Using the Machine Description Language LISA. In *Proc. of the ASPDAC/VLSI Design—Bangalore, India*, pages 239-244, January 2002.

[16] O. Schliebusch, A. Chattopadhyay, E. M. Witte, D. Kammler, G. Ascheid, R. Leupers, and H. Meyr. Optimization Techniques for ADL-driven RTL Processor Synthesis. In *RSP '05: Proc. of the 16th IEEE International Workshop on Rapid System Prototyping (RSP '05)*, pages 165-171, Washington, DC, USA, 2005, IEEE Computer Society.

[17] A. Chattopadhyay, D. Kammler, E. M. Witte, O. Schliebusch, H. Ishebabi, B. Geukes, R. Leupers, G. Ascheid, and H. Meyr. Automatic Low Power Optimizations during ADL-driven ASIP Design. In *IEEE International Symposium on VLSI Design, Automation and Test (VLSI-DAT)*, pages 1-4, Hsinchu, Taiwan, April 2006.

[18] A. Chattopadhyay, D. Zhang, D. Kammler, E. M. Witte, R. Leupers, G. Ascheid, and H. Meyr. Power-efficient Instruction Encoding Optimization for Embedded Processors. In *VLSI Design Conference*, pages 595-600, Bangalore, India, January 2007.

[19] O. Schliebusch, D. Kammler, A. Chattopadhyay, R. Leupers, G. Ascheid, and H. Meyr. JTAG Interface and Debug Mechanism Generation for Automated ASIP Design. In *GSPx 2004*, Santa Clara, CA, USA, September 2004.

[20] A. Aharon, D. Goodman, M. Levinger, Y. Lichtenstein, Y. Malka, C. Metzger, M. Molcho, and G. Shurek. Test Program Generation for Functional Verification of PowerPC Processors in IBM. In *DAC '95: Proc. of the 32nd ACM/IEEE Conference on Design Automation*, pages 279-285, NY, USA, ACM Press, 1995.

[21] S. Rubin, M. Levinger, R. R. Pratt, and W. P. Moore. Fast Construction of Test-program Generators for Digital Signal Processors. In *ICASSP '99: Proc. of the IEEE International Conference on Acoustics, Speech, and Signal Processing*, pages 1989–1992, Washington, DC, USA, 1999, IEEE Computer Society.

[22] O. Luethje. A Methodology for Automated Test Generation for LISA Processor Models. In *The Twelfth Workshop on Synthesis and System Integration of Mixed Information Technologies*, Kanazawa, Japan, pages 18–19, October 2004.

[23] A. Chattopadhyay, A. Sinha, D. Zhang, R. Leupers, G. Ascheid, and H. Meyr. Integrated Verification Approach during ADL-driven Processor Design. In *IEEE International Workshop on Rapid System Prototyping (RSP)*, pages 110–118, Chania, Crete, June 2006.

[24] A. Chattopadhyay, A. Sinha, D. Zhang, R. Leupers, G. Ascheid, and H. Meyr. ADL-driven Test Pattern Generation for Functional Verification of Embedded Processors. In *12th IEEE European Test Symposium*, May 2007.

[25] A. Wieferink, M. Doerper, R. Leupers, G. Ascheid, and H. Meyr (ISS, Aachen, Germany), T. Kogel, G. Braun, and A. Nohl (CoWare, Inc., USA). A System Level Processor/Communication Co-Exploration Methodology for Multi-processor System-on-chip Platforms. *IEE Proc. Computers & Digital Techniques*, 152(1):3–11, January 2005.

[26] E. Land and J. McCann. Lightness and Retinex Theory. *Journal of the Optical Society of America*, 61:1–11, 1971.

[27] G. D. Hines, Z. Rahman, D. J. Jobson, and G. A. Woodell. DSP Implementation of the Retinex Image Enhancement Algorithm. *Proc. of SPIE, Visual Information Processing XIII*, 5438, pages 13–24, 2004.

[28] M. Hohenauer, C. Schumacher, R. Leupers, G. Ascheid, H. Meyr, and H. van Someren. Retargetable Code Optimization with SIMD Instructions. In *CODES + ISSS '06: Proc. of the 4th International Conference on Hardware/Software Codesign and System Synthesis*, pages 148–153, NY, USA, ACM Press, 2006.

[29] N. Mitchell. Philips Trimedia: *A Digital Media Convergence Platform*, In *Proc. of WESCON/97 4–6*, pages 56–60, November 1997, DOI: 10.1109/WESCON.1997.632319, 0-7803-4303-4.

[30] CoWare/Processor Designer. *http://www.coware.com*. 15th February, 2008.

EXPRESSION: An ADL for Software Toolkit Generation, Exploration, and Validation of Programmable SOC Architectures

Prabhat Mishra and Nikil Dutt

EXPRESSION is an Architecture Description Language (ADL) for modeling, software toolkit generation, rapid prototyping, design space exploration, and functional verification of System-on-Chip (SOC) architectures. It was developed in the University of California, Irvine [1]. The EXPRESSION ADL follows a mixed level approach—it can capture both the structure and behavior supporting a natural specification of the programmable architectures consisting of processor cores, coprocessors, and memories. The EXPRESSION ADL was originally designed to capture processor/memory architectures and generate software toolkit (including compiler and simulator) to enable compiler-in-the-loop exploration of SOC architectures [1, 2]. The ADL has gone through various transformations primarily in two directions:

- The ADL is extended to capture a wide spectrum of processor and coprocessor architectures that represent many architectural domains, including RISC, DSP, VLIW, EPIC, and Superscalar. The ADL is also extended to support various novel memory configurations including partitioned register file, cache hierarchies, configurable scratch pad SRAM, stream buffer, and so on [3].

- The ADL is used to enable various design automation methodologies including retargetable compiler/simulator generation [4, 5], design space exploration [6], generation of hardware prototypes [7], and functional verification [8, 9] of programmable SOC architectures.

This chapter is organized as follows. Section 6.1 describes various features (syntax and semantics) of EXPRESSION ADL using illustrative examples. Section 6.2 describes software toolkit generation and exploration methodology. Section 6.3

133

presents a hardware generation approach, followed by the description of ADL-driven validation in Section 6.4. Finally, Section 6.5 concludes this chapter.

6.1 EXPRESSION ADL

EXPRESSION follows a syntax similar to LISP to ease specification and improve readability. This section describes the various features of EXPRESSION ADL using a simple example architecture. Fig. 6.1 shows an example architecture that can issue up to three operations (an ALU operation, a memory access operation, and a coprocessor operation) per cycle. The coprocessor supports vector arithmetic operations. In the figure, oval boxes denote units, dotted ovals indicate storages, bold edges are pipeline edges, and dotted edges are data-transfer edges. A data-transfer edge transfers data between units and storages. A pipeline edge transfers instruction (operation) between two units. A path from a root node (e.g., Fetch) to a leaf node (e.g., WriteBack) consisting of units and pipeline edges is called a *pipeline path*. For example, {*Fetch, Decode, ALU, WriteBack*} is a pipeline path. A path from a unit to the main memory or register file consisting of storages and data-transfer edges is called a *data-transfer path*. For example, {*MemCntrl, L1, L2, MainMemory*} is a data-transfer path. This section describes how the EXPRESSION ADL captures the structure and behavior of the architecture shown in Fig. 6.1 [9].

EXPRESSION description is composed of two primary sections: structure and behavior. Each of these primary sections is further subdivided into three

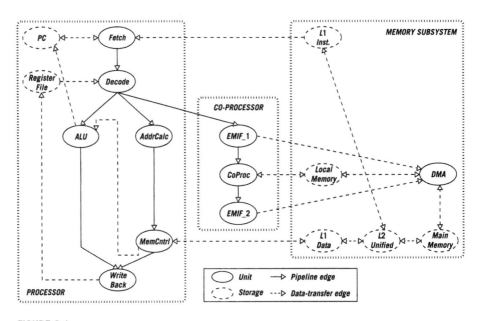

FIGURE 6.1

An example architecture.

subsections. For example, the structure is composed of three subsections: (i) components specification, (ii) pipeline and data-transfer paths description, and (iii) memory subsystem. Similarly, the behavior is composed of three subsections: (i) operations specification, (ii) instruction description, and (iii) operation mappings.

6.1.1 Structure

The structure of an architecture can be viewed as a netlist with the components as nodes and the connectivity as edges. The structure is composed of the following three subsections.

Components specification

This section describes each component and its attributes in the architecture. Typically, the list of components includes functional units, storage elements, ports, connections, and buses. It is also required to specify the timing information for multicycle or pipelined units. Each component can have a list of attributes. For example, the *ALU* unit has information regarding the number of instructions executed per cycle, timing of each instruction, supported opcodes, etc. The attributes are optional and can be any of the following:

- SUBCOMPONENTS: A compound component needs to describe all the individual subcomponents.

- LATCHES: If the component is a unit, it needs to list the input and output latches.

- PORTS: The list of ports attached to this component.

- CONNECTIONS: The list of connections to/from this component.

- OPCODES: The list (or group) of opcodes supported by this component.

- TIMING: In case of a multicycle unit or pipelined component, it needs to specify the timing behavior. The timing can be specified for each supported opcode (if necessary) or for a group of opcodes.

- CAPACITY: This indicates the number of operations that can be supported by this component in one clock cycle.

Fig. 6.2 shows the specification of some of the representative components of the example architecture in Fig. 6.1. For example, the *Fetch* unit can fetch three operations per cycle and can support all the operations.

Description of pipeline and data-transfer paths

This section describes the netlist of the processor consisting of components. The connectivity is established using the description of pipeline and data-transfer paths. Intuitively, pipeline paths describe the instruction flow through the pipeline stages, and data-transfer paths provide a mechanism for specifying valid data-transfers.

```
# Components specification
(SUBTYPE UNIT FetchUnit DecodeUnit ExecUnit CPunit ...)
(SUBTYPE PORT UnitPort Port)
(SUBTYPE CONNECTION MemoryConnection RegConnection)
(SUBTYPE LATCH PipelineLatch MemoryLatch)
( FetchUnit Fetch
  (capacity 3) (timing (all 1)) (opcodes all) ...) ...
)
( ExecUnit ALU
  (capacity 1) (timing (add 1) (sub 1) ...)
  (opcodes (add sub ...)) ...
)
( CPunit CoProc
  (capacity 1) (timing (vectAdd 4) ...)
  (opcodes (vectAdd vectMul))
)
```

FIGURE 6.2

Components specification.

```
# Pipeline and data-transfer paths
(pipeline Fetch Decode Execute WriteBack)
(Execute (parallel ALU LoadStore Coprocessor))
(LoadStore (pipeline AddrCalc MemCntrl))
(Coprocessor (pipeline EMIF_1 CoProc EMIF_2))
(dtpaths (WriteBack RegisterFile) (L1Data L2) ...)
```

FIGURE 6.3

Specification of pipeline and data-transfer paths.

Fig. 6.3 shows the specification of pipeline and data-transfer paths of the example architecture in Fig. 6.1.

Fig. 6.3 describes the four-stage pipeline as {*Fetch, Decode, Execute, Write-Back*}. The *Execute* stage is further described as three parallel execution paths: *ALU, LoadStore,* and *Coprocessor.* Furthermore, the *LoadStore* path is described using pipeline stages: *AddrCalc* and *MemCntrl.* Similarly, the coprocessor pipeline has three pipeline stages: *EMIF_1, CoProc,* and *EMIF_2.* The architecture has fifteen data-transfer paths. Seven of them are unidirectional paths. For example, the path {*WriteBack → RegisterFile*} transfers data in one direction, whereas the path {*MemCntrl ↔ L1Data*} transfers data in both directions.

```
# Storage section
( DCache L1Data
  (wordsize 64) (linesize 8) (associativity 2) (accessTime 1) ...
)
( ICache L1Inst
  (wordsize 64) (linesize 8) (associativity 2) (accessTime 1) ...
)
( DCache L2
  (wordsize 64) (linesize 16) (associativity 4) (accessTime 10) ...
)
( DRAM MainMemory
  (wordsize 64) (latency 100) ...
)
```

FIGURE 6.4

Specification of memory subsystem.

Memory subsystem

This section describes the components as well as the connectivity for specifying the memory subsystem. The memory subsystem structure is represented as a netlist of memory components. The memory components are described and attributed with their characteristics such as cache line size, replacement policy, write policy, and so on. Fig. 6.4 shows the specification of the memory subsystem of the example architecture in Fig. 6.1.

6.1.2 Behavior

The EXPRESSION ADL captures the behavior of the architecture as the description of the instruction set and provides a programmer's view of the architecture. The behavior is composed of the following three subsections.

Operations specification

This section describes the instruction set of the processor. The instruction set is organized into operation groups and each group contains a set of operations having common characteristics. Each operation is then described in terms of its opcode, operands, behavior, and instruction format. Each operand is classified either as the source or as destination. Furthermore, each operand is associated with a type that describes the type and size of the data it contains. The variable group description provides the facility for binding operands into a specific type of source or destination. Fig. 6.5 shows the specification of some of the representative operations in Fig. 6.1.

```
# Operations Specification
( opgroup aluOps (add, sub, ...) )
( opgroup memOps (load, store, ...) )
( opgroup cpOps (vectAdd, vectMul, ...) )
( opcode add
  (operands (s1 reg) (s2 reg/int16) (dst reg))
  (behavior dst = s1 + s2)
  (format 000101 dst(25-21) s1(21-16) s2(15-0))
)
( opcode store
  (operands (s1 reg) (s2 int16) (s3 reg))
  (behavior M[s1 + s2] = s3)
  (format 001101 s3(25-21) s1(21-16) s2(15-0))
)
( opcode vectMul
  (operands (s1 mem) (s2 mem) (dst mem) (length imm) )
  (behavior dst = s1 * s2)
)
( vargroup
  (reg RegisterFile)
  (mem L1Data — L1Inst | L2 — MainMemory)
  (imm int16 — int32)
)
```

FIGURE 6.5

Operations specification.

For example, the *aluOps* group includes all the operations supported by the *ALU* unit. Similarly, *memOps* and *cpOps* groups contain all the operations supported by the units *MemCntrl* and *CoProc*, respectively. The instruction format describes the fields of the instruction in both binary and assembly. Fig. 6.5 shows the description of *add*, *store*, and *vectMul* operations. Unlike normal instructions whose source and destination operands are register type (except load/store), the source and destination operands of *vectMul* are memory type. The *s1* and *s2* fields refer to the starting addresses of two source operands for the multiplication. Similarly *dst* refers to the starting address of the destination operand. The *length* field refers to the vector length of the operation that has immediate data type.

Instruction description

This section captures the parallelism available in the architecture. An instruction is viewed as a set of operations that can be executed in parallel. Each instruction

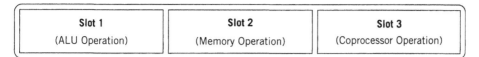

(a) Instruction for the Example Architecture

(**INSTR** (SLOTS (UNIT ALU) (UNIT AddrCalc) (UNIT EMIF_1)))

(a) Instruction Description in EXPRESSION

FIGURE 6.6

Instruction specification for the example architecture.

contains a list of slots (to be filled with operations). Each slot corresponds to a functional unit. For example, Fig. 6.6 shows the instruction description of the example architecture in Fig. 6.1. The instruction consists of three slots. The first slot is used to store the operation for the ALU unit. The second slot is used to store the operation for the AddrCalc unit (memory operations). The third slot is used to store coprocessor operations.

Operation mappings

This section is used to specify the information required by the compiler to perform various optimizations such as instruction selection and architecture-specific optimizations. This section also represents the mapping of an (sequence of) operation(s) to another (sequence of) operation(s). The mapping can be of two types: generic compiler operations to target processor operations, or target operations to target operations. The former is used by the instruction-selection algorithm whereas the latter is used by architecture-dependent optimizations.

Fig. 6.7 shows four representative operation mappings in EXPRESSION. The first one maps a generic integer addition operation to a target addition operation. The second describes the mapping of a multiply-by-2 operation by an addition operation. The third one maps the same add operation into a shift-left operation, and the last one maps multiply and addition operations into a MAC (multiply-and-accumulate) operation.

6.2 SOFTWARE TOOLKIT GENERATION AND EXPLORATION

This section describes software toolkit generation and exploration framework supported by EXPRESSION ADL. To enable rapid design space exploration, the framework generates a retargetable compiler (EXPRESS) and a cycle-accurate structural simulator (SIMPRESS). The framework also supports a Graphical User

```
# Operations Mappings
( OP_MAPPING
  (
    (GENERIC (IADD dst src1 src2))
    (TARGET (ADD dst src1 src2))
  )
  (
    (GENERIC (IMUL dst src1 #2))
    (TARGET (ADD dst src1 src1))
  )
  (
    (GENERIC (IMUL dst src1 #2))
    (TARGET (SHL dst src1 #1))
  )
  (
    (GENERIC ((IMUL tmpdst src1 src2) (IADD dst tmpdst src3))
    (TARGET (MAC dst src1 src2 src3))
  )
)
```

FIGURE 6.7

Operation mappings.

Interface (GUI) to enable visual specification and analysis. The EXPRESSION ADL is automatically generated from the GUI. A public release of this software is available at *http://www.ics.uci.edu/~express*. The remainder of this section describes compiler/simulator generation and design space exploration.

6.2.1 Retargetable Compiler Generation

In order to effectively compile for modern embedded processor architectures, the compiler needs to incorporate a large set of optimizations. These optimizations may target different aspects of the architecture, for example, conditional instructions or the application, SIMD. The EXPRESS retargetable compiler adopts a "toolbox" approach to incorporate both traditional and embedded systems specific compiler optimizations. However, in such an approach the phase ordering between the optimizations has a huge impact on the quality of generated code. The problem of determining the optimal phase ordering is further complicated by the fact that most applications have regions with different characteristics, for example, loop regions, if-block regions, etc., which require different optimization orderings. Statically determined phase orderings may not be able to satisfy the stringent constraints of performance, power, code size, etc. The compiler requires the ability to dynamically determine, based on the regions of interest, the best ordering of optimizations.

The EXPRESS compiler incorporates Transmutations, an approach that attempts to provide for dynamic ordering of the phases based on the program characteristics and available resources. The power of the EXPRESS compiler also stems from the fact that the operation reservation tables can be automatically generated from the ADL specification [2]. In the following paragraphs, we first present a brief overview of the EXPRESS compiler and then describe the Transmutations framework in detail.

The EXPRESS is an optimizing memory-aware Instruction-Level Parallelizing (ILP) compiler. EXPRESS uses the EXPRESSION ADL to retarget itself to a wide class of processor architectures and memory systems. Fig. 6.8 shows the EXPRESS compiler along with the Transmutations framework. The inputs to EXPRESS are the applications specified in C/C++ and the architecture specified in EXPRESSION. The front-end is GCC-based and performs some of the conventional optimizations. The core transformations in EXPRESS include resource-directed loop pipelining (RDLP) [10]—a loop pipelining technique, trailblazing percolation scheduling (TiPS)

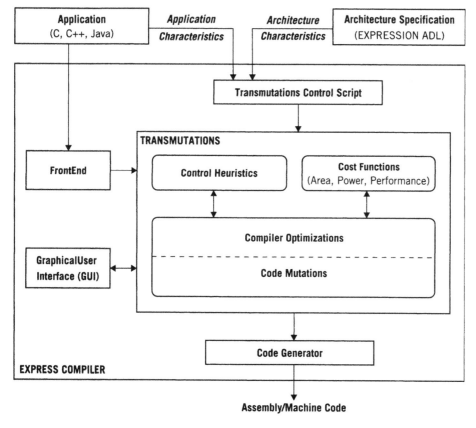

FIGURE 6.8

EXPRESS compiler framework.

[11]—a speculative code motion technique, instruction selection, register allocation, and if-conversion—a technique for architectures with predicated instruction sets. The back-end generates assembly code for the processor ISA.

Mutation Scheduling (MS) attempts to couple the phases of instruction selection and register allocation into the ILP scheduler by providing semantically equivalent computations of program values that have different resource usage patterns. The MS adopts a local view of the search space by only providing for mutations of values through algebraic transformations. Transmutations incorporates MS and also provides a framework for phase ordering between all transformations including the traditional compiler optimizations and memory optimizations. Furthermore, Transmutations attempts to customize the compiler for a wide variety of architecture styles including RISC, VLIW, and Superscalar.

Through the Transmutations framework the EXPRESS compiler is able to dynamically adapt both the program code and the order of transformations based on the resource availability and program region characteristics. Examples of code mutations [12] possible in EXPRESS include architecture-independent mutations such as tree-height reduction, and architecture-specific mutations such as strength reduction, synonyms, etc. Each code mutation has a cost function that determines its impact on performance, code size, memory access, etc. The heuristics in the Transmutations framework use this information in order to assign priorities for the mutations based on the resource availability. The heuristics also determine the ordering of the compiler phases. Transmutations also allows for user guidance through the transmutations control script, as shown in Fig. 6.8. In the script the user can specify new mutation transformation strategies for phase orderings and also specify the heuristics and cost functions. This allows the user to customize the compiler based on the application and processor domain.

The EXPRESSION ADL is used extensively for retargetable compiler generation in the context of compiler-in-the-loop exploration for embedded processors. Grun et al. have analyzed memory access patterns to perform access pattern-based memory and connectivity architecture exploration [13], memory miss traffic management [14], and memory-aware compilation through accurate timing extraction [15]. Similarly, Shrivastava et al. have studied compiler-aware exploration in the context of code size reduction using reduced bit-width ISAs [4], and efficient selection of partial bypassing in embedded processors [16].

6.2.2 Retargetable Simulator Generation

Simulators are critical components of the exploration toolkit for the system designer. They can be used to perform diverse tasks such as verifying the functionality and/or timing behavior of the system, and generating quantitative measurements (e.g., cache miss ratio) which can be used to aid Design Space Exploration (DSE). The SIMPRESS is a simulator generator which reads an EXPRESSION description and generates simulators for the target system. In the context of DSE, SIMPRESS can be used to generate a cycle-accurate structural simulator.

In addition to the simulator, SIMPRESS generates the hooks needed to connect the simulator to the VSAT GUI. It also incorporates statistics collection functionality into the simulator. The architecture model employed by SIMPRESS is generic enough to be able to incorporate a wide variety of processor/memory systems. The SIMPRESS assumes a two-phase clock cycle model with reads (from storage) occurring in the first phase and writes (to storage) occurring in the second phase. Furthermore, units function as active components that perform computation and request data from (or send to) other components. Storages are passive components which send data to (or receive from) active components.

Ports and connections are used to link units to storages and vice versa. The structure of a system can be viewed as a set of objects that communicate with each other via well-defined interfaces. Thus, an object-oriented language like C++ is well suited for describing a simulatable specification of the system, and we have chosen C++ for this very reason. The SIMPRESS reads in the EXPRESSION description and generates the structural model in C++. Other benefits of using C++ include encapsulation of information local to the components, hierarchical description of components, and ease of extensibility. Here, we present a brief overview of the organization of SIMPRESS.

SIMPRESS is organized as a collection of libraries (that can be classified as one of four types), as shown in Fig. 6.9.

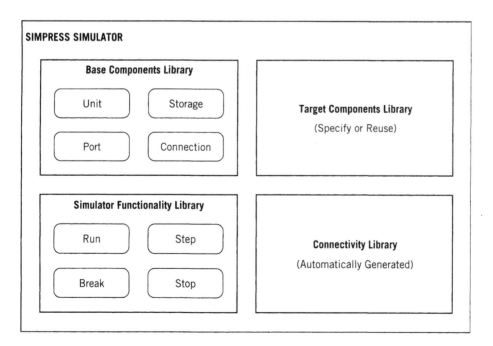

FIGURE 6.9

SIMPRESS simulator framework.

- **Base Components Library:** This library is the core of SIMPRESS. It defines each of the four component types in EXPRESSION (unit, storage, port, connection) as a class in the library. All other component types are derived from one of these classes. These four base classes define the public interface that governs the interaction between various components.

- **Target Components Library:** This library contains a description of each component class (derived from the base component classes) that is target specific (e.g., ALU, buses, etc). This library captures the functionality of the individual components that comprise the processor system and is used to capture soft, firm, and hard IP libraries.

- **Connectivity Library:** This library defines the connectivity as a netlist of components. Each component is an instantiation of a class defined in the Target Components Library.

- **Simulator Functionality Library:** This library defines functions (e.g., single-step, run, break, etc.) that can be used by the environment (e.g., GUI or compiler) to perform simulations and to control the simulator.

The *Base Components Library* and *Simulator Functionality Library* are target independent and hence do not vary from simulator to simulator. During generation of the simulator these libraries are linked with the relevant target components and connectivity libraries to form the simulator executable. The *Connectivity Library* is automatically generated from EXPRESSION. The target-specific components can either be built from the list of predefined components provided by SIMPRESS, or can be explicitly written (if necessary). In order to aid the designer, SIMPRESS provides a wide variety of parameterized components. These components are designed using functional abstraction [17].

SIMPRESS simulator is a cycle-accurate simulator and is based on interpretive simulation. Subsequent researches using EXPRESSION framework have developed fast and retargetable instruction-set simulators [18, 19]. Interpretive simulation is slow but flexible and allows easy debugging. Compiled simulation is fast but cannot be used in domains where the code can change during runtime. The Instruction-set Compiled Simulation (ISCS) technique [19] combines the benefits of interpretive (flexibility) and compiled (performance) simulation. The EXPRESSION ADL is used to generate retargetable instruction-set simulators [18] as well as cycle-accurate simulators [17].

6.2.3 Design Space Exploration

An EXPRESSION-based specification and software toolkit generation framework has been used in various design space exploration scenarios including processor exploration [20], coprocessor exploration [8, 21], memory exploration [8, 22–24], memory-aware compilation [14, 15], instruction-set exploration [25, 26], data forwarding (partial bypassing) exploration [16], and exploration of reduced instruction-set architectures (rISA) [4]. In this section, we present exploration experiments in two scenarios: coprocessor exploration and memory subsystem exploration.

Coprocessor exploration

This section describes the exploration experiments to study the impact of using coprocessors with TI C6x architecture. TI C6x [27] is an 8-way VLIW DSP processor with a novel memory subsystem (cache hierarchy, configurable SRAM, partitioned register file). The TI C6x processor has a deep pipeline, composed of four fetch stages (PG, PS, PR, PW), two decode stages (DP, DC), followed by the eight functional units. The fetch functionality consists of four stages, viz., program address generation, address send, wait, and receive. The architecture fetches one VLIW instruction (eight parallel operations) per cycle. The decode function decodes the VLIW word and dispatches up to eight operations per cycle to eight execution units. The functional units, L1, S1, M1, and D1, are connected to the "A" part of the partitioned register file whereas the remaining functional units, viz., L2, S2, M2, D2, are connected to the "B" of the register file. Two cross paths, viz., 1X and 2X, are used for transferring data from the other part of the partitioned register file.

Two configurations of the architecture are described using the EXPRESSION ADL. The first configuration of the TI C6x architecture has a functional unit for multiplication. The second configuration is a modification of the first one by adding a coprocessor (with DMA controller and local memory) that supports multiplication. The software toolkit is generated for both configurations. A set of DSPStone fixed point benchmarks are used to explore and evaluate the effects of adding a coprocessor.

Fig. 6.10 presents a subset of the experiments we performed to study the performance improvement due to the coprocessor. The light bar presents the number of execution cycles when the functional unit is used for the multiplication, whereas the dark bar presents the number of execution cycles when the coprocessor is used. We observe an average performance improvement of 22%. The performance improvement is due to the fact that the coprocessor is able to exploit the vector multiplications available in these benchmarks using its local memory. Moreover, functional units operate in register-to-register mode, whereas the coprocessor operates in memory–memory mode. As a result, the register pressure, and thereby the spilling, gets reduced in the presence of the coprocessor. However, the functional unit performs better when there are mostly scalar multiplications.

Memory subsystem exploration

Another important dimension for architectural exploration is the investigation of different memory configurations for a programmable architecture. This section explores various memory configurations for the TI C6x architecture with the goal of studying the trade-off between cost and performance. A set of benchmarks from the multimedia and DSP domains are used for the exploration experiments.

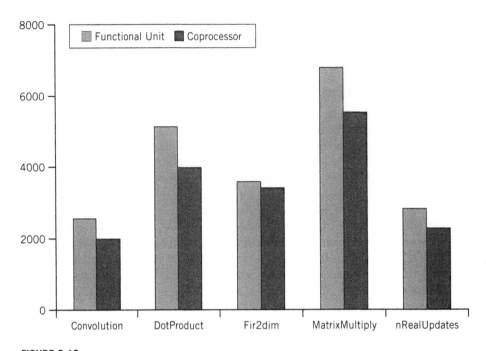

FIGURE 6.10

Functional unit versus coprocessor.

Table 6.1 presents various memory configurations. The numbers in Table 6.1 represent: the size of the memory module; the cache/stream buffer organizations *num_lines* × *num_ways* × *line_size* × *word_size*; the latency (in number of processor cycles); and the replacement policy (LRU or FIFO). Note that for the stream buffer, *num_ways* represents the number of FIFO queues present. The configurations are presented in the increasing order of cost in terms of area. The first configuration contains an L1 cache and a small stream buffer (256 bytes) to capitalize on the stream nature of the benchmarks. The second configuration contains the L1 cache and an on-chip direct mapped SRAM of 2K. A part of the arrays in the application is mapped to the SRAM. Due to the reduced control necessary for the SRAM, it has a small latency (of 1 cycle), and the area requirements are small. The third configuration contains L1 and L2 caches with FIFO replacement policy. Due to the control necessary for the L2 cache (of size 2K), the cost of this configuration is larger than Configuration 2. Configuration 4 contains an L1 cache, an L2 cache of size 1K, and a direct mapped SRAM of size 1K. Due to the extra buses to route the data to the caches and SRAM, this configuration has a larger cost than the previous one. The last configuration contains a large SRAM and has the largest area requirement. All the configurations contain the same off-chip DRAM module with a latency of 20 cycles.

Fig. 6.11 presents a subset of the experiments, showing the total cycle counts (including the time spent in the processor) for the set of benchmarks for different

Table 6.1   The memory subsystem configurations.

Config	L1 Cache	L2 Cache	SRAM	Stream Buffer	DRAM
1	4 x 2 x 4 x 4 latency = 1 (LRU)	—	—	4 x 4 x 4 x 4 latency = 4	latency = 20
2	4 x 2 x 4 x 4 latency = 1 (LRU)	—	2K latency = 1	—	latency = 20
3	4 x 2 x 4 x 4 latency = 1 (FIFO)	16 x 4 x 8 x 4 latency = 4 (FIFO)	—	—	latency = 20
4	4 x 2 x 4 x 4 latency = 1 (FIFO)	32 x 1 x 8 x 4 latency = 4 (FIFO)	1K latency = 1	—	latency = 20
5	—	—	8K latency = 1	—	latency = 20

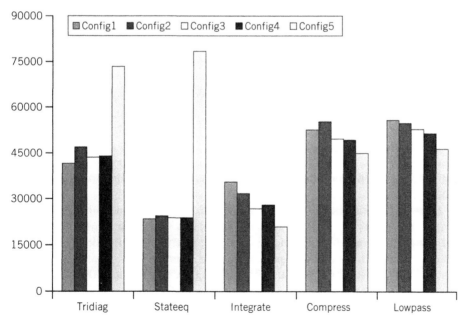

FIGURE 6.11

Cycle counts for the memory configurations.

memory configurations. Even though these benchmarks are kernels, we observed a significant variation in the trends shown by the different applications. For instance, in tridiag and stateeq, the first configuration (even though it has the lowest cost) performs better (lower cycle count means higher performance) due to the capability of the stream buffer to exploit the stream nature of the access patterns. Moreover, in these applications the most expensive configuration (Configuration 5) containing the large SRAM behaves poorly due to the fact that not all the arrays fit in the SRAM, and the lack of L1 cache to compensate the large latency of the DRAM creates its toll on the performance.

The expected trend of higher cost–higher performance was apparent in the applications integrate and lowpass. While the stream buffer in Configuration 1 has a comparable performance to the other configurations, Configuration 5 has the best behavior due to the low latency of the direct mapped on-chip SRAM. The complete study of the memory subsystem exploration can be found in [3].

6.3 ARCHITECTURE SYNTHESIS FOR RAPID PROTOTYPING

This section describes synthesizable HDL generation from the ADL specification. The generated hardware models can be used for both rapid prototyping (exploration) and functional validation of implementation. Section 6.3.2 describes various exploration experiments using the generated hardware models. The use of hardware models for implementation validation is discussed in Section 6.4.3.

6.3.1 Synthesizable HDL Generation

The hardware generation approach in EXPRESSION framework is similar to the cycle-accurate simulator generation technique using functional abstraction [17]. The only difference is that the generic components in the library are developed using HDL for hardware generation (instead of using C/C++ for simulator generation). This section briefly describes how to generate three major components of the processor: instruction decoder, datapath, and controller, using the generic HDL models [7].

A generic instruction decoder uses information regarding individual instruction format and opcode mapping for each functional unit to decode a given instruction. The instruction format information is available in the operation description. The decoder extracts information regarding opcode, operands, etc. from input instruction using the instruction format. The mapping section of the EXPRESSION ADL has the information regarding the mapping of opcodes to the functional units. The decoder uses this information to perform/initiate necessary functions (e.g., operand read) and decides where to send the instruction.

The implementation of a datapath consists of two parts. First, compose each component in the structure. Second, instantiate components (e.g., fetch, decode, ALU, LdSt, writeback, branch, caches, register files, memories, etc.) and establish connectivity using the appropriate number of latches, ports, and connections using the structural information available in the ADL. To compose each component in the structure we use the information available in the ADL regarding the functionality of the component and its parameters. For example, to compose an execution unit, it is necessary to instantiate all the opcode functionalities (e.g., ADD, SUB, etc., for an ALU) supported by that execution unit. Also, if the execution unit is supposed to read the operands, an appropriate number of operand read functionalities needs to be instantiated unless the same read functionality can be shared using multiplexors. Similarly, if this execution unit is supposed to write the data back to register file, the functionality for writing the result needs to be instantiated. The actual implementation of an execution unit might contain many more functionalities, for example, read latch, write latch, and insert/delete/modify reservation station (if applicable).

The controller is implemented in two parts. First, it generates a centralized controller (using generic controller function with appropriate parameters) that maintains the information regarding each functional unit, such as busy, stalled, etc. It also computes hazard information based on the list of instructions currently in the pipeline. Based on these bits and the information available in the ADL, it stalls/ flushes necessary units in the pipeline. Second, a local controller is maintained at each functional unit in the pipeline. This local controller generates certain control signals and sets necessary bits based on the input instruction. For example, the local controller in an execution unit will activate the add operation if the opcode is *add*, or it will set the busy bit in case of a multicycle operation.

6.3.2 Rapid Prototyping and Exploration

We have performed various exploration experiments using the generated hardware models for DLX processor based on silicon area, power, and clock frequency [7, 28]. We have used Synopsys Design Compiler [29] to synthesize the generated HDL description using LSI 10K technology libraries and obtained area, power, and clock frequency values. In this section we present three exploration experiments: pipeline path exploration, pipeline stage exploration, and instruction-set exploration. The reported area, power, and clock frequency numbers are for the execution units only. The numbers do not include the contributions from others components such as *Fetch*, *Decode*, *MEM*, and *WriteBack*.

Addition of functional units (pipeline paths)

Fig. 6.12 shows the exploration results due to addition of pipeline paths using an application program (kernel of the FFT benchmark). The first configuration has only one pipeline path consisting of *Fetch*, *Decode*, one execution unit (*Ex1*), *MEM*, and *WriteBack*. The *Ex1* unit supports five operations: *sin, cos*, $+$, $-$, and $\times$. The second configuration is exactly the same as the first configuration, except that it has one more execution unit (*Ex2*) parallel to *Ex1*. The *Ex2* unit supports three operations: $+$, $-$, and $\times$. Similarly, the third configuration has three parallel execution units: *Ex1*

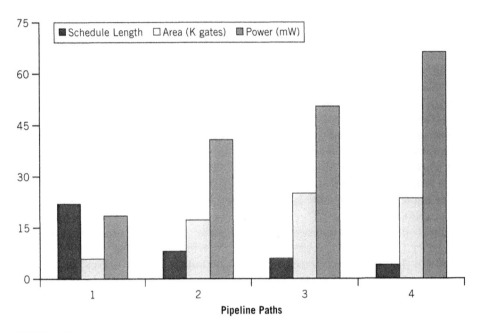

FIGURE 6.12

Pipeline path exploration.

$(+, -, \times)$, $Ex2$ $(+, -, \times)$, and $Ex3$ $(sin, cos, +, -, \text{and} \times)$. Finally, the fourth configuration has four parallel execution units: $Ex1$ (sin, cos), $Ex2$ $(+, -, \text{MAC})$[1], $Ex3$, and $Ex4$, where $Ex3$ and $Ex4$ are customized functional units that perform $a \times b + c \times d$.

The application program requires a fewer number of cycles (schedule length) due to the addition of pipeline paths, whereas the area and power requirement increases. The fourth configuration is interesting since both area and schedule length decrease due to the addition of specialized hardware and removal of operations from other execution units.

Addition of pipeline stages

Fig. 6.13 presents exploration experiments due to the addition of pipeline stages in the multiplier unit. The first configuration is a one-stage multicycle multiplier. The second, third, and fourth configurations use multipliers with two, three, and four stages, respectively. The clock frequency (speed) is improved due to addition of pipeline stages. The fourth configuration generated 30% speed improvement at the cost of 13% area increase over the third configuration.

---

[1]MAC performs multiply-and-accumulate of the form $a \times b + c$.

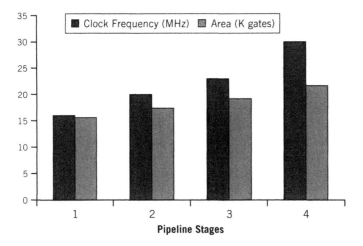

FIGURE 6.13

Pipeline stage exploration.

Addition of operations

Fig. 6.14 presents exploration results for addition of opcodes using three processor configurations. The three configurations are shown in Fig. 6.14. The first configuration has four parallel execution units: *FU1*, *FU2*, *FU3*, and *FU4*. The *FU1* supports three operations: $+$, $-$, and $\times$. The *FU2*, *FU3*, and *FU4* support $(+, -, \times)$, (*and, or*), and (*sin, cos*), respectively. The second configuration is obtained by adding a *cos* operation in the *FU3* of the first configuration. This generated reduction of schedule length of the application program at the cost of increase in area. The third configuration is obtained by adding multipliers both in *FU3* and *FU4* of the second configuration. This generated the best possible (using $+, -, \times, sin$, and *cos*) schedule length for the application program.

Each iteration in our exploration framework is in the order of hours to days depending on the amount of modification needed in the ADL and the synthesis time. However, each iteration will be in the order of weeks to months for manual or semi-automatic development of HDL models. The reduction of exploration time is at least an order of magnitude.

6.4 **FUNCTIONAL VERIFICATION**

The EXPRESSION ADL has been used to enable top-down validation of programmable SOC architectures [8]. This section describes three primary components in ADL-driven validation: validation of ADL specification, functional test generation,

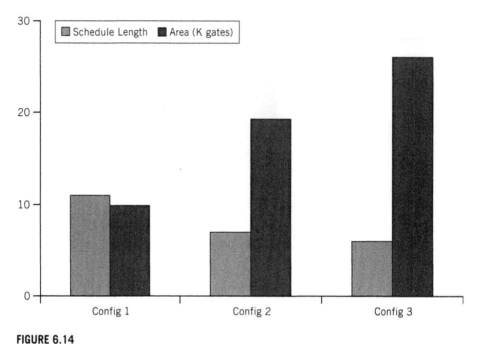

FIGURE 6.14

Instruction-set exploration.

and implementation validation using a combination of symbolic simulation and equivalence checking.

6.4.1 Specification Validation

One of the major challenges in validating the ADL specification is to verify the pipeline behavior in the presence of hazards and multiple exceptions. There are many important properties that need to be verified to validate the pipeline behavior. For example, it is necessary to verify that each operation in the instruction set can execute correctly in the processor pipeline. It is also necessary to ensure that execution of each operation is completed in a finite amount of time. Similarly, it is important to verify the execution style of the architecture. These properties are by no means complete to prove the correctness of the specification. Additional properties can be easily added and integrated into our validation framework.

We have developed validation techniques to ensure that the architectural specification is well formed by analyzing both static and dynamic behaviors of the specified architecture. Fig. 6.15 shows the flow for verifying the ADL specification. The graph model as well as the FSM model of the architecture are generated from the specification. We have developed algorithms to verify several architectural properties,

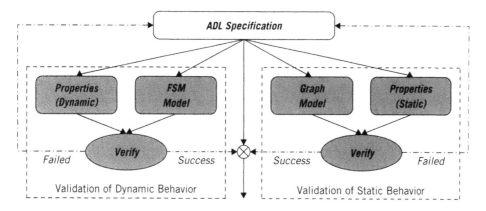

FIGURE 6.15

Validation of specification.

such as connectedness, false pipeline, data-transfer paths, completeness, and finiteness [9, 30, 31]. The dynamic behavior is verified by analyzing the instruction flow in the pipeline using a finite-state machine (FSM)-based model to validate several important architectural properties such as determinism and in-order execution in the presence of hazards and multiple exceptions [32–34]. The property checking framework determines if all the necessary properties are satisfied. In case of a failure, it generates traces so that a designer can modify the specification of the architecture. The validated ADL specification is used as a golden reference model for generating various executable models such as simulators and hardware implementations.

6.4.2 Test Generation Using Model Checking

Algorithm 1 shows three major steps in ADL-driven test generation using model checking. We use SMV model checker [35] in our framework. The first step produces one property (negated version) for each fault. The second step generates the SMV model of the processor architecture. The final step applies each property and the processor model to the model checker to generate the counterexample. The counterexample is analyzed to generate the test program consisting of instruction sequences.

For example, to generate a testcase for assigning a value 5 to a register R_7, the property states that "$R_7 \, != 5$". The model checker produces a counterexample which is converted to a test program. The conversion is straightforward in our framework since we model the design at the cycle-accurate level and instructions are modeled as a structure consisting of opcode, source, and destination operands. As a result, the counterexample consists of several instructions at different clock cycles that can be translated into an actual test by a simple text analysis. Based on the coverage report additional properties can be added or the fault model can be modified.

The model checking-based test generation (Algorithm 1) is a promising approach for automated generation of directed microarchitectural tests to exercise various intricate interactions and corner cases. Since the complete processor is used during model checking, this approach is limited by the capacity restrictions of the model checking tool. As a result, this approach is not suitable for today's pipelined processors since the time and memory requirements can be prohibitively large in many test generation scenarios. In fact, test generation may not be possible in various instances due to state space explosion. We proposed a test generation approach using decomposition of the processor model and properties to make the ADL-driven test generation applicable in practice.

Fig. 6.16 shows our test generation framework using decompositional model checking. The basic idea is to break one processor-level property into multiple module-level properties and apply them to the respective modules. In case the property was applied to an internal module, the generated counterexample is used to extract the output requirements for the parent module(s). This iteration continues until the primary input assignments (e.g., assignment to register file or instruction memory) are obtained. These primary input assignments are converted into test programs consisting of instruction sequences. Since the model checker is applied only at the module level, this approach can handle larger designs and reduce the test generation time. It is important to note that various constraints have to be captured in an efficient manner to generate a useful test. We model global input and environmental constraints as part of the design. We model local input constraints as part of the property during decomposition to ensure that model checker always generates correct partial counterexamples.

Algorithm 1: *Model Checking based Test Generation Algorithm*
Inputs: 1. Graph Model of the architecture (G).
        2. Functional fault model (F).
Output: Test programs for detecting all the faults in the fault model.
begin /*** *PropertyList* = {} ***/
    Step 1: **for** each fault f in the fault model F
            $prop_f$ = GenerateProperty(f); /*SMV version*/
            PropertyList = *PropertyList* $\cup$ *prop_f*;
        endfor
    Step 2: *design* = GenerateSMVmodel(G)
    Step 3: *TestProgramList* = {}
        for each property p in the *PropertyList*
        $test_p$ = ApplyModelChecking(p, *design*);
        TestProgramList = *TestProgramList* $\cup$ *test_p*;
        endfor
    return *TestProgramList*.
end

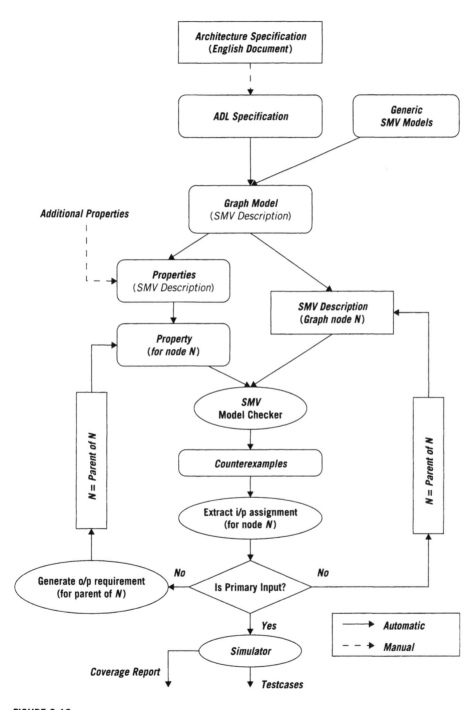

FIGURE 6.16

Test generation using decompositional model checking.

Test generations using module-level decompositions is very useful, but this approach may not be able to generate tests where module-level decompositions are not possible. We have developed novel design and property decomposition techniques [36] as well as SAT-based bounded model checking techniques [37] to address the state space explosion problem in test generation. Our initial study using a Freescale e500 processor [38] showed promising results [39] in applying the decompositional model checking-based test generation on industrial microprocessors.

6.4.3 Implementation Validation

Fig. 6.17 shows our top-down validation methodology. Logic designers implement the architecture at register-transfer level (*RTL design*). The structure and behavior of

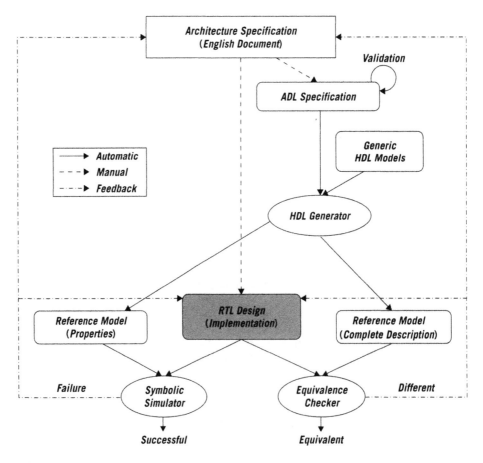

FIGURE 6.17

Implementation validation using formal methods.

the processor are captured using an ADL. The ADL description is validated to ensure that it specifies a well-formed architecture [30, 34], as described in Section 6.4.1. The reference model (HDL description) is automatically generated from the ADL specification [7], as described in Section 6.3. To verify that the implementation (*RTL design*) satisfies certain properties, our framework generates behaviors for the intended properties. We use the Versys2 symbolic simulator [40] to perform property checking. A counterexample is generated if a property fails in the *RTL design*. The feedback is used to modify the *RTL design*. Our framework also generates the complete description of the processor to enable equivalence checking using Formality [41]. In case of failure, the feedback is used to modify the *RTL design*. In case of an ambiguity in the original description that leads to the mismatch, the architecture specification needs to be updated.

Property checking using symbolic simulation

Symbolic simulation combines traditional simulation with formal symbolic manipulation [42]. Each symbolic value represents a signal value for different operating conditions, parameterized in terms of a set of symbolic Boolean variables. By this encoding, a single symbolic simulation run can cover many conditions that would require multiple runs of a traditional simulator. Fig. 6.18(a) shows a simple n-input *and* gate. Exhaustive simulation of the *and* gate requires 2^n binary test vectors. However, the ternary simulation (uses 0, 1, and x) requires $(n + 1)$ test vectors for the *and* gate. Fig. 6.18(b) shows the vectors: n vectors with one input set to "1" and the remaining inputs set to 'x', and one vector with all inputs set to "1". Finally, symbolic simulation [42] requires only one vector using n symbols $(S_1, S_2, ..., S_n)$ as shown in Fig. 6.18(c).

To verify that the implementation (*RTL design*) satisfies certain properties, our framework generates behaviors for the intended properties instead of generating the complete reference design. We use the Versys2 [40] that uses symbolic trajectory

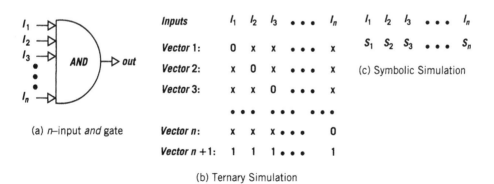

(a) n–input *and* gate

(b) Ternary Simulation

(c) Symbolic Simulation

FIGURE 6.18

Test vectors for validation of an *and* gate.

evaluation to perform property checking. It is necessary to manually specify the state mappings between the reference model and the implementation. This involves mapping of both latches and bit cells by specifying their names. The assertions are automatically generated from the reference model [43]. Versys2 symbolically simulates the implementation (*RTL design*) by using the generated assertions to ensure that the implementation satisfies the reference model. A counterexample is generated if an assertion fails in the *RTL design*. The feedback is used to modify the *RTL design*.

Equivalence checking

Equivalence checking is a branch of static verification that employs formal techniques to prove that two versions of a design are either functionally equivalent or not. The equivalence checking flow consists of four stages: *read*, *match*, *verification*, and *debug*. The match and verification stages are those most impacted by design transformations. During the *read* stage, both versions of the design are read by the equivalence checking tool and segmented into manageable sections called logic cones. Logic cones are groups of logic bordered by registers, ports, or black boxes. In *match* phase, the tool attempts to match, or map, compare points from the reference (golden) design to their corresponding compare point within the implementation design [44]. Two types of matching techniques are used: nonfunction (name-based) and function-based (signature analysis). During the *verification* stage, each matched compare point is proven either functionally equivalent or nonequivalent [45, 46].

Our framework generates the complete description of the processor to enable equivalence checking using Synopsys Formality [41]. The tool reads both reference design and the implementation, and attempts to match the compare points between them. The unmatched compare points need to be mapped manually. The tool tries to establish equivalence for each matched compare point. In case of failure, the failing compare points are analyzed to verify whether they are actual failures or not. The feedback is used to perform additional setup (in case of a false negative), or to modify the implementation (*RTL design*).

6.5 CONCLUSIONS

This chapter presented EXPRESSION ADL and its associated design automation methodologies. The powerful constructs in EXPRESSION allow specification of a wide variety of processor, coprocessor, and memory architectures. Automatic generation of a compiler and simulator enables fast and efficient design space exploration of architectures. The elegant formalism in EXPRESSION also enables top-down validation of complex architectures to complement existing bottom-up verification techniques. The EXPRESSION is successfully used to specify various architectures and perform early exploration, rapid prototyping, and functional verification of programmable SOC architectures.

REFERENCES

[1] A. Halambi, P. Grun, V. Ganesh, A. Khare, N. Dutt, and A. Nicolau. EXPRESSION: A language for architecture exploration through compiler/simulator retargetability. In *Proc. of Design Automation and Test in Europe (DATE)*, pages 485–490, 1999.

[2] P. Grun, A. Halambi, N. Dutt, and A. Nicolau. RTGEN: An algorithm for automatic generation of reservation tables from architectural descriptions. *IEEE Transactions on Very Large Scale Integration (VLSI) Systems*, 11(4):731–737, August 2003.

[3] P. Mishra, M. Mamidipaka, and N. Dutt. Processor-memory co-exploration using an architecture description language. *ACM Transactions on Embedded Computing Systems (TECS)*, 3(1):140–162, 2004.

[4] A. Shrivastava, P. Biswas, A. Halambi, N. Dutt, and A. Nicolau. Compilation framework for code size reduction using reduced bit-width ISAs (rISAs). *ACM Transactions on Design Automation of Electronic Systems (TODAES)*, 11(1):123–146, January 2006.

[5] M. Reshadi, P. Mishra, and N. Dutt. A retargetable framework for instruction-set architecture simulation. *To appear in ACM Transactions on Embedded Computing Systems (TECS)*, 5(2):431–452, 2006.

[6] P. Mishra, A. Shrivastava, and N. Dutt. Architecture description language (ADL)-driven software toolkit generation for architectural exploration of programmable SoCs. *ACM Transactions on Design Automation of Electronic Systems (TODAES)*, 11(3):1–33, July 2006.

[7] P. Mishra, A. Kejariwal, and N. Dutt. Synthesis-driven exploration of pipelined embedded processors. In *Proc. of International Conference on VLSI Design*, pages 921–926, 2004.

[8] P. Mishra and N. Dutt. *Functional Verification of Programmable Embedded Architectures: A Top-Down Approach*, Springer, July 2005.

[9] P. Mishra and N. Dutt. Automatic modeling and validation of pipeline specifications. *ACM Transactions on Embedded Computing Systems (TECS)*, 3(1):114–139, 2004.

[10] S. Novack, A. Nicolau, and N. Dutt. Resource directed loop pipelining: Exposing just enough parallelism. *The Computer Journal*, 40(6):311–321, 1997.

[11] A. Nicolau and S. Novack. Trailblazing: A hierarchical approach to percolation scheduling. In *Proc. of International Conference on Parallel Processing (ICPP)*, pages 120–124, 1993.

[12] S. Novack, A. Nicolau, and N. Dutt. A unified code generation approach using mutation scheduling. In *Code Generation for Embedded Processors*, pages 203–218, Kluwer, 1994.

[13] P. Grun, N. Dutt, and A. Nicolau. Access pattern-based memory and connectivity architecture exploration. *ACM Transactions on Embedded Computing Systems (TECS)*, 2(1):33–73, 2003.

[14] P. Grun, N. Dutt, and A. Nicolau. Mist: An algorithm for memory miss traffic management. In *Proc. of International Conference on Computer-Aided Design (ICCAD)*, pages 431–437, 2000.

[15] P. Grun, N. Dutt, and A. Nicolau. Memory aware compilation through accurate timing extraction. In *Proc. of Design Automation Conference (DAC)*, pages 316–321, 2000.

[16] A. Shrivastava, N. Dutt, A. Nicolau, and E. Earlie. PBExplore: A framework for compiler-in-the-loop exploration of partial bypassing in embedded processors. In *Proc. of Design Automation and Test in Europe (DATE)*, pages 1264–1269, 2005.

[17] P. Mishra, N. Dutt, and A. Nicolau. Functional abstraction driven design space exploration of heterogeneous programmable architectures. In *Proc. of International Symposium on System Synthesis (ISSS)*, pages 256–261, 2001.

[18] M. Reshadi, N. Bansal, P. Mishra, and N. Dutt. An efficient retargetable framework for instruction-set simulation. In *Proc. of International Symposium on Hardware/Software Codesign and System Synthesis (CODES+ISSS)*, pages 13–18, 2003.

[19] M. Reshadi, P. Mishra, and N. Dutt. Instruction set compiled simulation: A technique for fast and flexible instruction set simulation. In *Proc. of Design Automation Conference (DAC)*, pages 758–763, 2003.

[20] A. Khare, N. Savoiu, A. Halambi, P. Grun, N. Dutt, and A. Nicolau. V-SAT: A visual specification and analysis tool for system-on-chip exploration. In *Proc. of EUROMICRO Conference*, pages 1196–1203, 1999.

[21] P. Mishra, F. Rousseau, N. Dutt, and A. Nicolau. Architecture description language driven design space exploration in the presence of coprocessors. In *Proc. of Synthesis and System Integration of Mixed Technologies (SASIMI)*, 2001.

[22] P. Grun, N. Dutt, and A. Nicolau. Apex: Access pattern based memory architecture customization. In *Proc. of International Symposium on System Synthesis (ISSS)*, pages 25–32, 2001.

[23] P. Grun, N. Dutt, and A. Nicolau. Memory system connectivity exploration. In *Proc. of Design Automation and Test in Europe (DATE)*, pages 894–901, 2002.

[24] P. Mishra, P. Grun, N. Dutt, and A. Nicolau. Processor-memory co-exploration driven by an architectural description language. In *Proc. of International Conference on VLSI Design*, pages 70–75, 2001.

[25] P. Biswas and N. Dutt. Reducing code size for heterogeneous-connectivity-based VLIW DSPs through synthesis of instruction set extensions. In *Proc. of Compilers, Architectures, Synthesis for Embedded Systems (CASES)*, pages 104–112, 2003.

[26] S. Pasricha, P. Biswas, P. Mishra, A. Shrivastava, A. Mandal, N. Dutt, and A. Nicolau. A framework for GUI-driven design space exploration of a MIPS4K-like processor. Technical Report CECS 03-17, University of California, Irvine, 2003.

[27] Texas Instruments. *TMS320C6201 CPU and Instruction Set Reference Guide*, 1998.

[28] P. Mishra, A. Kejariwal, and N. Dutt. Rapid exploration of pipelined processors through automatic generation of synthesizable RTL models. In *Proc. of Rapid System Prototyping (RSP)*, pages 226–232, 2003.

[29] Synopsys, Inc. *http://www.synopsys.com*, February 20, 2008.

[30] P. Mishra, H. Tomiyama, A. Halambi, P. Grun, N. Dutt, and A. Nicolau. Automatic modeling and validation of pipeline specifications driven by an architecture description language. In *Proc. of Asia South Pacific Design Automation Conference (ASPDAC) / International Conference on VLSI Design*, pages 458–463, 2002.

[31] P. Mishra, N. Dutt, and A. Nicolau. Automatic validation of pipeline specifications. In *Proc. of High Level Design Validation and Test (HLDVT)*, pages 9–13, 2001.

[32] P. Mishra, H. Tomiyama, N. Dutt, and A. Nicolau. Automatic verification of in-order execution in microprocessors with fragmented pipelines and multicycle functional units. In *Proc. of Design Automation and Test in Europe (DATE)*, pages 36–43, 2002.

[33] P. Mishra and N. Dutt. Modeling and verification of pipelined embedded processors in the presence of hazards and exceptions. In *Proc. of Distributed and Parallel Embedded Systems (DIPES)*, pages 81-90, 2002.

[34] P. Mishra, N. Dutt, and H. Tomiyama. Towards automatic validation of dynamic behavior in pipelined processor specifications. *Kluwer Design Automation for Embedded Systems (DAES)*, 8(2-3):249-265, June-September 2003.

[35] http://www.cs.cmu.edu/~ modelcheck. *Symbolic Model Verifier*, February 20, 2008.

[36] H. Koo and P. Mishra. Functional test generation using property decompositions for validation of pipelined processors. In *Proc. of Design Automation and Test in Europe (DATE)*, pages 1240-1245, 2006.

[37] H. Koo and P. Mishra. Test generation using SAT-based bounded model checking for validation of pipelined processors. In *Proc. of ACM Great Lakes Symposium on VLSI (GLSVLSI)*, pages 362-365, 2006.

[38] http://www.freescale.com/files/32bit/doc/ref_manual/e500CORERM.pdf. *PowerPC^{TM} e500 Core Family Reference Manual*, November 15, 2007.

[39] H.-M. Koo, P. Mishra, J. Bhadra, and M. Abadir. Directed microarchitectural test generation for an industrial processor: A case study. In *Proc. of Microprocessor Test and Verification (MTV)*, pages 33-36, 2006.

[40] N. Krishnamurthy, M. Abadir, A. Martin, and J. Abraham. Design and development paradigm for industrial formal verification tools. *IEEE Design & Test of Computers*, 18(4):26-35, July-August 2001.

[41] Synopsys Formality. *http://www.synopsys.com*, February 20, 2008.

[42] R. Bryant. Symbolic simulation—techniques and applications. In *Proc. of Design Automation Conference (DAC)*, pages 517-521, 1990.

[43] L. Wang, M. Abadir, and N. Krishnamurthy. Automatic generation of assertions for formal verification of PowerPC microprocessor arrays using symbolic trajectory evaluation. In *Proc. of Design Automation Conference (DAC)*, pages 534-537, 1998.

[44] D. Anastasakis, R. Damiano, H. Ma, and T. Stanion. A practical and efficient method for compare-point matching. In *Proc. of Design Automation Conference (DAC)*, pages 305-310, 2002.

[45] C. Eijk. Sequential equivalence checking without state space traversal. In *Proc. of Design Automation and Test in Europe (DATE)*, pages 618-623, 1998.

[46] J. Marques-Silva and T. Glass. Combinational equivalence checking using satisfiability and recursive learning. In *Proc. of Design Automation and Test in Europe (DATE)*, pages 145-149, 1999.

ASIP Meister

Yuki Kobayashi, Yoshinori Takeuchi, and Masaharu Imai

7

The ASIP (Application Specific Instruction-set Processors) Meister is an ASIP design environment that provides designers with an interactive environment to specify, design, and generate an ASIP and possibly legacy off-the-shelf processor cores based on excellent implementation methodology. The ASIP Meister generates not only an HDL description of a target ASIP, but also application program development tools including an assembler, a compiler, and an instruction-set simulator. Designers enter the specifications of a target processor using a GUI dedicated to ASIP design that enables greater design productivity, as designers do not need to be concerned with ADL syntax. The ASIP Meister supports a flexible pipeline model where fetch or decode stages can be changed along with execution stages whose modification ADLs usually support. Then designers can perform a wide range of design space exploration in a short time. An extension to VLIW architectures has also been developed in recent years. The ASIP Meister, which has been used in universities as well as companies, contributes to state-of-the-art processor research and education by providing easy prototyping environments of configurable processors.

This chapter is organized as follows. Section 7.1 describes an overview of ASIP Meister. An architecture model of generated processors is introduced in Section 7.2. Section 7.3 describes processor specifications for ASIP Meister. Section 7.4 describes an HDL generation method employed in ASIP Meister, and Section 7.5 describes various case studies generated with ASIP Meister. Finally, Section 7.6 concludes this chapter.

7.1 OVERVIEW OF ASIP MEISTER

The ASIP Meister is an interactive design environment of ASIPs. Designers can specify processor architecture and obtain synthesis-oriented HDL description, simulation-oriented HDL description, and application program (AP) development tools in a short time. Once designers enter a high-level specification of a target processor, modifying or extending the processor architecture is very easy. In this way, ASIP Meister improves the design productivity of ASIPs.

163

7.1.1 Framework Overview

The ASIP Meister combines the advantages of both the *fixed base processor core approach* and the *ADL approach* by providing designers with the configurability of base processors in an easy to use interactive environment and the flexibility of specifying a new processor core behavior. A dedicated Graphical User Interface (GUI) of ASIP Meister provides designers with the easy configuration of a target processor. At the same time, since the behavior of the instructions is fully configurable using microoperation description, it achieves the high flexibility of processor architecture.

Fig. 7.1 shows an overview of ASIP Meister design methodology. Designers enter the target processor specifications using Architecture Design Entry System (GUI). Based on the specification, the processor synthesizer generates synthesis-oriented and simulation-oriented HDL descriptions, and then the AP Development Tool Generator generates such software tools as a compiler and an assembler. Using the simulation-oriented HDL description and the AP development tools, designers can evaluate the target processor performance through hardware/software co-simulation, and using the synthesis-oriented HDL description, designers can perform logic synthesis for implementation on FPGA or LSI.

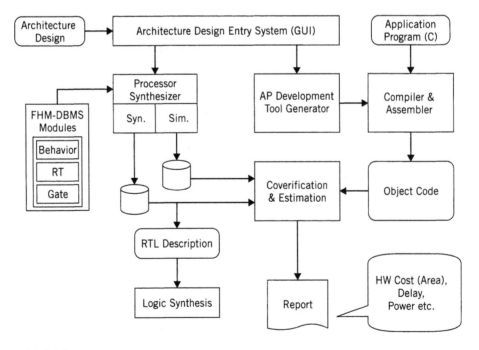

FIGURE 7.1

Overview of ASIP Meister.

Processor core design in ASIP Meister [1, 2] mainly depends on *design reuse* and *cycle-based specification*. Design reuse is demonstrated by a parameterized resource (i.e., hardware component) model called the Flexible Hardware Model (FHM) [3]. The parameters include the design abstraction levels (such as behavior, RT, and gate) of the target architecture, the hardware algorithm, and the bit width. Designers can select suitable resources from the FHM database for their processor design. The FHM Data Base Management System (FHM-DBMS) is utilized in the framework to obtain an instance of resources from the FHM database. The FHM-DBMS can also estimate the design quality (area, delay, and power consumption) of the selected resource. Designers can evaluate the quality of the target processor in an early stage of design, that is, before detailed design and HDL generation, hence, the design iteration cost is reduced. Since FHM can contain different design abstraction levels, the processor synthesizer can generate both synthesis-oriented and simulation-oriented HDL descriptions. In ASIP Meister, designers develop their ASIPs with higher-level abstraction language called microoperation description language. In this language, designers specify the instruction behavior of each pipeline stage independent of other instructions. The benefit of this language is that designers do not have to consider pipeline registers, control logic, and datapath selectors. Furthermore, since specifications of instruction are independent of each other, designers can easily modify the target processor, delete unnecessary instructions, or rapidly add new effective instructions.

Application system designers need software tools, such as an assembler, a compiler, and a simulator, to develop application programs for each designed processor. Therefore, software development tools play an important role in ASIP design. ASIP Meister's meta-assembler is a retargetable assembler that reads the target architecture specifications including instruction format, storage format, and addressing modes; then it performs assembly. The ASIP Meister's compiler generator [4] produces a target compiler from the instruction-set, storage, and structural specifications. A cycle-based Instruction-Set level Simulator (ISS) [5] is also generated from the specifications.

Processor design flow using ASIP Meister

Fig. 7.2 shows the processor design flow using ASIP Meister, where a processor is designed step by step. First, the design goal and pipeline architecture information are set. Then the processor outline is specified, which includes a declaration of resources, a definition of the input and output ports of the processor, and a definition of the instruction format. With such information, designers can perform a first-cut estimation. In the microoperation description step, designers specify the behavior of instruction in a pipeline stage level. The datapath of the processor is constructed based on the specification. Then ASIP Meister generates an HDL description for the entire processor that consists of the datapath and control logic.

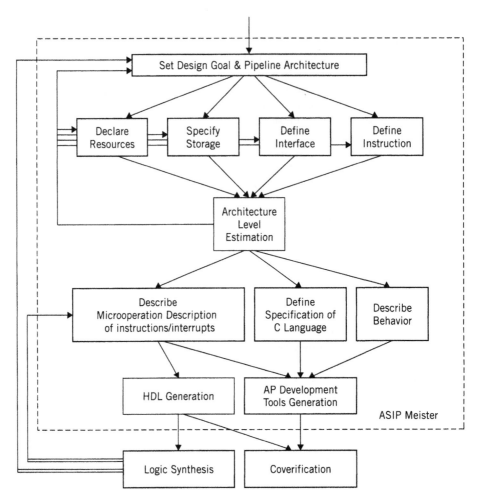

FIGURE 7.2

Design flow of ASIP Meister.

Designers also specify the behavior of instruction in an instruction level in the behavior description step. Then AP development tools are also generated based on the information.

Input and output

The input and output of ASIP Meister are summarized as follows:

- Input
  - Processor specification entered with GUI
  - Additional model for FHM-DB (optional)

- Output
  - Synthesis-oriented HDL description of target processor [6]
  - Simulation-oriented HDL description of target processor [6]
  - Assembler
  - Instruction-set simulator (ISS) [5]
  - Compiler [4]

7.1.2 Features

The following are the main features of ASIP Meister:

- Flexibility of target architecture
  - Number of pipeline stages can be changed
  - All pipeline stages including fetch stage are fully customizable
  - Hardware resources are flexible due to FHM
  - Multicycle execution of resources can be handled
  - Delayed branch with a variable number of delayed slots is supported
  - External and internal interrupts (exceptions) can be handled
  - Instruction format and processor interface ports are fully customizable
- Automatic generation of pipelined datapath and control logic
  - Automatic insertion of pipeline registers
  - Automatic insertion of datapath selectors
  - Resolution of structural hazard
  - Pipeline stall due to multicycle execution
  - Pipeline flush due to control hazard (e.g., jump/branch)
- User friendly GUI for rapid prototyping
- Design reuse using FHM
- Early estimation of design quality

Advantage of our approach

The ASIP Meister utilizes dedicated GUI for design entry. Since designers do not need to be concerned about ADL format or syntax, this benefits the productivity of ASIPs and frees designers from tedious matters.

Approaches based on a fixed base processor usually do not have such a flexible method, limiting the design flexibility of the target processor. On the contrary, flexible ADL-based approaches tend to be error prone since they typically require detailed specifications for pipeline control logic, and so on. The ASIP Meister employs a very flexible language and microoperation description for representing instruction behavior. In the ASIP Meister paradigm, designers can configure almost any kind of behavior that they imagine. Furthermore, designers do not need to specify pipeline registers and datapath selectors since they are automatically generated in ASIP Meister. The FHM contributes not only design productivity due to design reuse, but also flexibility; once designers implement a specific hardware component for their target processor, by registering it as FHM, it can be easily integrated into the processor with proper control logic added by ASIP Meister.

7.1.3 A Short History

The ASIP Meister was formerly called the PEAS-III system, which was developed in collaboration with the Toyohashi University of Technology, Osaka and Shizuoka Universities, the Tsuruoka National College of Technology, and the Semiconductor Technology Academic Research Center (STARC), Japan. PEAS-III research was launched in 1995, and PEAS-III was rewritten and released in 2002 as ASIP Meister. The ASIP Meister had been distributed among academia from 2002 to 2005, and the number of registered users was 180 from 37 countries/regions, as of January 2006. ASIP Meister is widely recognized all over the world as a processor development tool that can change various architecture parameters including pipeline structure and even the behavior of fetch and decode, as well as execution stages. Since April 2005, ASIP Solutions Inc. has taken over the maintenance and development of ASIP Meister and released ASIP Meister. Several companies have already started using it for their commercial products.

7.1.4 ASIP Meister Usage in Academia

Many academic research groups have used ASIP Meister for implementing and confirming their state-of-the-art architecture. Due to ASIP Meister's rapid design and its flexible processor model, their research succeeded. Fei et al., who changed fetch and decode stages to monitor code integrity, used ASIP Meister to implement an ASIP that supports program code integrity monitoring, since the feature requires a very flexible pipeline model for ASIP design tools [7]. Ragel et al. proposed a hardware technique for fault detection [8, 9] for security threats in ASIPs. The proposed approaches employ ASIP Meister to tweak an ASIP to detect attacks by adding comparators that check code integrity. Radhakrishnan et al. used ASIP Meister in research on multi-pipeline processors [10, 11]. They generated each pipeline using ASIP Meister and then built a multi-pipeline processor by integrating the pipelines. Due to the easiness of ASIP design, many processor instances can be obtained in a short time. Many universities also have adopted ASIP Meister as an education tool in teaching and lab classes. The ASIP Meister is used in courses in Universität Karlsruhe in Germany and Kinki, Shizuoka, and Osaka Universities in Japan.

7.2 ARCHITECTURE MODEL

This section describes an architecture model of a processor in ASIP Meister, which targets a pipeline processor. Fig. 7.3 shows a processor model in ASIP Meister. A processor consists of a datapath and a controller. A processor also consists of multiple pipeline stages.

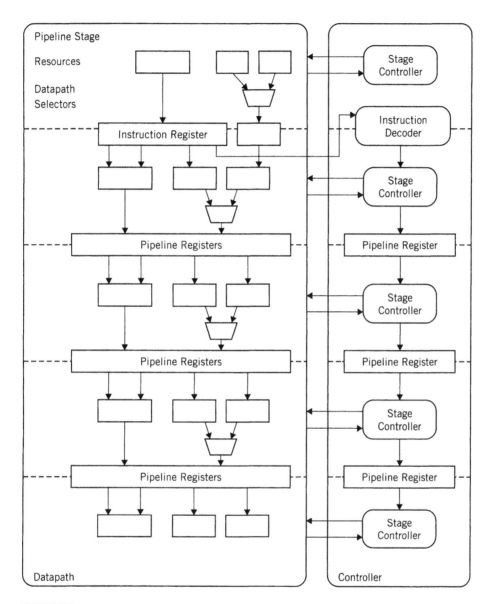

FIGURE 7.3

Processor model.

Fig. 7.4 shows a model of a pipeline stage in ASIP Meister. Each pipeline stage consists of datapath and controller parts. The datapath part, which consists of resources, pipeline registers, and datapath selectors, is controlled by the controller part.

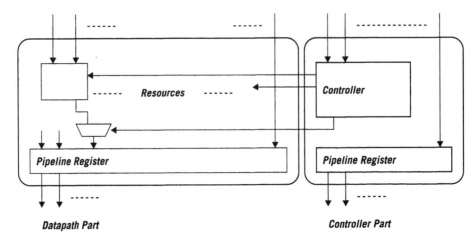

Datapath Part   **Controller Part**

FIGURE 7.4

Pipeline stage model.

Functional execution in the microoperation description is performed by a resource, and an assignment is implemented as a connection between resources. The datapath selector is used to resolve signal conflicts. The results of resource computation are transferred to the next stage by pipeline registers.

7.3 ADL IN ASIP MEISTER

This section introduces the ADL used in ASIP Meister.

7.3.1 Overview of ADL Structure

The ADL structure in ASIP Meister is shown here:

- Pipeline architecture
- Design goal
- Resource declaration
- Storage definition
- Instruction type declaration
- Instruction declaration
- Interface definition
- Type specification definition
- Interrupt definition
- Instruction definition (microoperation description)

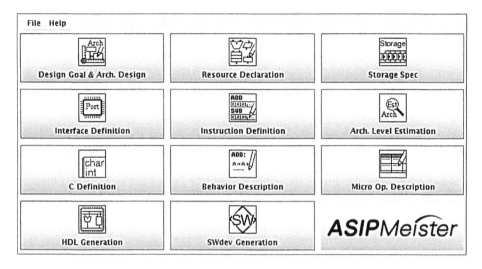

FIGURE 7.5

Main window of ASIP Meister.

7.3.2 Specification Entry Using GUI

Designers enter the specification of their target processor using GUI in ASIP Meister. Fig. 7.5 shows a snapshot of the main window of ASIP Meister. From this window, designers invoke a subwindow corresponding to each design step shown in Fig. 7.2.

For instance, the window shown in Fig. 7.6 is a subwindow for the design step of instruction type and instruction definition. In this window, designers declare instruction types and then define instructions derived from the instruction type. As discussed earlier, since the GUI clarifies the required fields and omits syntax errors, design productivity is improved. Also, a check box for each instruction or instruction type easily disables the instruction or type, which is useful during design space exploration where many parameters are being managed.

Fig. 7.7 shows a FHM browser window, where designers can add hardware resources to their target processor by choosing a proper model from the FHM database. In ASIP Meister, designers easily modify the parameterized model depending on their application, and instantiate it for the processor.

Fig. 7.8 shows a microoperation description window, where designers describe microoperation for each pipeline stage for a selected instruction in the left column. Section 7.3.3 introduces the details of microoperation description.

7.3.3 Microoperation Description

In microoperation description, designers define clock-based instruction and interrupt behaviors. In the microoperation description of instruction, instruction

File　Edit　Search　Sort　　　　　　　　　　　　　　　　　　　　　　Help

☑ Complete

| Instruction type/Instruction | Exception |

Instruction Type Definition　　　　　　　　　　　　　　　　　　New Type

Valid	Name	#	MSB	LSB	Field Type	Field Attr	Name/Value	Addr mode	Operand Name	element	reg class
			31	26	OP-code	name	opecode				
			25	21	Operand	name	rs0	Reg Direct	rs0	Resource	GPR
☑	R_R	1	20	16	Operand	name	rs1	Reg Direct	rs1	Resource	GPR
			15	11	Operand	name	rd	Reg Direct	rd	Resource	GPR
			10	0	OP-code	name	func				
			31	26	OP-code	name	opecode				
			25	21	Operand	name	rs0	Reg Direct	rs0	Resource	GPR
☑	R_I	1	20	16	Operand	name	rd	Reg Direct	rd	Resource	GPR
			15	0	Operand	name	const	Immediate data	const	Immediate	
			31	26	OP-code	name	opecode				
☑	L_S	1	25	21	Operand	name	rs0	Reg Indirect with Disp	addr	Resource	GPR

Instruction Field Definition　　　　　　　　　　　　　　　　　　New Instruction

Valid	Name	Type	#	Field	Format
☑	ADD	R_R	1	0,0,0,0,0,0 r,s,0 r,s,1 r,d 0,0,0,0,0,1,0,0,0,0,0	ADD rs0 rs1 rd
☑	ADDU	R_R	1	0,0,0,0,0,0 r,s,0 r,s,1 r,d 0,0,0,0,0,1,0,0,0,0,1	ADDU rs0 rs1 rd
☑	ADDI	R_I	1	0,0,1,0,0,0 r,s,0 r,d c,o,n,s,t	ADDI rs0 const rd
☑	ADDUI	R_I	1	0,0,1,0,0,1 r,s,0 r,d c,o,n,s,t	ADDUI rs0 const rd
☑	SUB	R_R	1	0,0,0,0,0,0 r,s,0 r,s,1 r,d 0,0,0,0,0,1,0,0,0,1,0	SUB rs0 rs1 rd
☑	SUBU	R_R	1	0,0,0,0,0,0 r,s,0 r,s,1 r,d 0,0,0,0,0,1,0,0,0,1,1	SUBU rs0 rs1 rd
☑	SUBI	R_I	1	0,0,1,0,1,0 r,s,0 r,d c,o,n,s,t	SUBI rs0 const rd
☑	SUBUI	R_I	1	0,0,1,0,1,1 r,s,0 r,d c,o,n,s,t	SUBUI rs0 const rd
☑	MULT	R_R	1	0,0,0,0,0,0 r,s,0 r,s,1 r,d 0,0,0,0,0,0,1,1,0,0,0	MULT rs0 rs1 rd

FIGURE 7.6

Instruction definition window.

operations are described for each pipeline stage, such as reading values from the register file, execution on ALU, and writing back the execution result to the register file. In the microoperation description of interrupts, interrupt operations are described, such as setting specific values to status registers and jumping to the interrupt handler routine.

Microoperation consists of three kinds of statements: (i) operation executed by a resource, e.g., arithmetic and logical operation, read from/write to register file, (ii) data transfer between resources, and (iii) conditional execution of (i) and (ii).

Fig. 7.9 shows an example of the microoperation description of ADD instruction. In Fig. 7.9, the behaviors in pipeline Stages 1, 2, 3, and 5 are described. First, the keyword *wire* declares three 32-bit variables, *src0*, *src1*, and *res*, as global variables (lines 2–4). When a global variable is accessed from different stages, pipeline registers are inserted between stages. There are three other 5-bit global variables, *rs0*, *rs1*, and *rd* (lines 5–7) for the operand of the ADD instruction that are automatically inserted by ASIP Meister, based on the specification of the instruction format. Note that designers only describe lines marked with * in Fig. 7.9. In Stage 1, an instruction address is read from the Program Counter (PC) using function *read( )* (line 13). The address is then fed into the Instruction Memory InterFace Unit (IMIFU), which returns an instruction corresponding to the address (line 14). The fetched instruction is stored in the Instruction Register (IR), so that the instruction decoder decodes the instruction (line 15). PC is simultaneously incremented to proceed to

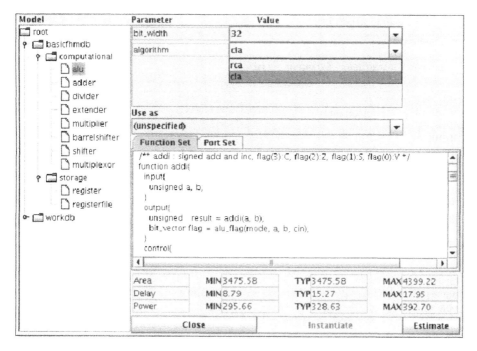

FIGURE 7.7

FHM browser window.

the next instruction (line 16). In Stage 2, ASIP Meister first generates a description that reads the operand from IR. Then, values for operands *src0* and *src1* are read from the general purpose register *GPR* (lines 26 and 27). In Stage 3, *src0* and *src1* are summed by *ALU*, and the result is stored into *res* (line 33). In Stage 4, there is no description in the ADD instruction. The result is simply conveyed through the global variable and is written back into *GPR* in Stage 5.

Note that a common behavior through instructions, such as instruction fetch in Stage 1, does not need to be described repeatedly. The ASIP Meister provides a macro feature to improve design productivity, which inserts a user-defined description for multiple instructions.

The information of connections between resources is extracted from the microoperation description. Fig. 7.10 shows a Data Flow Graph (DFG) of ADD instruction generated from the microoperation description in Fig. 7.9. In a DFG of Fig. 7.10, the output port of PC is connected to the input port of IMIFU, where the instruction address is conveyed. A port of the resource is determined using resource information registered in the resource's FHM. By specifying a resource function, corresponding input and output ports are determined. For instance, the register file with two read ports, functions *read0( )* and *read1( )*, corresponds to a pair of ports, (*r_sel0*, *data_out0*) and (*r_sel1*, *data_out1*), respectively.

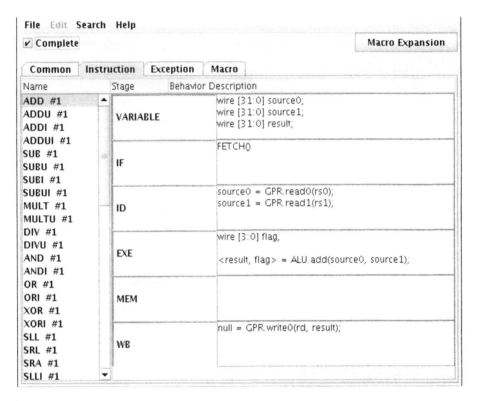

FIGURE 7.8

Microoperation description window.

As described earlier, this behavior information for instruction is used to generate a processor's datapath and its control logic in the HDL generation phase. The detailed algorithm is discussed in Section 7.4.

7.3.4 **VLIW Extension**

Although an HDL generation algorithm of ASIP Meister targets a scalar processor (i.e., with one issue slot), an extended algorithm for a Very Long Instruction Word (VLIW) processor with multiple issue slots is also proposed [2]. The algorithm also employs microoperation description for representing operation behavior. Additional information on a dispatching model enables HDL generation of the VLIW processor. This section introduces a VLIW processor model [5] used by the VLIW processor generation method. This model can represent various kinds of architectures of VLIW processors.

A VLIW instruction consists of multiple operations that are executed simultaneously. Dispatching is a managing process to assign issued operations to appropriate

```
 1: micro_operation ADD
* 2:   wire [31:0] src0;
* 3:   wire [31:0] src1;
* 4:   wire [31:0] result;
  5:   wire [4:0] rs0;
  6:   wire [4:0] rs1;
  7:   wire [4:0] rd;
  8:
  9:   stage 1 {
*10:     wire [31:0] current_pc;
*11:     wire [31:0] inst;
*12:
*13:     current_pc = PC.read();
*14:     inst = IMIFU.read(current_pc);
*15:     null = IR.write(inst);
*16:     null = PC.inc();
 17:   };
 18:
 19:   stage 2 {
 20:     wire [31:0] tmp_ir;
 21:
 22:     tmp_ir = IR.read();
 23:     rs0 = tmp_ir[25:21];
 24:     rs1 = tmp_ir[20:16];
 25:     rd = tmp_ir[15:11];
*26:     src0 = GPR.read0(rs0);
*27:     src1 = GPR.read1(rs1);
 28:   };
 29:
 30:   stage 3 {
*31:     wire [3:0] flag;
*32:
*33:     <result,flag> = ALU0.add(src0,src1);
 34:   };
 35:
 36:   stage 4 {
 37:   };
 38:
 39:   stage 5 {
*40:     null = GPR.write0(rd, result);
 41:   };
 42: };
```

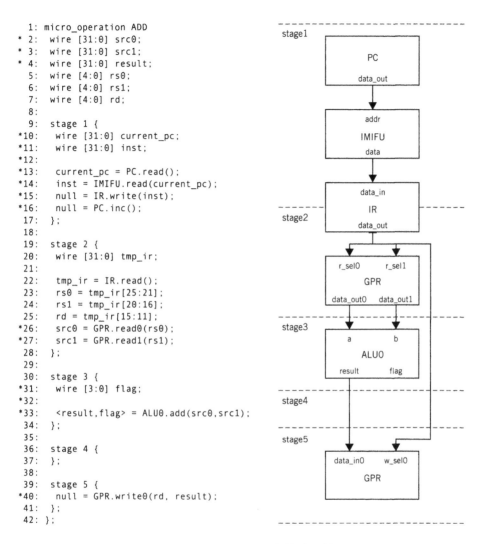

FIGURE 7.9

Microoperation description of ADD instruction (designers need to describe only lines marked with *).

FIGURE 7.10

DFG of ADD instruction generated from corresponding microoperation description of Fig. 7.9.

hardware resources. A slot is a unit to issue an operation. A VLIW processor has one or more slots and issues multiple operations from the slots in parallel. Note that in this model a VLIW processor with one slot is equivalent to a scalar processor.

Fig. 7.11(a) shows the dispatching model of Reference [5]. To represent the complex dispatching rules in a simple description, two concepts are introduced:

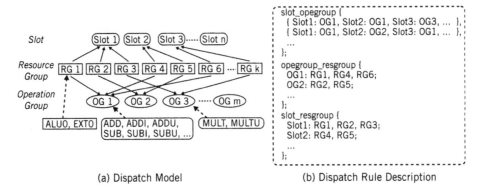

(a) Dispatch Model   (b) Dispatch Rule Description

FIGURE 7.11

Example of the VLIW dispatching model.

operation group and *resource group*. An *operation group* is a set of operations that have the same dispatching characteristic; for instance, a member of an operation group can be executed on the same kind of resources. A *resource group* is a set of resources used when a certain operation is executed in a certain slot. Note that a resource group belongs to one slot and one operation group. In Fig. 7.11(a), operations ADD, ADDI, and so on are members of operation group OG1. Resource group RG1 consists of resources ALU0 and EXT0. RG1 belongs to OG1 and Slot1. A resource can belong to one or more resource groups, which means that a shared resource is represented by belonging to multiple resource groups. In this way, dispatching rules are described using three relations: between slots and operation groups, between slots and resource groups, and between operation groups and resource groups. Fig. 7.11(b) shows a dispatching rule description of the above model. Microoperation is described for a pair of a resource group and an operation. This method allows the representation of as wide a range of dispatching rules as designers intend to make.

7.4 GENERATION OF HDL DESCRIPTION

A processor consists of a number of hardware resources, including pipeline registers, datapath selectors, and a controller. Hardware resources are basically extracted from FHM-DBMS. Pipeline registers and datapath selectors are generated by ASIP Meister as well as a controller.

Fig. 7.12 depicts the overall flow of HDL generation in ASIP Meister [6]. Microoperation description is first analyzed, and a DFG is generated for each instruction. The generated DFGs are then integrated into one DFG for the processor datapath. In the merged DFG, some input ports have a signal conflict where several edges go to one input port. The ASIP Meister inserts multiplexers into the point as datapath

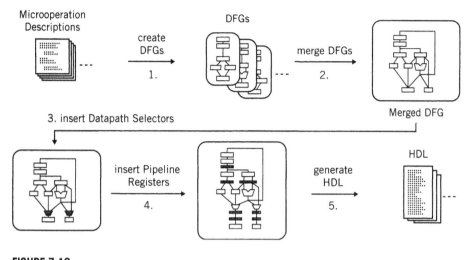

FIGURE 7.12

Overview of HDL generation in ASIP Meister.

selectors to resolve signal conflicts. Then pipeline registers are inserted on stage boundaries to form a pipelined datapath. Finally, synthesizable HDL description is generated from the DFG.

7.4.1 DFG Construction from Microoperation Description

In Reference [6]'s method, a DFG corresponding to an instruction is derived from a microoperation description representing behavior in each pipeline stage of instruction, and then DFGs corresponding to all instructions are merged into one datapath that represents an entire processor. Fig. 7.13 shows an example of the microoperation description of ADD instruction and shift left logical with immediate (SLLI) instruction. Note that this figure only shows the essential part of the instructions; we omitted the instruction fetch and operand decode part for simplification. Fig. 7.14(a) and (b) shows DFGs corresponding to ADD and SLLI instructions, respectively, where SFT is a barrel shifter. Fig. 7.14(c) shows an example of merging DFGs. By identifying the same resource from DFGs, a datapath for an entire processor can be generated. In Fig. 7.14(a) and (b), GPR is a common resource in the two DFGs, and then the DFGs are merged sharing GPR as a common neighbor of ALU and SFT.

7.4.2 Pipeline Registers and Datapath Selectors Insertion

When an input port of a certain resource is shared by the destination of multiple connections in DFG, a signal conflict occurs at the port for the connections.

```
micro_operation ADD {              micro_operation SLLI {
 wire [31:0] src0;                  wire [31:0] src0;
 wire [31:0] src1;                  wire [31:0] res;
 wire [31:0] res;                   stage 2 {
 stage 2 {                           src0 = GPR.read0(rs0);
  src0 = GPR.read0(rs0);            };
  src1 = GPR.read1(rs1);            stage 3 {
 };                                  res  = SFT.add (src0,imm);
 stage 3 {                          };
  wire [3:0] flag;                  stage 5 {
  <res,flag> = ALU.add(src0,src1);   null = GPR.write0(rd,res);
 };                                  };
 stage 5 {                         };
  null = GPR.write0(rd,res);
 };
};
        (a) ADD Instruction                (b) SLLI Instruction
```

FIGURE 7.13

Examples of microoperation description: (a) ADD instruction and (b) SLLI instruction.

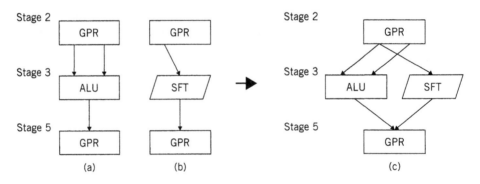

FIGURE 7.14

Example of merging DFGs.

For such a port, a datapath selector is required to solve the conflict. Fig. 7.15 shows an inserting datapath selector. In this example, a conflict on the input port of GPR is resolved by inserting a selector. By choosing the left input for the selector, instructions using ALU can be executed, while instructions using SFT can be executed by choosing the right input port.

When data is transferred over a pipeline stage boundary, a pipeline register is required to transfer the results of the previous stage to the next stage. Fig. 7.16 shows inserting pipeline registers. By inserting the registers, each pipeline stage can be operated independently.

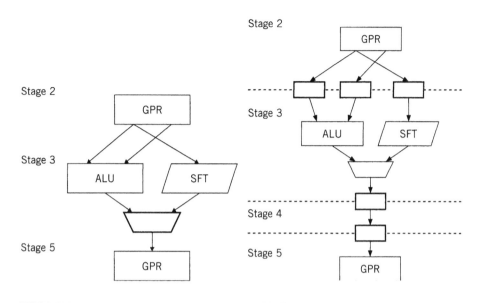

FIGURE 7.15

Example of inserting datapath selector.

FIGURE 7.16

Example of inserting pipeline registers.

7.5 CASE STUDY

The ASIP Meister can be successfully used to generate various ASIP cores with different design specifications. The following subsections introduces some case studies designed using ASIP Meister.

7.5.1 MIPS and DLX

This case study shows that creating a new processor and changing architecture parameters is easy with ASIP Meister due to its GUI interface.

We first designed MIPS subset architecture, PEAS R3K, from scratch. It only took 8 hours to implement a 32-bit 5-stage pipeline processor with 52 instructions including arithmetic, logical, shift, multiply, divide, load, store, branch, and jump instructions.

Then we modified the model of the architecture and designed a subset of the DLX processor architecture. The processor, PEAS DLX, has 51 instructions, and it took just 3.5 hours to complete the design of PEAS DLX with ASIP Meister.

Table 7.1 shows the detailed results. In this case study, we confirmed that the design productivity of ASIP Meister is much higher than manual designs.

Note that this case study was designed by one of the developers of ASIP Meister, that is, an ASIP Meister expert.

Table 7.1 Design productivity of PEAS R3K and PEAS DLX.

		PEAS R3K	PEAS DLX
Area	(K gate)	36.7	72.5
Manpower	(hour)	8.0	3.5
Design Productivity	(K gate/hour)	4.6	20.7

(90-nm CMOS library)

Table 7.2 Synthesis result of M32R.

		M32R
Area	(K gate)	52.6
Maximum Frequency	(MHz)	746

(90-nm CMOS library)

7.5.2 **M32R**

We designed a processor with the instruction set of an M32R processor, a product of the Renesas Technology Corporation. The processor, a 32-bit RISC type, took 36 hours to implement the design with ASIP Meister by an ASIP Meister expert. Table 7.2 shows a synthesis result of the processor, where logic synthesis is performed in 90-nm process technology.

7.5.3 **MeP**

We designed a processor with the instruction set of MeP architecture, which is a Toshiba product. MeP is a configurable processor core that can be configured and optimized for specific applications. Once the architecture is developed on ASIP Meister, it is very easy to tweak various architecture parameters, not only by adding some user-custom instructions or custom DSP units, but also by enhancing the fetch stage or the pipeline architecture itself.

We implemented an instruction-compatible processor based on MeP-c3 core architecture that took 210 hours to implement using ASIP Meister. Note that the manhours include time for thinking implementation, while the previous case studies in PEAS DLX and PEAS R3K only show the time for design implementation. Therefore, this example reveals how little effort is needed during the design phases; that is, a realistic processor can be designed in comparable time even when thinking time is included.

Table 7.3 Synthesis result of MeP.

		MeP
Area	(K gate)	140
Maximum Frequency	(MHz)	230

(90-nm CMOS library)

The generated processor is synthesized in 90-nm CMOS technology, its hardware cost is about 140 K gate, and the maximum frequency is about 230 MHz, as shown in Table 7.3.

7.6 CONCLUSION

The design productivity of ASIPs can be remarkably improved using ASIP Meister, and architecture optimization can be done very easily because designers can explore the design space in a short time. The following are the key features of ASIP Meister:

1. Supports various architecture parameters:

   - Variable numbers of pipeline stages
   - Parameterized hardware resources instantiated from FHM-DBMS
   - Supports multicycle instructions, delayed branches, and handling of external and internal interrupts.

2. Designers can input and change architecture parameters using a Graphical User Interface (GUI).

3. Estimation of such design quality as area, delay, and power consumption of target ASIP at an early stage of design phase.

4. Both the datapath and control logic of ASIP are automatically generated from the architectural parameters and microoperation description of instructions.

7.7 ACKNOWLEDGMENTS

We would like to thank the Semiconductor Technology Academic Research Center (STARC) for their financial and technical support, especially in the early stage of this research project. We would also like to thank all the members of the PEAS project and ASIP Solutions, Inc.

REFERENCES

[1] M. Itoh, S. Higaki, J. Sato, A. Shiomi, Y. Takeuchi, A. Kitajima, and M. Imai. PEAS-III: An ASIP design environment. In *Proc. IEEE International Conference on Computer Design: VLSI in Computers & Processors (ICCD 2000)*, pages 430-436, September 2000.

[2] Y. Kobayashi, S. Kobayashi, K. Okuda, K. Sakanushi, Y. Takeuchi, and M. Imai. Synthesizable HDL generation method for configurable VLIW processors. In *Proc. Asia and South Pacific Design Automation Conference (ASP-DAC 2004)*, pages 843-846, January 2004.

[3] M. Imai, A. Shiomi, Y. Takeuchi, and J. Sato. Integrated circuit design method, database apparatus for designing integrated circuit and integrated circuit design support apparatus. US Patent #6,026,228, 2000.

[4] S. Kobayashi, Y. Takeuchi, A. Kitajima, and M. Imai. Compiler generation in PEAS-III: An ASIP development system. *SCOPES 2001*, St. Goal, Germany, March 2001.

[5] K. Okuda, S. Kobayashi, Y. Takeuchi, and M. Imai. A simulator generator based on config- urable VLIW model considering synthesizable HW description and SW tools generation. In *Proc. the Workshop on Synthesis And System Integration of Mixed Information Tech- nologies (SASIMI 2003)*, pages 152-159, April 2003.

[6] M. Itoh, Y. Takeuchi, M. Imai, and A. Shiomi. Synthesizable HDL generation for pipelined processors from a micro-operation description. *IEICE Trans. on Fundamentals of Electro- nics Communications and Computer Sciences*, E83-A(3):394-400, March 2000.

[7] Y. Fei and Z. J. Shi. Microarchitectural support for program code integrity monitoring in application-specific instruction set processors. In *Proc. of Design, Automation and Test in Europe (DATE 2007)*, pages 815-826, 2007.

[8] R. G. Ragel and S. Parameswaran. Hardware assisted pre-emptive control flow checking for embedded processors to improve reliability. In *Proc. of International Conference on Hardware/Software Co-Design and System Synthesis (CODES+ISSS 2006)*, pages 100-105, 2006.

[9] R. G. Ragel, S. Parameswaran, and S. M. Kia. Micro embedded monitoring for security in application specific instruction-set processors. In *Proc. of International Conference on Com- pilers, Architecture, and Synthesis for Embedded Systems (CASES 2005)*, pages 304-314, 2005.

[10] S. Radhakrishnan, H. Guo, and S. Parameswaran. Dual-pipeline heterogeneous ASIP design. In *Proc. of International Conference on Hardware/Software Co-Design and System Synthesis (CODES+ISSS 2004)*, pages 12-17, 2004.

[11] S. Radhakrishnan, H. Guo, S. Parameswaran, and A. Ignjatovic. Application specific for- warding network and instruction encoding for multi-pipe ASIPS. In *Proc. of International Conference on Hardware/Software Co-Design and System Synthesis (CODES+ISSS 2006)*, pages 241-246, 2006.

TIE: An ADL for Designing Application-specific Instruction Set Extensions

8

Himanshu A. Sanghavi and Nupur B. Andrews

8.1 INTRODUCTION

This chapter describes the Tensilica Instruction Extension (TIE) language and its use for creating application-specific instruction extensions for the Xtensa microprocessor core. It provides a detailed description of the syntax and semantics of the TIE language. It also describes the TIE compiler tool, which automates the hardware implementation and integration of the instruction extensions, as well as the complete software tool chain of the customized processor. The chapter starts with a brief explanation of the advantage of customizing the processor for a specific application by extending the instruction set, and a description of the design flow for creating such extensions. It then describes the TIE language, starting with a simple instruction definition followed by the language semantics for creating custom register files and complex datapaths. Advanced TIE language constructs used to create VLIW machines and new external interfaces for the microprocessor core are also described. The chapter includes a discussion of the hardware implementation and verification issues involved in adding instruction set extensions, and the software tool support required to make the resulting machine easy to program. The chapter concludes with a case study on the design of a 24-bit fixed point Audio DSP core. This commercially available DSP supports a variety of audio codecs with attractive area, power, code size, and performance characteristics.

8.1.1 Adapting the Processor to the Application

A typical system-on-chip (SOC) design consists of a combination of programmable processor cores, memories, and hardware blocks. The benefits of programmability make processor-core-based designs appealing, but a general purpose processor may not have the computation power and data bandwidth to match the needs of complex applications. On the other hand, hardwired RTL-based designs may provide an efficient implementation, but they do not provide the flexibility of a programmable solution. As a trade-off between these two options, it is common in SOC designs to implement the performance critical or datapath intensive sections in hardware, while control-oriented functions (and functions likely to require frequent changes) are targeted to the processors.

An innovative solution to the problem is to have a processor whose instruction set can be customized by the designer for a specific application or class of applications [1]. Such a customized processor can provide application performance that is significantly higher than a general purpose processor. A performance improvement of 2X, 5X, and in some cases orders of magnitude can be achieved, with a lower hardware area and lower energy dissipation [2]. The goal of configurable processors is to provide the flexibility of a programmable solution, with implementation efficiency comparable to RTL-based solutions [3]. Another benefit of using pre-verified IP cores over RTL-based designs is that they significantly reduce the design and verification time.

8.1.2 Tensilica Instruction Extension Language and Compiler

The Tensilica Instruction Extension (TIE) language is used to describe instruction set extensions for the Xtensa family of processor cores. The language provides a simple and intuitive syntax to describe new instructions, new register files and data types, and new interfaces between the Xtensa processor and external devices [4, 5]. These instructions are not micro-coded, but are implemented in hardware, in a manner very similar to the base instructions of the Xtensa processor. Further, these instructions are supported by the full software tool chain including an Instruction Set Simulator (ISS), a C/C++ compiler, assembler, and debugger.

The TIE compiler is a tool that automates the implementation of instruction set extensions defined using the TIE language. The TIE compiler reads the description of the instruction extensions defined by the designer, and generates both the hardware implementation and software tools for the new customized processor.

In this chapter, the TIE language is discussed in the context of extending an Xtensa processor core with application-specific instructions. The base Xtensa core, with a typical RISC instruction set, is assumed to exist. The TIE language is capable of describing the base instruction set of the Xtensa processor. In fact, the entire instruction set, pipeline, and exception semantics of the Xtensa processor are defined in the TIE language. Only some units such as instruction fetch, load/store, and bus interface are designed directly in RTL. However, the focus of this chapter is on instruction set extensions and thus some of the TIE language features used to describe the base Xtensa core will not be discussed.

8.2 DESIGN METHODOLOGY AND TOOLS

8.2.1 Designing Application-specific Instructions

Instruction set extensions are used to accelerate data intensive and compute intensive functions, called "hot-spots," in application code. Because these functions are heavily exercised by the program, the efficiency of their implementation can have a significant impact on the efficiency of the whole program. It is possible to increase the performance of the application by a factor of two, five, or even an order of magnitude by adding a few carefully selected instructions. Such gains do come at the cost of additional hardware, but in most cases it is possible to achieve a solution that offers significant performance gain for an acceptable hardware cost.

Fig. 8.1 is a flowchart of a design methodology to create application-specific instruction set extensions. It highlights the important phases of the instruction set design process; identification of instructions, functional description, verification, and hardware optimization. TIE development starts once a base Xtensa processor has been selected for the application.

As illustrated in Fig. 8.1, a careful study of the execution profile of the application is a good way to identify the functions that are the hot spots of an application. Next,

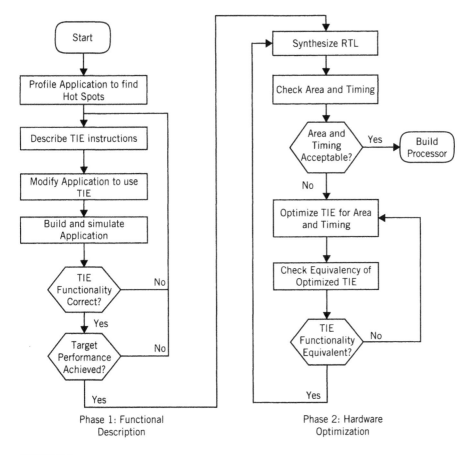

FIGURE 8.1

Flowchart of TIE development.

the behavior of these functions is mapped to a set of TIE instructions, by describing the functionality of these instructions in the TIE language. The TIE compiler tool is used to generate simulation libraries for these new instructions, and the application code modified to replace the original functions with TIE intrinsics. The modified application code is then simulated to verify that the results match the original code. Once the application is functionally verified to be correct, the execution profile is checked again to evaluate performance. The designer may iterate through this process a few times, modifying the TIE instructions or creating new ones until the desired performance is achieved.

Once the basic functionality of the TIE instructions has been defined and verified, the next step is to optimize the hardware implementation of these new instructions. The TIE compiler tool is once again used to generate the hardware implementation of these instructions as synthesizable Verilog. This RTL description can be synthesized to obtain the area and timing information. This information has to be reviewed to evaluate whether the area increase is acceptable, and the design meets the timing (frequency or MHz) goals of the project. There are several hardware optimization techniques that can be used to reduce the area and improve timing if the initial design does not meet the requirement. These techniques are described in Section 8.6. Once again, as illustrated in Fig. 8.1, the designer may iterate through this process, optimizing the hardware implementation until the design goals are met. In some situations, it may be necessary to revisit the definition of one or more TIE instructions and perhaps modify them in order to meet the area and timing goals of the design. Once TIE development is complete, the final processor is built and the resulting hardware and software tools are ready for integration with the rest of the system.

Although the focus here is the acceleration of the computation, in many systems the data bandwidth and memory subsystem performance are often a bigger bottleneck. Using the TIE methodology, significant system performance gain may be achieved by optimizing the data bandwidth in the design. This can be done by designing custom data channels using the data interfaces described in Section 8.7, which allow high-speed data transfer between a processor(s) and other blocks.

8.2.2 Design Automation with the TIE Compiler

Extending a processor's instruction set goes well beyond writing the RTL code of the new instruction hardware. Integrating the new hardware into the processor requires in-depth knowledge of the processor pipeline and microarchitecture, to ensure that the hardware works correctly under all conditions such as data hazards, branches, and exceptions. Similarly, the task of modifying all the software tools to understand the extensions is nontrivial. The ability to modify the software tools quickly is crucial to designer productivity. It enables a quick profile of their impact on application performance, thus allowing the designer to iterate over different design options in a short amount of time.

The TIE compiler is a processor synthesis tool that automates the process of extending the Xtensa processor with instruction set extensions. The TIE developer simply describes the functional behavior of the new instructions, without regard to the microarchitecture of the processor. The TIE compiler takes this description and automatically generates the complete hardware implementation of these instructions, and updates all the software tools to recognize these extensions. The hardware implementation is generated as synthesizable Verilog RTL, along with synthesis and place and route scripts. Test programs to verify the microarchitectural implementation of these instructions are also automatically generated. The software tools are updated so that the assembler can assemble these new instructions and the C/C++ compiler recognizes these instruction intrinsics and can schedule them efficiently in compiled code. The Instruction Set Simulator (ISS) can simulate these new instructions, and the debugger is made aware of any new state that is added to the processor, so that its value can be examined when debugging programs.

8.3 BASICS OF TIE LANGUAGE

The fundamental concepts of the TIE language are introduced in this section, starting with a simple example to illustrate a basic TIE instruction. Note that these instructions are added to a predefined core Xtensa ISA, consisting of a 32-bit, 16-entry register file, and a RISC instruction set consisting of load/store, ALU, shift and branch instructions.

8.3.1 A basic TIE Acceleration Example

Consider an example where the application code calculates a dot product from two arrays of unsigned 16-bit data. The product is right-shifted by 8 bits before accumulating it into a 32-bit unsigned accumulator.

```
unsigned int i, acc;
short *sample, *coeff;
for (i=0; i<N; i++) {
  acc += (sample[i] * coeff[i]) >> 8;
}
```

On a base Xtensa processor with a single cycle multiplier, each iteration of the loop will take at least three cycles to execute, excluding the loads and stores of data and result, and updating the pointers to memory. This computation can be accelerated by creating a single TIE operation that is commonly known as a "fusion." The operation combines the multiplication, shifting, and accumulation as:

```
operation dotprod {inout AR acc, in AR sample, in AR coeff} { } {
  assign acc = acc + ((sample[15:0] * coeff[15:0]) >> 8);
}
```

In the aforementioned example TIE code, *operation* is a TIE language keyword indicating the definition of a new instruction. The name or mnemonic of the instruction is dotprod. This instruction has three operands from the 32-bit *AR* register file of the Xtensa processor core. The two input operands sample and coeff are read from the register file, while the third operand acc is both read and written by this instruction.

The TIE instruction dotprod is used in C code as illustrated here:

```
# include <xtensa/tie/dotprod.h>
unsigned int i, acc;
short *sample, *coeff;
for (i=0; i<N; i++) {
    dotprod(acc, sample[i], coeff[i]);
}
```

The header file dotprod.h generated by the TIE compiler defines the prototype of the intrinsic dotprod as:

```
void dotprod(unsigned int *acc, unsigned int sample,
            unsigned int coeff);
```

This C code can be compiled by the Xtensa C compiler and simulated by the instruction set simulator. The inner loop of the computation goes from three cycles to a single cycle as a result of the TIE instruction. If the inner loop is in the critical section of an application, this acceleration will improve the performance significantly.

8.3.2 Defining a Basic TIE Instruction

The most fundamental TIE construct is *operation*. It specifies the name, arguments, and the behavior of a TIE instruction. The syntax of the *operation* construct is show here:

```
operation <name> {<argument list>} {<state interface list>}
        {<datapath description>}
```

The *operation name* is the instruction mnemonic, and its definition is followed by the *argument list*. This is a comma-separated list of the register file and immediate operands of the instruction. Each argument in this list has three components. The first is a direction identifier which can be *in*, *out*, or *inout*, specifying that the instruction reads, writes, or read-modify-writes the operand, respectively. The second component specifies the type of operand; either the name of the register file that this operand references, or the identifier for an immediate operand. The third component is the name of the operand, which is used in a similar manner as argument variable names in a C function. These names are used in the description of the operation. Operands listed in the *argument list* are encoded in the instruction word and appear in the assembly-language-syntax of the instruction.

The *argument list* is followed by the *state interface list*, which is a list of the implicit operands of the instruction. Implicit operands are not encoded in the instruction word and are not listed in the assembly language syntax of the instruction. A TIE state refers to a single entry register file. Interfaces are special resources used in the definition of load/store instructions to access memory, or to create specialized instructions that directly connect the Xtensa processor to external devices.

The *state interface list* is followed by the *datapath description*, which is the main body of the *operation* definition. It describes the computation that is performed by the instruction as a list of one or more assignments.

Immediate operands

Immediate operands of an instruction are defined using the *immediate_range* TIE construct, which defines a range of legal values that the operand can take. It has the following syntax:

```
immediate_range <name> <low value> <high value> <step size>
```

name is the identifier used to reference the immediate range in an *operation* definition, *low value* and *high value* are signed integers representing the lowest and highest value the operand can take, respectively, and *step size* is the amount by which the successive values in this range increment.

8.3.3 Instruction Encodings

The dotprod operation described in Section 8.3.1 does not specify the encoding of the instruction. The TIE compiler is capable of automatically generating encodings for instructions; this is just another aspect of the automation it provides. It is, however, possible for the designer to provide the encoding if desired. In this section we briefly describe the TIE constructs used to specify instruction encodings, and show an example encoding for the dotprod instruction.

The dotprod instruction has three operands that reference the *AR* register file of the Xtensa processor. This register file has 16 entries, thus each of these operands will require 4 bits to be encoded in the instruction word. The remaining bits of the instruction word can be used to specify the opcode encoding of this instruction.

The base Xtensa processor has a 24-bit instruction word, referred to by the TIE keyword *Inst*. The TIE language allows the designer to create bit fields in this instruction word, and to specify operand and opcode definitions using these bit fields. The following example illustrates how to use the *field, operand*, and *opcode* constructs of the TIE language to specify one possible encoding for the dotprod instruction.

```
// Define 4-bit bit-fields of Inst
field r Inst[ 3: 0]
```

```
field s Inst[ 7: 4]
field t Inst[11: 8]

// Create operands that reference the AR register file
operand ars s { AR[s] }
operand arr r { AR[r] }
operand art t { AR[t] }

// Specify the opcode encoding for the instruction
field opc Inst[23:12]
opcode dotprod opc=12'h0A2

// Alternative syntax for instruction definition, when using

// explicit operand definitions
iclass dp { dotprod } { inout arr, in ars, in art }
reference dotprod {
    assign arr = arr + ((ars[15:0]*art[15:0]) >> 8);
}
```

In the aforementioned description, *field* is a TIE keyword used to define three bit fields named r, s, and t, respectively. Each of these fields is 4-bits wide, and is used to create the operands arr, ars, and art using the TIE keyword *operand*. These operands reference the *AR* register file, and the *operand* syntax indicates that the *field* provides the address or index into the register file. The operands consume 12 bits out of the 24-bit instruction word. The remaining 12 bits are used to define another field, opc, and the opcode encoding of the instruction dotprod is defined (using the TIE keyword *opcode*) as the value of this field being equal to the hex value 0x0A2. The *iclass* definition specifies the input and output operands of an instruction, and the *reference* describes the instruction behavior or datapath.

8.3.4 Instruction Datapath

The datapath of a TIE instruction is described in the *operation* definition as a list of assignment statements. The assignment statement has the following syntax:

```
assign <name> = <expression>;
```

name is the name of an *out* or *inout* operand of the *operation* or an intermediate *wire* that has earlier been declared in the computation. Expressions in the TIE language follow the same syntax as combinational logic in Verilog, and can be any expression using TIE operators and the names of *in* or *inout* operands or intermediate *wires*. A list of TIE operators is shown in Table 8.1. These operators are essentially identical to Verilog-95 HDL and follow the same rules of precedence. All operators treat the data as unsigned. Thus the rules governing the computed width of expressions in TIE are the same as Verilog rules for unsigned data type.

Table 8.1   TIE operators.

Operator Type	Operator Symbol
Sub range	a[n:m], where n >= m
Arithmetic	+, −, *
Logical	!, &&, ||
Relational	>, <, >=, <=, ==, !=
Bitwise	~,&, |,^,^~, ~^
Reduction	&, |,^,^~, ~^
Shift	>>, <<
Concatenation	{}
Replication	{n{}}
Conditional	?:

Wires and assignment

For all but the simplest of TIE instructions, it becomes impractical to describe the entire datapath using only the input and output operands of the instruction. Wires are temporary "scratchpad" variables and are assigned the intermediate results of a computation. The syntax for declaring a wire is:

```
wire [n:m] <name> [= <expression>];
```

where $[n:m]$ specifies the width of the wire. By default, a wire is 1-bit wide if the width is omitted. The assignment to the wire can be specified in a separate assign statement or in the declaration itself. As an example, the dotprod operation introduced in Section 8.3.1 can be rewritten as:

```
operation dotprod {inout AR acc, in AR sample, in AR coeff} {} {
    wire [15:0] s = sample[15:0];
    wire [15:0] c = coeff[15:0];
    assign acc = acc + ((s * c) >> 8);
}
```

Note that the lexical ordering of the assign statements in an operation's description may not bear any relationship with the actual order in which the statements are executed. All TIE assignments are evaluated for dependencies and computed in the order of dependency. All input arguments in a statement are read before any output arguments are computed. For an *inout* operand, the input value is read before any computation is performed, and the output value written after all computations are performed.

Conditional writes to output operand

The TIE language provides a mechanism to predicate the result of a TIE operation by specifying the condition under which the write to an output operand is suppressed or killed. For every output operand *<name>*, the TIE compiler automatically generates a signal *<name>_kill*, which can be assigned in the TIE operation. If this signal is true, the write is killed. An example of a conditional move operation is shown here:

```
operation MOVCOND { out AR d, in AR a } {} {
    assign d = a;
    assign d_kill = (a == 32'b0);
}
```

In the aforementioned example, the move is predicated when the input is zero. Note that although the operand is an output of the move, the Xtensa compiler will be aware of the predication.

Built-in computation modules

The TIE language contains a rich set of built-in computation modules that can be used in expressions. These modules are used to specify computations that are cumbersome to describe. Further, the hardware implementation of these modules has been pre-optimized for area and timing, and therefore results in better hardware across a cross section of standard cell libraries. A list of these modules is shown in Table 8.2.

8.4 ADDING PROCESSOR STATE

The instruction set extensions described so far read and write the predefined 32-bit *AR* register file of the Xtensa processor. However, in many designs there is a need to customize the data size of operands. For example, the designer may want to use a 24-bit data type for fixed point audio processing, or a 128-bit data type to represent eight 16-bit values for SIMD operations. The TIE language provides constructs to add new processor states and register files that can then be used as operands of TIE instructions.

8.4.1 Defining a State Register

In the TIE language, the term *state* is used to refer to a single entry register file. TIE states are useful for maintaining frequently accessed data in the processor instead

Table 8.2 TIE built-in modules.

Format	Description	Result Definition
TIEadd(a, b, cin)	Add with carry-in	$a + b + \text{cin}$
TIEaddn(a_0, a_1, ..., an)	N number addition	$a_0 + a_1 + \cdots + a_n$
TIEcmp(a, b, sign)	Signed and unsigned comparison	$\{a < b,\ a <= b,\ a == b,\ a >= b,\ a > b\}$
TIEcsa(a, b, c)	Carry save adder	$\{\text{carry, sum}\}$ $= \{a\,\&\,b\,|\,a\,\&\,c\,|\,b\,\&\,c,\ a\char`^b\char`^c\}$
TIEmac(a, b, c, sign, negate)	Multiply accumulate, sign specifies how a and b are sign extended	negate ? $c - a^*b : c + a^*b$
TIEmul(a, b, sign)	Signed and unsigned multiplication	$\{\{m\{a[n-1]\,\&\,\text{sign}\}\}, a\}^*$ $\{\{n\{b[m-1]\,\&\,\text{sign}\}\}, b\}$, where n is the size of a and m is the size of b
TIEmulpp(a, b, sign, negate)	Partial product multiply	$\{p0, p1\} = \text{result}$ $p0 + p1 = \text{negate?} - (a^*b) :$ (a^*b)
TIEmux(s, d_0, d_1, ..., d_{n-1})	n-way multiplexer	$s==0?d_0 : s==1?d_1 : \ldots s==$ $n-2 : d_{n-2} : d_{n-1}$
TIEpsel(s_0, d_0, s_1, d_1, ...s_{n-1}, d_{n-1})	n-way priority selector	$s_0?d_0 : s_1?d_1 : \ldots s_{n-1}?d_{n-1} : 0$
TIEsel(s_0, d_0, s_1, d_1, ...s_{n-1}, d_{n-1})	n-way 1 hot selector	$(\text{size}\,\{s_0\}\,\&\,d_0)\,|$ $(\text{size}\,\{s_1\}\,\&\,d_1)\ldots|$ $(\text{size}\,\{s_{n-1}\}\,\&\,d_{n-1})$

of accessing it through memory. The TIE states can be of arbitrary width, and once defined, can be used as an operand of any TIE instruction. The syntax for defining a state is

```
state <name> <width> [<reset value>] [<add_read_write>] [<export>]
```

name is the unique identifier for referencing the state and *width* is the bit-width of the state. Optional parameters of a state definition include a *reset value* that specifies the value that the state will have when the processor comes out of reset. The *add_read_write* keyword directs the TIE compiler to automatically create instructions to transfer data between the TIE state and the AR register file, as described in the next example. Finally, the *export* keyword specifies that the value of this state is to be made visible outside the processor as a new top-level interface. This is described in detail in Section 8.7.1.

Consider a modified version of the dotprod instruction of Section 8.3.1, in which a 40-bit accumulator is used. The accumulator can no longer be stored in the *AR* register file, which is only 32-bits wide. The following example illustrates the use of TIE state to create such an instruction. It also uses the built-in module *TIEmac*, described in Section 8.3.4, to perform the multiply and accumulate operation.

```
state ACC 40 add_read_write
operation dotprod {in AR sample, in AR coeff} {inout ACC} {
    assign ACC = TIEmac(sample[15:0], coeff[15:0], ACC, 1'b0, 1'b1);
}
```

General purpose software access to TIE state

In addition to allowing the use of the TIE state as an operand for any arbitrary TIE instruction, it is useful to have direct read and write access to TIE state in a program. This mechanism also allows a debugger to view the value of the state, or an operating system to save and restore the value of the state across a context switch.

The TIE compiler can automatically generate an *RUR* (read user register) instruction that reads the value of the TIE state into an *AR* register and a *WUR* (write user register) instruction that writes the value of an *AR* register to a TIE state. If the state is wider than 32 bits, multiple *RUR* and *WUR* instructions are generated, each with a numbered suffix that begins with 0 for the least significant word of the state. The automatic generation of these instructions is enabled by the use of the *add_read_write* keyword in the state declaration. In the TIE example in Section 8.4.1, instructions RUR.ACC_0 and WUR.ACC_0 are generated to read and write bits [31 : 0] of the 40-bit state ACC, respectively, while instructions RUR.ACC_1 and WUR.ACC_1 are generated to access bits [39 : 32] of the state.

8.4.2 Defining a Register File

A TIE state is well suited for storing a single variable, but register files are for a more general purpose. A TIE register file is a custom set of addressable registers for use by TIE instructions. Many algorithms might require data wider than 32 bits, and programming it on a 32-bit datapath machine becomes cumbersome. A custom register file that matches the natural size of the data is much more efficient. The TIE construct for defining a register file is

```
regfile <name> <width> <depth> <short name>
```

The *width* is the width of each register, while *depth* indicates the number of registers in the register file. The short name is used by the assembler and debugger to reference the register file. When a register file is defined, its name can be used as an operand type in a TIE operation. An example of a 64-bit wide, 32 entry register file and an XOR operation that operates on this register file is shown here:

```
regfile myreg 64 32 mr
operation widexor {out myreg o, in myreg i0, in myreg i1} {} {
    assign o = i0^i1;
}
```

Load store operations and memory interface

Every register file definition is accompanied by load, store, and move instructions for that register file. The Xtensa C/C++ compiler uses these instructions as described in Section 8.4.3. These basic instructions are typically generated by the TIE compiler automatically unless specified by the designer. An example of these instructions for the myreg register file is shown here:

```
immediate_range imm4 0 120 8
operation ld.myreg {out myreg d, in myreg *addr, in imm4 offset}
            {out VAddr, in MemDataIn64} {
    assign VAddr = addr + offset;
    assign d = MemDataIn64;
}
operation st.myreg {in myreg d, in myreg *addr, in imm4 offset}
            {out VAddr, out MemDataOut64} {
    assign VAddr = addr + offset;
    assign MemDataOut64 = d;
}
operation mv.myreg {out myreg b, in myreg b} {} {
    assign b = a;
}
```

The instruction ld.myreg performs a 64-bit load from memory with the effective virtual address provided by the pointer operand addr, plus an immediate offset. The operand addr is specified as a pointer with the * to indicate to the compiler to expect a pointer to a data type that resides in the register file myreg. Since the 64-bit load operation requires the address aligned to a 64-bit boundary, the step size of the offset is 4 (bytes). The load/store operations send the virtual address to the load-store unit of the Xtensa processor in the execution stage. The load data is received from the load-store unit in the memory stage while the store data is sent in the memory stage. This is done using a set of standard interfaces to the load-store unit of the processor. A list of these memory interfaces is shown in Table 8.3.

The designer can also write load and store instructions with a variety of addressing modes. For example, updating load instructions is useful to efficiently access an array of data values in a loop. Similarly, bit-reversed addressing is useful for DSP applications.

Table 8.3 Load store memory interface signals.

Name	Width	Direction	Purpose
Vaddr	32	Out[1]	Load/store address
MemDataIn {128,64,32,16,8}	128,64,32,16,8	In	Load data
MemDataOut {128,64,32,16,8}	128,64,32,16,8	Out	Store data
LoadByteDisable	16	Out	Byte disable signal
StoreByteDisable	16	Out	Byte disable signal

8.4.3 Data Type and Compiler Support

The TIE compiler automatically creates a new data type for every register file defined by the TIE designer. The data type has the same name as the register file, and is available for use as the "type" of a variable in C/C++ programs. The Xtensa C/C++ compiler understands that variables of this type reside in the custom register file, and can also perform register allocation for these variables. The load, store and move instructions described in the previous TIE example are used by the C/C++ compiler to save/restore register values when performing register allocation and during a context change for multi-threaded applications.

The C programming language does not support constant values wider than 32 bits. Thus, initialization of data types wider than 32 bits is done indirectly, as illustrated in the example here for the myreg data type generated for the same register file:

```
# define align_by_8 __attribute__ ((aligned)8)
unsigned int data[4] align_by_8 = { 0x0, 0xffff, 0x0, 0xabcd };
myreg i1, *p1, i2, *p2, op;
p1 = (myreg *)&data[0]; p2 = (myreg *)&data[2];
i1 = *p1; i2 = *p2;
op = widexor(i1, i2);
```

In this example, variables i1 and i2 are of type myreg and are initialized by the pointer assignments to a memory location. The compiler uses the appropriate load/store instructions corresponding to the register file when initializing variables. Note that the data values should be aligned to an 8-byte boundary in memory for the 64-bit load/store operations to function correctly, as specified by the attribute pragma in the code.

---

[1]"In" signals go from Xtensa core to TIE logic, "out" signals go from TIE logic to Xtensa core.

Multiple data types

The TIE language provides constructs to define multiple data types that reside in a single register file, and to perform type conversions between them. The myreg register file can hold a 40-bit data type in addition to the 64-bit type, described in Section 8.4.3. This is done using the *ctype* TIE construct as illustrated here:

```
ctype myreg 64 64 myreg default
ctype my40 40 64 myreg
```

The syntax of the *ctype* declaration provides the width and memory alignment of the data, and the register file it resides in. In this description, the second data type my40 has a width of 40 bits. Both data types are aligned to a 64-bit boundary in memory. The keyword *default* in a *ctype* declaration indicates that unless otherwise specified, it is the type of a variable used by any instruction that references the register file.

The Xtensa C/C++ compiler requires load, store, and move instructions corresponding to each *ctype* of a register file. The *proto* construct of the TIE language is used to indicate to the Xtensa C/C++ compiler the load, store, and move instruction corresponding to each *ctype* as illustrated here:

```
proto myreg_loadi {out myreg d, in myreg *p, in immediate o} {} {
    ld.myreg d,p,o;
}
proto myreg_storei {in myreg d, in myreg *p, in immediate o} {} {
    st.myreg d,p,o;
}
proto myreg_move {out myreg d, in myreg a} {} {
    mv.myreg d,a;
}
proto my40_loadi {out my40 d, in my40 *p, in immediate o} {} {
    ld.my40 d,p,o;
}
proto my40_storei {in my40 d, in my40 *p, in immediate o} {} {
    st.my40 d,p,o;
}
proto my40_move {out my40 d, in my40 a} {} {
    mv.my40 d,a;
}
```

The *proto* uses stylized names of the form *<ctype>_loadi* and *<ctype>_storei* to define the instruction sequence to load and store a variable of type *<ctype>* from memory. The proto *<ctype>_move* defines a register to register move. The load/store instructions defined in the previous TIE example are used to define the *proto* for the *ctype* myreg. Similar instructions for the 40-bit type my40 can be defined using only the lower 40 bits of the memory interfaces *MemDataIn64* and *MemDataOut64*. In some cases, multiple instructions may be needed in the *proto*, such as loading a register file whose width is greater than the maximum allowable data memory width of 128 bits.

The *proto* construct can also be used to specify type conversion from one *ctype* to another. For example, conversion from the fixed point data type my40 to myreg involves sign extension, while the reverse conversion involves truncation with saturation as shown:

```
operation mr40to64 {out myreg o, in myreg i} {} {
    assign o = {{24{i[39]}}, i[39:0]};
}
operation mr64to40 {out myreg o, in myreg i} {} {
    assign o = (i[63:40] == {24{i[39]}}) ? i[39:0] :
            {i[63], {39{~ i[63]}}};
}
proto my40_rtor_myreg {out myreg o, in my40 i} {} {
    mr40to64 o, i; }
proto myreg_rtor_my40 {out my40 o, in myreg i} {} {
    mr64to40 o, i; }
```

The *proto* definition follows a stylized name of the type *<ctype1>_rtor_<ctype2>* and gives the instruction sequence required to convert a variable of type ctype1 to a variable of type ctype2. These *protos* are used by the C/C++ compiler when variables of the different data types are assigned to each other.

The C intrinsic for all operations referencing the register file myreg will automatically use the default 64-bit *ctype* myreg since it is the default *ctype*. If any operation should use the 40-bit data type my40, this can be specified by writing a *proto* as shown:

```
operation add40 { out myreg o, in myreg i1, in myreg i2 } {
    assign o = { 24'h0, TIEadd(i1[39:0], i2[39:0], 1'b1) };
}
proto add40 { out my40 o, in my40 d1, in my40 d2} {} {
    add40 o, d1, d2; }
```

The *proto* add40 specifies that the intrinsic for the operation add40 uses the data type my40.

8.4.4 Data Parallelism and SIMD

The ability to use custom register files allows the designer to create new machines targeted for a wide variety of data processing tasks. For example, the TIE language has been used to create a set of floating point extensions for the Xtensa processor core. Many DSP algorithms that demand a high performance share common characteristics—the same sequence of operations is performed repetitively on a stream of data operands. Applications such as audio processing, video compression, and error-correction-coding fit this computation model. These algorithms can achieve large performance benefits from a single instruction multiple data (SIMD) technique, which is easy to design with custom register files.

In the example given later, the average of two arrays is computed. In an iteration of the loop, two short values are added and the result is shifted by 1 bit, which requires two Xtensa instructions as well as load/store instructions.

```
unsigned short *a, *b, *c;
for ( i=0; i<N; i++ ) {
    c[i] = (a[i] + b[i]) >> 1;
}
```

If the aforementioned loop is in the critical section of the application, creating a TIE fusion operation to combine the add and shift computations will accelerate the performance. However, since the same computation is performed in every iteration of the loop, this is an ideal candidate for a SIMD or vector instruction. A SIMD instruction that performs four averages in parallel can be written as shown:

```
regfile VEC 64 8 v
operation VAVERAGE {out VEC res, in VEC input0, in VEC input1} {} {
    wire [16:0] res0 = input0[15: 0]  + input1[15: 0];
    wire [16:0] res1 = input0[31:16] + input1[31:16];
    wire [16:0] res2 = input0[47:32] + input1[47:32];
    wire [16:0] res3 = input0[63:48] + input1[63:48];
    assign res = {res3[16:1], res2[16:1], res1[16:1], res0[16:1]};
}
```

Compared to a scalar fusion of add and shift, which takes one 16-bit adder, the SIMD average instruction will require four 16-bit adders. Also, since the vector data is 64 bits, a new 64-bit register file will be needed.

8.5 VLIW MACHINE DESIGN

Very Long Instruction Word (VLIW) processors are well known for speeding application performance by taking advantage of instruction-level parallelism. This approach of combining multiple independent operations into an instruction word to execute them simultaneously is supported in the TIE language using Flexible Length Instruction Extension (FLIX) technology.

Although the base Xtensa processor has a 24-bit instruction word, 32- and 64-bit instructions can be created using the TIE language. Instructions of different sizes can be freely mixed without any mode bit, because the processor identifies, decodes, and executes any size instruction word coming from the instruction stream. The 32- and 64-bit instructions can be divided into multiple instruction issue slots, with independent operations placed in each slot. A VLIW instruction consists of any combination of operations allowed in each issue slot. The Xtensa C/C++ compiler recognizes the parallelism available through the definition of the VLIW formats and automatically does software pipelining to achieve this performance in compiled code.

8.5.1 Defining a VLIW Instruction

A VLIW instruction can be created using two TIE constructs. The first is *format*, which defines the length of the instruction and the number of issue slots. The second is *slot_opcodes*, which defines all the possible operations that can belong to each slot of a format. The syntax is shown here:

```
format <name> <length> <list of slots>
slot_opcodes <slot_name> <list of operations>
```

name is a unique identifier for the *format*. *length* defines the length of the instruction and can be either 32 or 64. The list of slots contains slot names in the order that they appear in the instruction word. For each slot name defined in the list of slots, there must be a corresponding *slot_opcodes* statement that declares the list of operations available in that slot. This list can consist of any of the Xtensa ISA instructions and any of the designer-defined TIE instructions.

The size of each slot is automatically determined by the size of the operations defined in that slot. Slots can be of different sizes. Further, the encoding of each operation in each of the slots is also automatically determined by the TIE compiler. It is possible for the designer to explicitly specify the instruction bits corresponding to each slot, and the encoding of the instructions within each slot using syntax similar to the one described in Section 8.3.3.

Consider the inner loop in the application code in Section 8.4.4. On the base Xtensa processor, this requires the arithmetic instructions 32-bit add (ADD) and a shift-right (SRAI) to compute the average, two 16-bit load (L16I) and one 16-bit store (S16I) instructions to load the input data and store the result, and three add immediate (ADDI) instructions to update the pointers to memory. The execution of the inner loop would thus take eight cycles. To accelerate this code, a simple 3-issue VLIW processor can be created as follows:

```
format vliw3 64 {slot0, slot1, slot2}
slot_opcodes slot0 {L16I, S16I}
slot_opcodes slot1 { ADDI }
slot_opcodes slot2 {ADD, SRAI}
```

Each 64-bit instruction of the vliw3 format consists of three issue slots, each of which can issue any instructions from its *slot_opcodes* list. slot0 can issue a 16-bit load or store instruction, slot1 can issue the "add with immediate" instruction, and slot2 can issue either a 32-bit add or a "shift right arithmetic" instruction. All the operations of this example are predefined Xtensa ISA instructions, so the description of these operations is not required in the TIE code. With this simple description, the inner loop of this code can be reduced to three instructions.

8.5.2 Hardware Cost of VLIW

An important aspect of VLIW design is estimating and optimizing the hardware cost. To replicate operations in multiple slots, multiple copies of the hardware execution units are instantiated and this can cause a substantial increase in gate count. To create hardware for each slot, the slots are grouped by slot index, where the index refers to the slot position in a format, starting at the index 0 for the least significant bit. The slots with the same slot index share the execution units. The single 24-bit slot *Inst* of the base processor has an index 0. This is an important consideration for creating VLIW instructions with base processor operations. For the VLIW format in Section 8.5.1, slot0, slot1, and slot2 will have indices 0, 1, and 2, respectively. The load store operations in slot0 will therefore not cost any extra hardware, since they will share the hardware with the base processor load/store operations. The ADD and ADDI instructions in slot1 and slot2 will result in replicating the hardware for the arithmetic units. It is also important to note that load/store instructions can only be replicated if the base processor supports multiple load-store units. The size and complexity of the register file also increase with VLIW. The input and output operands of an operation are automatically mapped to read and write ports of a register file. When an operation is replicated in multiple slots, the read and write ports are also replicated, thus causing the register file to have increased hardware.

8.6 LANGUAGE CONSTRUCTS FOR EFFICIENT HARDWARE IMPLEMENTATION

The design of TIE extensions involves a trade-off between accelerated application performance and the cost of the added hardware. The TIE language provides several constructs to enable efficient hardware implementation, and thus minimize the hardware cost of TIE extensions.

8.6.1 Sharing Hardware between Instructions

There are many situations in which hardware execution units can be shared between two or more instructions. For example, the same adder can be used to compute the sum and difference of two operands, and multiply-add and multiply-subtract instructions can share the same multiplier. The TIE language includes a construct named *semantic*, which is used to describe the hardware implementation of one or more earlier defined instructions. The following example illustrates a *semantic* that implements a 32-bit add and subtract instruction:

```
operation ADD {out AR res, in AR in0, in AR in1} {
    assign res = in0 + in1;
}
operation SUB {out AR res, in AR in0, in AR in1} {
    assign res = in0 − in1;
}
semantic addsub {ADD, SUB} {
    wire carry = SUB;
    wire [31:0] tmp = ADD ? in1 : ~ in1;
    assign res = TIEadd(in0, tmp, carry);
}
```

The semantic addsub implements the instructions ADD and SUB. It uses the simple principle of 2's complement arithmetic that (a − b) is equal to (a + ~b + 1). For every instruction implemented in a TIE semantic, a 1-hot signal with the same name as the instruction is available in the semantic. This signal is true when the corresponding instruction is being executed, and false otherwise. In the aforementioned example, the 1-hot signal for the subtract instruction is used to set the value of the 1-bit "carry" and the 1-hot signal for the add instruction is used to select between the original value and bit inverted value of the second operand.

In the absence of the addsub semantic, the ADD and SUB instructions would be individually implemented, resulting in two 32-bit adders in the hardware. In the presence of a semantic, the semantic defines the hardware implementation for all the instructions implemented in the semantic. Thus, the addsub semantic implements both instructions using a single 32-bit add with carry and a 32-bit multiplexor.

8.6.2 TIE Functions

TIE functions, like Verilog functions, encapsulate some computation that can then be instantiated in other TIE descriptions. They can be used to make the code modular, and to optionally share the hardware between multiple instantiations of the function. A TIE function has the following syntax:

```
function [n:m] <name> ([input_list])[<shared/slot_shared>]
                      {<function statements>}
```

The *input_list* of a function lists the input arguments of the function. The function has a single output with the same name as the name of the function and the width of this output is $[n : m]$ bits. An optional argument declares the function to be either *shared* or *slot_shared*. A shared function is one that has exactly one instance in the hardware, and is shared by all instructions that use it. A *slot_shared* function has one instance per slot index of a VLIW design; all the instructions within a slot index share this hardware, but it is not shared with instructions in other slots. If neither of these attributes is specified, the function is not shared and its purpose is to encapsulate the computation for modular code development. The following example demonstrates the use of shared functions for the ADD and SUB instruction of Section 8.6.1.

```
function [31:0] addsub([31:0]in0, [31:0]in1, cin) shared {
    assign addsub = TIEadd(in0, in1, cin);
}
operation ADD {out AR res, in AR input0, in AR input1 } {
    assign res = addsub(input0, input1, 1'b0);
}
operation SUB {out AR res, in AR input0, in AR input1 } {
    assign res = addsub(input0, ~ input1, 1'b1);
}
```

Instead of writing a *semantic* for the two instructions to share the 32-bit adder, the adder is encapsulated in a shared function. The TIE compiler will automatically generate the control logic required to multiplex the inputs from the operations into the single copy of the hardware for function addsub, as well as to resolve any resource hazard between instructions using the same function.

8.6.3 Defining Multicycle Instructions

The Xtensa processor has a 5-stage pipeline with stages instruction fetch (I), instruction decode and register read (R), execution (E), memory access (M), and write back (W), which is very similar to the DLX pipeline described in [6]. By default, all TIE operations perform the computation in a single clock cycle during the execution (E) stage of the pipeline, that is, the output operands are written at the end of the same cycle in which the input operands are read. Depending on the nature of the computation and the target frequency of the processor, it may not be possible to finish the computation in one clock cycle. If this is true, the computation needs to be distributed over multiple clock cycles. The TIE language provides the *schedule* construct to define multicycle instructions, and in a more general sense, to allow the designer control over the pipeline stage in which input operands are read, and output operands are computed.

The syntax of the *schedule* construct is as shown:

```
schedule <name> <instruction list> {
    use <operand> <stage>; . . . use <operand> <stage>;
    def <operand> <stage>; . . . def <operand> <stage>;
}
```

The *instruction list* is a list of all instructions to which the schedule is applicable. A *use* schedule can be applied to one or more input operands of an instruction, while a *def* schedule can be applied to output operands as well as the intermediate wires of a computation. The *use* schedule defines the pipeline stage at the beginning of which the input operand is available to the execution unit. The *def* schedule defines the pipeline stage at the end of which the output operand is expected to be available from the execution unit. The stage parameter of the schedule is defined in terms of the processor's pipeline stages, such as *Estage*, *Mstage*, *Wstage*, and so on.

The following example illustrates the use of the schedule construct to define a multicycle TIE instruction:

```
state PROD 64 add_read_write
operation MUL32 {in AR m0, in AR m1} {out PROD} {
    assign PROD = m0 * m1;
}
schedule mul {MUL32} {
    use m0 Estage; use m1 Estage; def PROD Mstage;
}
```

The MUL32 instruction multiplies two 32-bit numbers to generate a 64-bit product. The schedule construct specifies that the input operands are read in the *Estage* and the result is available in the *Mstage*. Thus the multiply computation is spread over two clock cycles.

Multicycle TIE instructions are always implemented in a fully pipelined manner, so that it is possible to issue a new instruction at every clock cycle, even though the instruction takes more than one cycle to complete. Thus, the increased latency does not affect performance unless the computation result is immediately needed in the following instruction. The TIE compiler automatically generates the control logic for detecting data hazards in the processor pipeline. This control logic inserts bubbles in the pipeline to prevent data hazards. It also does appropriate data forwarding so that the instruction held in the pipeline can proceed as soon as the earlier instruction completes its computation, without waiting for the result to be written to the register file.

The Xtensa C/C++ compiler is aware of the schedule of TIE instructions, and performs intelligent scheduling of instructions to avoid data hazard stalls in hardware as far as possible.

Note that while the Xtensa processor pipeline is five stages, the execution of the TIE operation can extend beyond the final write back (W) stage. In this case, the execution pipeline will be extended, and the write back to the TIE register file will occur at the end of this extended pipeline.

Iterative use of shared hardware

The TIE language allows designers to create iterative instructions, which are defined as instructions that use the same piece of hardware multiple times, in different clock cycles. This can be viewed as a way of implementing non-pipelined multicycle instructions. The primary advantage of this implementation is the reduced hardware cost, and the disadvantage is the fact that the same instruction cannot be issued at every cycle.

The following example illustrates a vector dot product computation implemented using a single multiplier:

```
function [31:0] mul16 ([15:0] a, [15:0] b, signed) shared {
    assign mul16 = TIEmul(a,b,signed);
}
operation VDOTPROD {out AR acc, in AR m0, in AR m1} {
  wire [31:0] prod0 = mul16(m0[15: 0], m1[15: 0], 1'b1);
  wire [31:0] prod1 = mul16(m0[31:16], m1[31:16], 1'b1);
  assign acc = prod0 + prod1;
}
schedule dotprod {VDOTPROD} { def acc Estage + 2; }
```

In the aforementioned example, the VDOTPROD instruction performs two 16×16 multiply operations using the shared function mul16. Because there is only one copy of the mul16 function in the hardware, this instruction requires at least two cycles to execute, as specified by the schedule.

8.7 CUSTOM DATA INTERFACES

Over the last several years, processors have become significantly faster and more powerful. However, memory subsystems have not kept pace with processors in this regard, making the data bandwidth in and out of the processor the bottleneck in many systems [7]. For such systems, reducing the latency and increasing the bandwidth of the connection between the processor and the rest of the system are the key to improving system performance. A traditional processor communicates with system memory and other peripheral devices through a bus that takes several cycles to perform one data transfer. The bus can become a bottleneck in systems where a large amount of data needs to be transferred between the processor and external devices.

The TIE language provides a mechanism to create new interfaces between the Xtensa processor and external devices. These interfaces allow direct, point to point connections between the Xtensa processor and other design blocks, including other Xtensa processors. Any number of such interfaces can be defined, and they can be of arbitrary bit-width. These interfaces allow high bandwidth and low latency communication between the processor and external devices, without using the memory subsystem or the processor bus. A rich set of communication protocols is supported using these interfaces to exchange status and control information between processors and devices, read and write data to a standard FIFO, and perform a memory lookup.

8.7.1 Import Wire and Export State

The TIE export state and import wire constructs are used to define interfaces used to transfer status or control information between an Xtensa processor and external devices.

Export state

Section 8.4.1 describes how to define a new processor state in the TIE language. The value of any TIE state can be made visible outside the processor via a top-level port. Such a state is known as an exported state. A state is exported by adding a keyword *export* to the state declaration. The architectural copy of an exported state is available as a primary output port of the Xtensa processor. Speculative writes to a TIE state are not visible on this interface—the value on the interface is updated only after the instruction writing to the state has committed.

Import wire

Import wires allow TIE instructions to read inputs from top-level ports of the Xtensa processor. This port is driven by the external device that generates the data value. For example, the export state of one processor can be connected to the import wire of another processor. An import wire is defined in the TIE language as:

```
import_wire <name> <width>
```

Import wires can be of arbitrary width, and can be used as an input operand of any TIE instruction. The following example illustrates a 20-bit value being read from an import wire, sign extended to 32 bits, and assigned to an AR register:

```
import_wire IMP_WIRE 20
operation READ_WIRE {out AR reg} {in IMP_WIRE} {
    assign reg = {{12{IMP_WIRE[19]}}, IMP_WIRE};
}
```

The value of IMP_WIRE is sampled when the READ_WIRE instruction executes, and this sampling happens before the instruction has committed. The read of the import wire interface is thus speculative in nature, that is, the data sampled may not be consumed if the instruction does not commit due to a branch, exception, or interrupt. Thus the read must be free of any side effect, because there is no handshake with the external device. Because of this property, import wires are suited for passing status or control information, rather than data transfer between devices.

8.7.2 TIE Queue

In a typical data flow system, it is a common practice to use a FIFO to create a point to point buffered data channel between two blocks, such as a producer–consumer scenario. In traditional processor-based systems, the FIFO channels are implemented in shared memory [8]. Such an implementation suffers from data bandwidth and a latency limitation imposed by the shared memory, and requires synchronization overhead.

The TIE language provides the *queue* construct to add new interfaces to the Xtensa processor that directly connect to an external FIFO. TIE instructions can read data from an input FIFO or write data to an output FIFO using the *queue* interface. Synchronization is achieved by stalling on an empty or full FIFO. Thus, TIE queues provide a natural and efficient way to map data flow applications to Xtensa processor-based systems.

The syntax for defining a queue interface is:

```
queue <name> <width> <in | out>
```

The width of the *queue* interface can be arbitrary, and the keyword *in* denotes an input queue interface from which data can be read while the keyword *out* denotes an output queue interface to which data can be written. Once a TIE queue has been defined, it can be used as an operand in any TIE instruction. This is illustrated in the following example:

```
queue IPQ0 40 in
queue IPQ1 40 in
queue OPQ 40 out
operation QADD {} {in IPQ0, in IPQ1, out OPQ} {
    assign OPQ = IPQ0 + IPQ1;
}
```

This example defines two 40-bit input queues and one 40-bit output queue. The instruction QADD reads from the two input queues, computes the sum of their data values, and writes the sum to the output queue. The QADD instruction stalls if either of the input queues is empty, or the output queue is full.

For each queue defined in the TIE description, the TIE compiler creates the ports necessary to interface the Xtensa processor to an external synchronous FIFO, as illustrated in Fig. 8.2. The hardware to read and write the FIFO using the pop/push interface and to stall on an empty/full FIFO is also automatically generated.

When an instruction that reads data from an input queue executes, data is popped from the external queue before the instruction has committed. The read is thus speculative, that is, the data popped from the external queue may not be consumed if the instruction does not commit due to an interrupt or exception. In this event, the processor buffers this data internally, so that when the instruction is re-executed, the data from the internal buffer is used without popping the external queue.

A FIFO connected to an output queue interface is written to only after the instruction performing the write has committed. This is achieved by buffering the write data

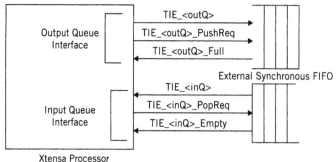

FIGURE 8.2

Top-level pins for TIE queue.

inside the processor until the instruction commits. Thus, writes to output queues are not speculative.

Nonblocking queue access

The operation QADD in Section 8.7.2 stalls the processor if either the input queues are empty or the output queue is full. In some cases, it is useful to check the status of a queue before issuing the instruction if a stall is not desirable. The TIE compiler automatically generates a signal for each queue called *<queue_name>_NOTRDY*, which, when asserted, indicates that the next access to the queue will stall. The signal is based on the state of the internal buffer as well as the external queue. This signal can be read as an input operand of a TIE operation, to implement "test and branch" instruction sequences, such that a queue read or write access is guaranteed to be successful.

The TIE compiler also creates an interface *<queue_name>_KILL* for each TIE queue, which can be written as an output operand of a TIE instruction that reads or writes the queue. If this signal is asserted, the queue access is cancelled. If the processor is stalled due to the queue being full or empty, the stall is also released, and the instruction completes. The *KILL* signal can be assigned based on a run-time condition, and used to programmatically access or not access a certain queue.

Using the two interfaces *NOTRDY* and *KILL*, a nonblocking version of the QADD instruction in Section 8.7.2 can be defined as shown:

```
operation QADD_NOBLOCK {} {in IPQ0, in IPQ1, out OPQ, out BR b} {
    wire kill = IPQ0_NOTRDY | IPQ1_NOTRDY | OPQ_NOTRDY;
    assign OPQ = IPQ0 + IPQ1;
    assign IPQ0_KILL = kill; assign IPQ1_KILL = kill;
    assign OPQ_KILL = kill; assign b = ~ kill;
}
```

If either of the input queues are empty or the output queue is full, the instruction QADD_NOBLOCK does not stall, but the read and write operations are cancelled and the instruction completes after setting the status bit to 0, indicating an unsuccessful operation.

8.7.3 TIE Lookup

A TIE lookup is an interface designed to send a request or an address to an external device and get a response or data from the device one or more cycles later. This interface can be used for a variety of system design tasks, such as an interface to an external lookup table stored in a ROM, or to set up a request–response connection with any hardware blocks external to the processor. The external device can also stall the processor if it is not ready, thus making arbitration between processors and other devices possible.

The request can be sent out at any pipeline stage at or beyond the execution stage, and the response can be received a cycle or more later. The syntax for defining a lookup is:

```
lookup <name> {<out width>, <out stage>}
              {<in width>, <in stage>} [rdy]
```

The width and pipeline stage in which the address/request is sent out are specified using the parameters *out width* and *out stage*, respectively. The width and pipeline stage in which the data/response is read back are specified using the parameters *in width* and *in stage*, respectively. Thus, the latency of the external memory or device is (*<in stage>* - *<out stage>*). An optional argument *rdy* in the lookup definition indicates that the device should be able to stall the processor when it is not ready to accept a request. Note that once the device has accepted a request, the response must be guaranteed after the specified latency.

The TIE compiler automatically generates the hardware for the request–response protocol of a TIE lookup. The top-level ports generated for a TIE lookup are shown in Fig. 8.3. *TIE_<name>_Out* is the output or address of the lookup and can be connected to the address input of the memory. *TIE_<name>_In* is the input data for the lookup, and can be connected to the data port of the memory. The 1-bit output port *TIE_<name>_Req* is asserted for once clock cycle when the processor makes a lookup request. If the lookup is configured with *rdy*, the signal *TIE_<name>_Rdy* is a 1-bit input that can be used to signal that the device is not accepting requests, in which case the processor will stall the execution pipeline and present the request until it is accepted.

Once a lookup is defined, it can be used as an operand in any TIE instruction. The following example illustrates the use of a TIE lookup to access a coefficient table that is external to the processor.

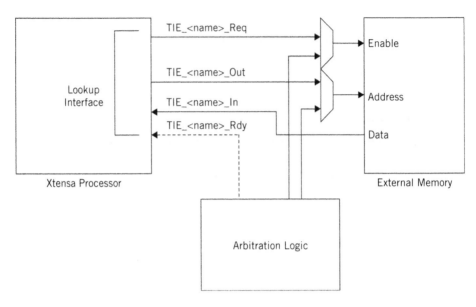

FIGURE 8.3

Top-level connections for a TIE lookup.

```
lookup coeff {8, Estage} {256, Mstage}
state ROM_DATA 256
operation ROM_LKUP {in AR index}
        {out coeff_Out, in coeff_In, out ROM_DATA } {
    assign coeff_Out = index;
    assign ROM_DATA = coeff_In;
}
```

8.8 HARDWARE VERIFICATION

The TIE language allows the designer to describe the instruction extensions at a high level of abstraction. The designer typically specifies only the operands of the instruction and the combinational relationship between the output and inputs. The designer may optionally specify the instruction set encoding and the pipeline stages in which input operands are read and output operands are written. The TIE compiler generates the RTL for implementing the datapath and control logic for integrating the new instructions into the Xtensa pipeline.

The designer is responsible for verifying that the combinational relationship between the input and output operands of the instruction is correct. This verification can be done by writing self-checking test programs for each TIE instruction. The general format of such test programs is to initialize the input operands of the instructions to a set of values, execute the instruction, and ensure that the output value matches the expected value of the result. This is done over a number of test vectors that cover all the corner case values and several typical values. Simulation of the application code using the TIE instrinsics is also part of this verification.

The control logic that integrates the new instructions into the Xtensa processor is automatically generated by the TIE compiler. This includes instruction decode logic, pipelining controls such as data hazard prevention and data forwarding, and appropriate interfaces to the instruction fetch and load-store units of the processor. The designer is not responsible for verifying these aspects of the instruction implementation, as this logic is considered to be "correct-by-construction". Microarchitectural test programs that verify the control logic are automatically generated by the TIE compiler and associated tools.

8.8.1 Microarchitectural Verification

A rich set of microarchitectural diagnostics is automatically generated for all TIE instructions to test the control logic of the TIE extensions. These diagnostics have no knowledge of the datapath of the instructions and do not depend on the datapath. They test specific microarchitectural features. For example, diagnostics are generated to test that the pipeline stalls due to data dependencies between TIE operations are correct. These depend entirely on the read and write stages of the operations that are described in TIE. The algorithm to generate an instruction sequence for such a diagnostic is shown here:

```
foreach reg in ( designer defined register_files ) {
  foreach wstage in ( write/def stages of reg ) {
   foreach rstage in ( read/use stages of reg ) {
           1) Find instruction I1 that writes reg in wstage
           2) Find instruction I2 that reads reg in rstage
           3) Allocate registers or select operands of I1 and I2
              with the constraint to generate data dependency, i.e.
              read operand of I2 = write operand of I1
           4) Calculate stall cycles = (rstgae – wstage) if (rstage
              > wstage)
           5) Generate instruction sequence as
              a. Read cycle count register into AR register a1
              b. Print instruction I1
              c. Print instruction I2
              d. Read cycle count register into AR register a2
              e. Execution cycles = a2 – a1
              f. Compare execution cycles with expected value and
                 generate error if not correct.
     }
    }
  }
```

The algorithm described here can be used to generate a self-checking diagnostic test program to check that the control logic appropriately handles read-after-write data hazards. This methodology can be used to generate an exhaustive set of diagnostics to verify specific characteristics of TIE extension, as described in [9].

8.9 CASE STUDY OF AN AUDIO DSP DESIGN

The TIE language has been used to design several complex extensions to the Xtensa processor core by Tensilica, its customers, and research institutions [10, 11]. This section presents a case study of an Audio DSP designed by Tensilica using the TIE language. The HiFi2 Audio Engine is a programmable, 24-bit fixed point Audio DSP designed to run a wide variety of audio codecs—from low bit-rate speech codec to high fidelity multichannel audio codecs for consumer living room applications.

8.9.1 Architecture and ISA Overview

The HiFi2 Audio Engine is a VLIW-SIMD design that exploits both instruction and data parallelism available in typical audio processing applications. In addition to the 16- and 24-bit instruction formats of the Xtensa processor, it supports a 64-bit, 2-slot VLIW instruction set defined using the TIE language. While most of the HiFi2 instructions are available in the VLIW instruction format, all the instructions in the first slot of the VLIW format are also available in the 24-bit format for better code density.

Fig. 8.4 shows the datapath of the HiFi2 design. In addition to the AR register file of the Xtensa core, it has two custom register files. The P register file is an 8-entry, 48-bit register file. Each 48-bit entry can be operated upon as two 24-bit values, making the HiFi2 a two-way SIMD engine. The Q register file is a four entry, 56-bit register file that serves as the accumulator for multiply-accumulate operations. The architecture also defines a few special purpose registers such as an overflow flag bit, a shift amount register, and a few registers for implementing efficient bit extraction, and manipulation hardware. The computation datapath features SIMD multiply accumulate, ALU, and Shift units that operate on the two elements of the P register file. A variety of scalar operations on the Q register file and on the AR register file of the Xtensa core are also defined.

Table 8.4 provides a high-level summary of the HiFi2 ISA, along with an indication of which slot of the VLIW format the instruction group is available in. It also lists the approximate number of instructions belonging to each group. There are a total of over 300 operations between the two slots. In addition to the Load/Store, MAC, ALU, and Shift instructions, the HiFi2 ISA supports several bit manipulation instructions that enable efficient parsing of bitstream data and variable length encode and decode operations common in audio bitstreams.

While it is possible to program the HiFi2 in assembly language, it is not necessary to do so. All the HiFi2 instructions are supported as intrinsics in C/C++ code, and

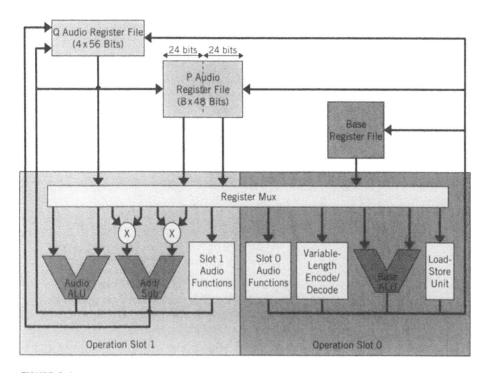

FIGURE 8.4

HiFi2 Audio Engine datapath.

variables for the P and Q register files can be declared using custom data types. The Xtensa C/C++ compiler does register allocation for these variables and appropriate scheduling of the intrinsics. All the codecs developed by Tensilica are written in C, and the performance numbers quoted in Table 8.5 are achieved without any assembly programming.

8.9.2 Implementation and Performance

The HiFi2 Audio Engine is available as synthesizable RTL, along with synthesis and place and scripts to target its implementation using any modern process technology. The synthesized netlist corresponds to about 67 K gates, of which approximately 45 K gates correspond to the TIE extensions and 22 K gates are for the base Xtensa processor core. In TSMC 90 nm process technology ("G" process), this corresponds to a die size area of 0.84 mm^2. The design achieves a clock frequency of 350 MHz, and its power dissipation at this frequency is 41.3 mW.

While the HiFi2 design can achieve a clock frequency of 350 MHz, the amount of cycles required for audio decoders and encoders is significantly lower. This allows headroom for other computations to be performed on the processor, or to target the design for lower clock frequency and lower power for better battery life in

Table 8.4   HiFi2 ISA summary.

Operation Type	Slot	Count	Description
Load/Store	0	28	Load with sign extension, store with saturation to/from the P and Q register files. SIMD and scalar loads supported for P register file. Addressing modes include immediate and index offset, with or without update.
Bit Manipulation	0	17	Bit extraction and variable length encode/decode operations.
Multiply & Multiply-Accumulate	1	175	24 × 24 to 56-bit signed single and dual MAC. 24 × 24 to 56-bit signed single MAC with saturation. 32 × 16 to 56-bit single and dual MAC. 16 × 16 to 32-bit signed single MAC with saturation. Different saturation and accumulation modes.
Arithmetic	1	22	Add, subtract, negate, and absolute value on P (element wise) and Q registers, with and without saturation. Minimum and maximum value computations on P and Q registers.
Logic	1	8	Bitwise AND, NAND, OR, XOR on P and Q registers.
Shift	1/0	20	Arithmetic and logical right shift on P (element-wise) and Q registers. Left shift with and without saturation. Immediate or variable shift amount (using special shift amount register).
Miscellaneous	1/0	42	Rounding, normalization, truncation, saturation, conditional, and unconditional moves.

mobile applications. More than 20 codecs have been ported to this platform, and Table 8.5 lists the performance of a few that are well known. The table lists the MCPS (millions of clocks/sec) required for real-time encode/decode, and the amount of program and data memory used by the codecs. The performance numbers illustrate the versatility and efficiency of the architecture in handling a wide variety of audio algorithms.

8.10   CONCLUSIONS

The phrase "design productivity gap" has been frequently used in the past few years to refer to the imbalance between the number of available transistors on a piece of silicon, and the ability of designers to make good use of these transistors. Designing

Table 8.5 HiFi2 codec performance.

Codec Name	Cycle Count (MCPS)	Program Memory Size (KB)	Data Memory Size (KB)
MP3 Decoder	5.7	20	41.6
Dolby Digital Plus (7.1 channel)	63	44.3	227
Dolby Digital Encoder (5.1 channel)	60	39	123
G.729 Speech Codec	13.7	36.4	13.2

SoCs at a higher level of abstraction has been proposed as a potential way to close this gap. A recently published book [12] describes the process of using application-specific processor IP as the building block for the next generation of highly integrated SoC designs. It describes how configurable microprocessor cores can serve as the "new NAND gate" for designs, and significantly improve designer productivity. The TIE language and compiler are at the heart of the methodology described in this book.

The goal of this chapter has been to give the reader a flavor of the TIE language and the automation provided by the TIE compiler. Additional information on the TIE language and its use in creating application-specific processors is available in the references and on Tensilica's website (*www.tensilica.com*). The authors wish to acknowledge the contribution of the entire technical staff of Tensilica Inc., whose work forms the basis for the content of this chapter.

REFERENCES

[1] R. Gonzales. Xtensa: A Configurable and Extensible Processor. *IEEE Micro Vol. 20, #2*, March–April 2000, pages 60–70.

[2] D. Chinnery and K. Keutzer. *Closing the Power Gap between ASIC and Custom: Tools and Techniques for Low Power Design*. Springer, 2007, Chapter 5. New York.

[3] A. Wang, E. Killian, D. Maydan, and C. Rowen. Hardware/Software Instruction Set Configurability for System-on-Chip Processors. *Proc. of Design Automation Conference*, June 2001, pages 184–190.

[4] Tensilica Instruction Extension (TIE) Language Reference Manual. Issue Date 11/2007, Tensilica, Inc., Santa Clara, CA, U.S.A.

[5] Tensilica Instruction Extension (TIE) Language User's Guide. Issue Date 11/2007, Tensilica Inc., Santa Clara, CA, U.S.A.

[6] J. L. Hennessey and D. A. Patterson. *Computer Architecture: A Quantitative Approach*, Morgan Kaufmann Publishers, San Francisco, California, 1990.

[7] D. Burger, J. Goodman, and A. Kagi. Limited Bandwidth to Affect Processor Design. *IEEE Micro*, November–December 1997, pages 55–62.

[8] M. J. Rutten, J. T. J. van Eijndhoven, E. G. T. Jaspers, P. van der Wolf, O. P. Gangwal, A. Timmer, E. J. D. Pol. A Heterogeneous Multiprocessor Architecture for Flexible Media Processing. *IEEE Design and Test of Computers*, July–August 2002, pp. 39–50.

[9] N. Bhattacharyya and A. Wang. Automatic Test Generation for Micro-Architectural Verification of Configurable Microprocessor Cores with User Extensions. *High-Level Design Validation and Test Workshop*, November 2001, pages 14–15.

[10] M. Carchia and A. Wang. Rapid Application Optimization Using Extensible Processors. *Hot Chips 13*, 2001.

[11] N. Cheung, J. Henkel, and S. Parameswaran. Rapid Configuration and Instruction Selection for an ASIP: A Case Study. *Proc. of the Conference on Design, Automation and Test in Europe*, March 2003, pages 10802.

[12] C. Rowen. *Engineering the Complex SOC: Fast, Flexible Design with Configurable Processors*, Prentice Hall, Upper Saddle River, New Jersey, 2004.

MADL—An ADL Based on a Formal and Flexible Concurrency Model[1]

9

Wei Qin, Subramanian Rajagopalan, and Sharad Malik

9.1 INTRODUCTION

Increasingly, complex system functionalities and shrinking process feature sizes are changing how electronic systems are implemented today. Hardwired Application Specific Integrated Circuit (ASIC) solutions are no longer attractive due to sharply rising Nonrecurring Engineering (NRE) costs to design and to manufacture the integrated circuits. The NRE costs have increased the break-even volume beyond which these hardwired parts are preferred over programmable solutions. Thus, increasingly we are seeing a shift toward systems implemented on programmable platforms. This shift is helped by the development of application/domain specific processors that attempt to reduce the power and performance overhead of programmability by providing microarchitectural features matched to domain specific computational requirements. However, there is relatively little available in terms of design tools for the software environment (e.g., simulators, compilers) for these processors and thus these tend to be hand crafted—a fairly low productivity task and especially limiting during processor design space exploration when a large number of design points need to be evaluated. Therefore it is highly desirable that the software design tools be synthesized automatically from high-level processor specifications.

Though microprocessors are diverse in their computation power and their underlying microarchitectures, they share some common general properties inherited from their common ancestor—the von Neumann computer. They fetch instruction streams from memory, decode instructions, read register and memory states, evaluate instructions as per their semantics, and then update register and memory states. As a result of this commonality, there exists a potential to develop high-level models to abstract a wide range of microprocessors. These abstract models can then be used to assist the development of design automation tools that prototype, synthesize, verify, and program microprocessors. One example of such an abstract data model is the register-transfer-list used by the compiler community for code generation.

---

[1] Portions of this paper were taken from W. Qin, S. Rajagopalan, S. Malik. A Formal Concurrency Model Based Architecture Description Language For Synthesis of Software Development Tools, *Proceedings of ACM 2004 Conference on Languages, Compilers, and Tools for Embedded Systems*, pages 47–56, June 2004.

Architecture Description Languages (ADLs) have been created to capture abstract processor models. Our initial survey of ADLs revealed that despite some common goals, their respective emphasis in supporting different tools and different architecture families made them highly diverse in their syntaxes and semantics. Further investigation showed that the quality of an ADL is largely determined by its underlying semantic model. Some existing ADLs have well-documented semantic models, while some solely rely on undocumented architecture templates. But there is a lack of a flexible processor model that could express precise data and control semantics at a high abstraction level. Therefore, no prior ADL could be used for synthesizing both cycle accurate simulators and compilers for a wide range of architectures. The lack of a good model became the first challenge that we faced when developing the Mescal Architecture Description Language (MADL).

After experimenting with various processor models and software development tools, we proposed the Operation State Machine (OSM) formalism [1] to fill the gap. The OSM model provides not only a high level, but also a flexible abstraction mechanism to model the data and control semantics of a microprocessor. Based on the model, we designed the two-layer MADL to support accurate modeling of a broad range of architectures. The main features of MADL are:

- A formal semantic model that allows flexible and rigorous modeling of a large family of processors.

- A two-layered design that separates the fundamental architectural properties from tool-dependent information.

MADL also utilizes the established AND-OR graph technique for instruction factoring first used in nML [2]. The technique enables users to factor descriptions to significantly reduce their length.

This chapter is organized as follows. Section 9.2 introduces the OSM model, including its static version, its dynamic version, and the scheduling algorithm. We describe the design of MADL in Section 9.3. Section 9.4 describes the generation of several software tools from MADL descriptions, and Section 9.5 presents the experiment results. Section 9.6 discusses related work in the field. Finally, Section 9.7 provides some concluding remarks.

9.2 OPERATION STATE MACHINE MODEL

In this section we discuss the two versions of the OSM model: the static OSM model, and the dynamic OSM model. The former is the theoretical foundation of the MADL framework. All formal analyses in the framework were performed on the static OSM model. However, the static model is inconvenient to describe since it does not benefit from the similarity among instructions. Therefore, the dynamic OSM model was later developed to enable shared description among similar instructions. The syntax of MADL was based on the dynamic OSM model and the AND-OR graph. The MADL compiler can convert a dynamic OSM model into an equivalent static one.

9.2.1 Static OSM Model

An architecture model largely determines the quality of the ADL built on top of it. It is important to define a solid computation model as the architecture model. The Operation State Machine (OSM) model is proposed for such a purpose [1]. It is based on the observation that all processors share the following common properties:

- They all interpret instructions. They inherit the common fetching, decoding, and executing behavior from their common ancestor—the von Neumann computer.

- They are logic circuits. They are composed of microarchitectural components that collectively facilitate the execution of instructions.

These two properties map into the two levels in the OSM model: the operation level and the hardware level. The operation level contains the ISA and the dynamic execution behavior of the operations. The hardware level represents the simplified microarchitecture as a result of the abstraction mechanisms used in the OSM model.

At the operation level, Extended Finite State Machines (EFSM) are used to model the execution of operations. An EFSM is a traditional Finite State Machine (FSM) with internal state variables [3]. It can be transformed to a traditional FSM by dividing each of its states into n states, where n is the size of the state space defined by the internal state variables. The use of these state variables compresses the state diagram and hides less relevant details of the state machine from its external view.

In an OSM-based processor model, one EFSM represents one operation, thus the name operation state machine is used for these special-purpose EFSMs. The states of an OSM represent the execution status of its corresponding operation, while its edges represent the valid execution steps. The states and the edges form the state diagram of the OSM, which must be a strongly connected graph so that there exists no dead-end state or isolated subcomponent.[2] The internal state variables of the EFSM are used to hold intermediate computation results of the operation. To control the state transition of the OSM, each edge of the OSM is associated with a condition, which represents the readiness of the operation to progress along the edge. Such readiness is expressed as the availability of execution resources, including structural resources, data resources, and artificial resources created for the purpose of modeling. Example resources include pipeline stages, reorder-buffer entries, and the availability of operands.

The execution resources are maintained in the hardware level of the OSM model. They are modeled as *tokens*. A token may optionally contain a data value. A *token*

---

[2] This is a requirement of the OSM model, not the EFSM.

manager controls a number of tokens of the same type. It assigns the tokens to the OSMs according to its token management policy and the requests from the OSMs. In essence, a token manager is an abstract implementation of a control policy in the processor.

To model the interaction between the operation level and the hardware level, we defined four types of token transactions—*allocate*, *inquire*, *release*, and *discard*. Transactions are requested by the OSMs, and are either accepted or denied by the token managers. The primitives help model control behaviors in a pipeline such as the handling of hazards. Additionally, to model the data flow in the processor, we introduced two data communication primitives—*read* and *write*. An OSM can read the value of a token either if it is *allocated* the token or if it successfully *inquires* about the token. It can write to a token only if it is *allocated* the token. An OSM can then exchange data values with the token managers through these two primitives.

As a simple example, Fig. 9.1 shows a 4-stage pipelined processor and a state machine modeling its "add" operation. The states of the state machine represent the execution status of the instruction. The state machine starts from the initial state *I*. At each control step (a clock cycle or phase), it will try to advance its state by one step. When it returns to state *I*, the instruction retires. In order to advance the state along an edge in the state diagram, the state machine must satisfy the transition

IF — fetch stage
ID — decode stage
EX — execution stage
WB — writeback stage
RF — register file

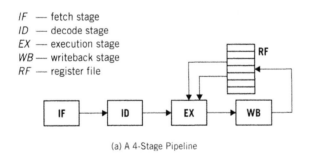

(a) A 4-Stage Pipeline

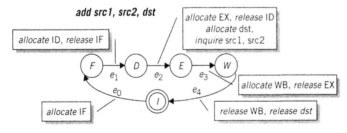

(b) OSM for the Add Instruction

FIGURE 9.1

An OSM model example.

condition on the edge. This condition is defined as the conjunction of the outcomes of a list of token transactions, such as the ones shown in the text boxes in Fig. 9.1(b).

In Fig. 9.1, the state machine reads source register values on edge e_2 and writes its destination register value on edge e_4. In addition, we also include as part of the OSM model a set of computation operators that can be used to evaluate instruction semantics. All token transactions and computation primitives are called *actions* and they are all bound to the edges of the state machines.

Modern processors are typically pipelined and hence may have more than one operation executing simultaneously. In the simulation model of such a processor, multiple OSMs may be alive at the same time, each representing an operation in the pipeline. They compete for resources through their token transaction requests and transition their states concurrently. They constitute the operation level of the model.

The hardware level of the processor example includes 4 token managers modeling the pipeline stages, each controlling one token that represents the ownership of the corresponding pipeline stage. In addition, we model the register file as a token manager containing one token for each register. The token managers respond to the token transactions required by the state machines. All managers contain a set of interface functions corresponding to all types of transactions. These interface functions implement the resource assignment policies of the token manager and control the execution flow of the state machines.

The OSM is a highly flexible model capable of modeling arbitrarily connected execution paths of an operation, including those of superscalar processors with out-of-order issuing. For example, Fig. 9.2 shows part of a state machine that can be used to model the "add" operation in an out-of-order processor. In this example, "RS" represents a token manager modeling a reservation station and "RB" is a token manager for the reorder buffer. After the operation has been decoded at state D,

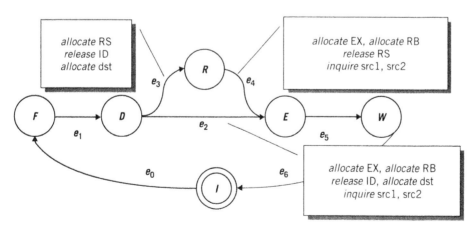

FIGURE 9.2

OSM example for out-of-order issuing.

the state machine has two options to proceed. It can either proceed to state E—the execution stage—if an execution unit, a reorder buffer entry, and the source operands are available, or enter state R—the reservation station stage—to wait for these resources. Such out-of-order issuing behavior cannot be modeled by other pipeline models such as the pipeline diagram, but can be easily modeled by the OSM. The flexibility of the operation level as well as the extensibility of the hardware level allow the OSM model to capture the precise execution semantics of microprocessors for cycle accurate simulator generation.

The OSM model has well-defined semantics and therefore can be directly used for simulating microprocessors. From the point of view of compiler developers, it is possible to analyze the state diagram and the *actions* to extract useful properties such as the operation semantics and the resource usage of operations. Automatic extraction of this information is very useful for the development of retargetable compilers.

To summarize, the OSM model provides a flexible yet formal means to model microprocessors. It is suitable as the underlying semantic model of a machine description language. It serves as the theoretical foundation of MADL.

9.2.2 Dynamic OSM Model

The OSM example in Fig. 9.1 models one operation. To model a full instruction set, we need a bigger finite state machine so that every operation can define its own actions. Fig. 9.3 illustrates such an OSM. The OSM is comprehensive in that it is capable of representing all types of operations in the ISA. In its state diagram, states I, F, and D are still shared among all the operations. After D, the state diagram splits into multiple paths. Each path models the execution behavior of one type of operation.

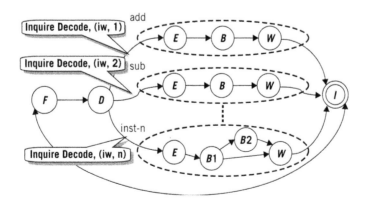

FIGURE 9.3

Comprehensive OSM for an instruction set.

To steer an OSM to the right execution path matching the operation that it represents, an artificial token manager named Decoder can be used. As a pure abstraction of the steering mechanism, Decoder contains no token. It responds to inquiry requests with token identifiers in the form of (integer, integer). The first integer contains the instruction word, and the second contains an index denoting the type of an operation. If the instruction word decodes to the type as specified by the index, the inquiry request is approved by the token manager. Otherwise, it is rejected.

In Fig. 9.3, an inquiry action to decode exists on each output edge from state D. The instruction word value in iw and an index value unique to the operation type corresponding to the edge are used as the token identifier. In this way, the OSM can only proceed along the edge that matches the actual operation type.

The large size of the comprehensive FSM poses a challenge to the description writers. Since a real-world ISA may contain several hundred operations, the state diagram often contains hundreds of states and edges. This implies that specifying the OSM can be a rather laborious task unless some simplification means are used.

Fortunately, most of the real-world processors are designed with regularity. Their instruction sets are usually organized as classes. Operations from the same class share many common properties. Most instruction sets are also designed to be orthogonal. The operand fields are usually separate from one another and from the opcode field in the binary encoding, allowing the addressing mode of each operand to be changed independently. Such regular organization of the instruction set reduces the complexity of the processor implementation. It also provides opportunities to simplify the description of the OSMs in a similar hierarchical and orthogonal way.

To take advantage of this, the OSM model is adapted to its dynamic version. The key idea is to replace the static binding relationship between the state diagram and the actions with a dynamic one. A dynamic OSM starts with a small set of actions so that it is able to enter the pipeline and fetch the operation from the instruction memory. This set of actions should be common to all operation types. Once the operation is decoded, the actions corresponding to the actual operation type are then bound to the state diagram, allowing the OSM to continue its execution.

For the dynamic OSM, the state diagram can be significantly reduced compared to the one in Fig. 9.3. It is no longer necessary to have a separate branch for each type of operation. Similar operations can share the same states and edges, and possibly some common actions among them. Their specific actions will be bound onto the state diagram after the instruction is decoded. This reduces the complexity of the description of the state diagram. An example of the dynamic OSM-based description is given in Section 9.3.

Given that the dynamic OSM model is more compact to describe, it forms the underlying semantic model of MADL. The MADL simulator is essentially an interpreter of the dynamic OSM model, with some run-time overhead of action binding. In all practical situations, a dynamic OSM model can be converted to a static

OSM model. The MADL compiler is thus capable of converting a dynamic OSM into a static OSM, which can be used for property analysis and also for compiled simulation [4].

9.2.3 Scheduling of the OSM Model

In a processor model, multiple OSMs attempt to obtain required resources and advance their states. When resources are limited, competition exists. As many token managers respond to the resource requests in a first come-first serve fashion, the activation order of the OSMs can affect the priorities of the OSMs to get the resources. This section explains the details of the scheduling algorithms used.

Fig. 9.4 shows the overall picture of the OSM-based modeling scheme. The entities in the scheme include the OSMs, the microarchitectural components (hardware units in the figure), and the token managers. The token managers are wrapped around by the hardware units, which communicate via the Discrete-Event (DE) computation model. They implement the resource management policy of the corresponding microarchitecture components. At the operation level, the OSMs follow their own scheduler. They do not communicate with each other, but with the token managers via the token transaction protocol. Therefore, DE-based communication at the hardware level and token transactions across the two levels are the basic means of communication in the model.

At the operation level, the OSMs are sorted according to the time when they last leave the initial state. The earliest one has the highest priority. The ones that remain

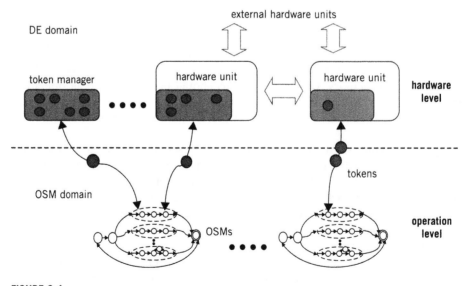

FIGURE 9.4

Overall structure of the OSM model.

```
Scheduler::control_step()
{
  update(OSMList);
  OSM = OSMList.head;
  while ((OSM = OSM.next) != OSM.tail)
  {
    EdgeList = OSM.currentOutEdges();
    foreach edge in EdgeList
    {
      result = OSM.requestTransactions(edge);
      if (result == satisfied)
      {
        OSM.commitTransactions(edge);
        OSM.updateState(edge);
        OSMList.remove(OSM);
        OSM = OSMList.head;
        break;
      }
    }
  }
}
```

Algorithm 1: OSM scheduling algorithm.

at the initial state have the lowest priority. Their order does not matter since they do not represent any operations and therefore all are equal.

Algorithm 9.1 shows the pseudo-code of the algorithm used for scheduling in each control step. The OSMList contains sorted OSMs, the oldest one first. The scheduler activates the OSMs in the list order. When an OSM successfully transitions, it is removed from the list so that it will not be scheduled for state transition again in the same control step. Stalled OSMs are kept in the list to await future opportunities to transition. A transitioning OSM may free up resources useful to some stalled OSMs that have higher priorities. To allow them to obtain the resources in the same control step, the outer-loop is restarted from the remaining OSM with the highest priority. When the OSMList becomes empty or when no OSM can change its state, the scheduler terminates and the control step finishes.

A control step in the operation level is an atomic step from the point of view of the hardware level. It occurs synchronously at the edge of the triggering clock signal. Between two control steps, the hardware components communicate with one another in the DE domain. Such communication affects the internal states of the components. As the token managers share states with their wrapper components, the communication result may also affect their responses to the actions in

```
nextEdge = 0.0;
eventQueue.insert(clock_event(nextEdge));
while (!eventQueue.empty())
{
  event = eventQueue.pop();
  if (event.timeStamp >= nextEdge)
  {
    director.control_step();
    nextEdge += regularInterval;
    eventQueue.insert(clock_event(nextEdge));
    break;
  }
  event.run();
}
```

Algorithm 2: Adapted DE scheduler for OSM.

the following control step. At a triggering clock edge, a control step is executed during which the token managers interact with the OSMs at the operation level. The results of such an interaction affect the internal states of the token managers. However, as the control step is indivisible in the DE domain, the state changes of a token manager cannot generate any event in the DE domain and are kept to the token manager itself within the control step. Only after the control step finishes and the DE scheduler resumes can its wrapper component communicate its new states to other components. In other words, during a control step, the token managers act independently. They may only "collaborate" during the interval between two control steps. Such scheduling can be explained by the pseudo-code in Algorithm 9.2. The pseudo-code iteratively generates regular clock edge events. It performs the book-keeping tasks of a standard DE scheduler such as ordering events by time and firing them in order. A control step of the OSM domain is activated at each clock edge.

9.3 MADL

Based on the OSM model introduced in Section 9.2, we designed MADL, an ADL that can be used to describe a broad range of microprocessors including scalar, super-scalar, VLIW, and multithreaded architectures. The goal of the description language is to assist the generation of software tools such as simulators and compiler optimizers.

To make MADL simple yet extensible, we adopted a two-layer description structure for microprocessor modeling. The first layer is the core language that describes the operation layer of the OSM model. It contains specifications of instruction semantics (the *actions*), binary encoding, assembly syntax, execution timing, and resource consumption. This layer forms the major part of a processor description. In Section 9.3.2, we describe how we can achieve concise descriptions in this layer by combining the OSM semantic model with an effective syntax model, namely the AND-OR graph.

The second layer of MADL describes information that is relevant to specific software tools. It supplements the core description and assists the software tools to analyze and extract processor properties. This layer is called the annotation layer. The annotation syntax is generic and hence flexible and extensible to use. Unlike the core layer, its semantics are subject to interpretation by the software tools that use the information.

9.3.1 The AND-OR Graph: A Review

A good syntax model is important to the usability of an ADL. In this section, we review the AND-OR graph model that is commonly used to describe instruction encoding, assembly syntax, and evaluation semantics. It appears in different forms in ADLs including nML [2], ISDL [5], and LISA [6]. An AND-OR graph is a directed acyclic graph with only one source node. It is composed of a number of *nodes*. A *node* is either an AND-rule(·) and all its OR-rule(+) children, or a leaf AND-node. Fig. 9.5(a) shows an example of an AND-OR graph with five *nodes*. An *expansion* of an AND-OR graph can be obtained by short circuiting each OR-rule with an edge from its parent to one of its children. Fig. 9.5(b) shows all possible *expansions* of the graph, each of which corresponds to an instruction. Compared to enumerating instructions in their expanded forms, the AND-OR graph model is much more compact as it factors common properties into the upper levels, thereby minimizing redundancy.

Most, if not all, instruction set architectures (ISA) organize instructions into hierarchical classes. All instructions in a class share properties such as encoding format, assembly syntax, etc. Hence, the AND-OR graph model is a natural choice to model

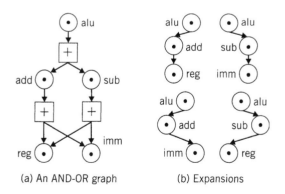

(a) An AND-OR graph   (b) Expansions

FIGURE 9.5

AND-OR graph example.

ISAs, especially orthogonal ISAs such as RISC. For the toy instruction set shown in Fig. 9.5, since the operands and the opcodes are orthogonal, they can be separated into different nodes. The common nodes among different instructions are then merged to form the AND-OR graph.

9.3.2 The Core Layer

The core layer uses the dynamic OSM model as its semantic model, and the AND-OR graph as its syntax model. Thus, the task of designing the layer is to combine these two to create a description language that is expressive and efficient for describing architectural details. These include the encoding, assembly syntax, semantics, and timing information for the instructions, as well as their interactions with the microarchitecture.

Note that the focus of the core layer is restricted to the operation level and does not include the hardware level (mainly the token managers). The execution models of the token managers are currently organized as a C++ template library in the current implementation of MADL. We are investigating the proper syntax for describing token managers in descendants of MADL [7]. A clean description of the token managers may allow the ADL compiler to potentially analyze the token managers for verification purposes and for extracting other useful information.

In the rest of this section, we describe the core layer of MADL using an example of the 4-stage pipelined processor of Fig. 9.1 with the toy instruction set shown in Fig. 9.5.

As mentioned in Section 9.2, the edges of a static OSM are statically bound with the *actions*. Since different types of operations differ in their *actions*, it would be a natural thought to adopt separate state machines to model different operation types. However, the actual type of an operation cannot be resolved until it is fetched and decoded, which occurs during the execution of the state machine. On the other hand, we cannot decide which type of state machine to execute until we know the actual type of the operation. Such mutual dependency creates a bootstrapping problem.

To solve the problem, we use a polymorphic state machine that can model all types of operations. Fig. 9.6 shows one such state machine for the example processor. In such a state machine, each path from state F to state I represents the execution path of one type of operation. The state machine is capable of fetching and decoding the operation and choosing to execute along the right path that corresponds to the actual type of the operation. An artificial token manager can be implemented to help steer the execution path. It should be noted that although the state machine is polymorphic, it still represents only one operation at a time. Multiple such state machines are needed to model the operations in a pipelined processor.

In general, the paths of the polymorphic state machine may have heterogeneous topologies. For real-world instruction set architectures with hundreds of operations, the state machine may be very large and cumbersome to specify. Moreover, although

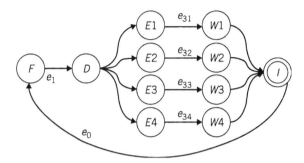

FIGURE 9.6

Polymorphic state machine for the toy ISA.

we may still apply the AND-OR graph to describe the *actions* of all operations, it is not directly usable in specifying the static binding relationships between the *actions* and the edges if most of the edges are not sharable across operations.

To avoid the potential combinatorial explosion in descriptions, we adopted the dynamic OSM model. As explained in Section 9.2, in the dynamic OSM model, the *actions* are no longer statically bound to the state diagram. Instead, they are bound to the edges dynamically during the execution of the state machine. In such a dynamic model, two different operations can refer to the same set of edges in the *action* description. As long as the syntax ensures that no conflict may occur during binding, the same state diagram can be shared by all operations in the description. This makes it possible for us to fully apply the AND-OR graph model to specify the *actions*.

In the dynamic OSM model, the state diagram is simply a directed graph. We call such a bare diagram a *skeleton*. To apply the AND-OR graph model, we describe operations based on *syntax operations*, each of which corresponds to one node in the AND-OR graph. All *syntax operations* in one *expansion* of the AND-OR graph form one operation. At run time, the *actions* of the *syntax operations* in the *expansion* will bind to the *skeleton* to form a finite state machine that models the operations.

Fig. 9.7 shows an example based on such a description scheme for the 4-instruction toy processor. The first USING statement indicates that the following *syntax operations* are based on the *skeleton* named "diagram_a," whose structure is the same as the state diagram shown in Fig. 9.1. A *skeleton* definition (not shown here) contains a list of states and edges, and a list of variables that can be accessed by all the *syntax operations* bound to it. The rest of the example defines a set of OPERATIONs, each corresponding to a *syntax operation*. An OPERATION contains sections such as VAR where local variables are defined, CODING where binary encoding is specified, EVAL where the initialization of local variables can be performed, and TRANS where *actions* are defined with respect to the edges in the *skeleton*. In

```
#this skeleton defines
#variables rdst, v_src1, v_src2, v_dst,
#fbuf, dbuf, ebuf, wbuf, rbuf
USING diagram_a;

OPERATION alu
    VAR oper : {add, sub};                      #an OR-node
        iw   : uint<14>;                        #i-word

    TRANS e0  : { fbuf = IF [] };               #allocate IF
          e1  : { dbuf = ID [],                 #allocate ID
                  iw  = *fbuf,                   #read i-word
                  ! fbuf }                       #release IF
                  +oper = iw;                    #decode & bind
          e2  : { ebuf = EX [],                  #allocate EX
                  rbuf = RF [rdst],              #allocate dst
                  ! dbuf };                       #release ID
          e3  : { wbuf = WB[],                    #allocate WB
                  ! ebuf };                       #release EX
          e4  : {*dst_buf = v_dst,               #write dst
                  ! rbuf,                         #release dst
                  ! wbuf };                       #release WB

OPERATION add
    VAR rs1  : uint<4>;                         #assigned from CODING
        src2 : { imm, reg };                    #an OR-node
    CODING 1 rdst rs1 src2;
    SYNTAX "add R"^rs1 "," src2 ", R"^rdst;     #assembly syntax
    EVAL   +src2;                                #decode per bit 0-4 of CODING
    TRANS e2 : {v_src1 = *RF [rs1] } ;           #inquire&read src1
          e3 : v_dst = v_src1 + v_src2;              #compute addition

OPERATION imm
    VAR v_im : uint<4> ;                         #assigned from CODING
    CODING 1 v_im ;
    SYNTAX v_im;
    TRANS id_ex : v_src2 = (uint<32>) v_im;

OPERATION reg
    VAR rs : uint<4>;                            #assigned from CODING
    CODING 0 rs;
    SYNTAX "R"^rs;
    TRANS e2 : { v_src2 = *RF [rs] };            #inquire&read src2

OPERATION sub

    ......
```

FIGURE 9.7

Example MADL description.

a VAR section, besides variables of normal arithmetic data types, a special type of variable that corresponds to the OR-rule in the AND-OR graph can also be defined. An example of such a variable is the "oper" variable in "alu," which corresponds to the top-most OR-rule in Fig. 9.5(a). In the TRANS sections, control and data *actions* are enclosed within curly braces, while computations are not.

In order to understand the process of dynamic binding of *actions* onto edges, consider the *syntax operations* shown in Fig. 9.7 and the *skeleton* shown in Fig. 9.1. During run time, the top-level *syntax operation* "alu" will be associated with the *skeleton* first. It will bind its *actions* specified in the TRANS section onto the corresponding edges of the *skeleton*. In this simple example, the *actions* include reading the instruction word from the token manager "IF," which models the instruction fetching unit, when leaving state *F*, and then decoding the OR-node variable "oper" with the instruction word. Decoding is done by matching the instruction word against the CODING of the *syntax operations* pointed to by "oper." Depending on the value of the instruction word, decoding will resolve "oper" to one of "add" and "sub." If the result is "add," then "add" will also bind its *actions* to the edges of the same *skeleton*. Meanwhile it will further decode its OR-node variable "src2" based on the lowest order 5 bits of its encoding value. If the result is "imm," then "imm" will bind its *action* onto the *skeleton*. The combined result of "alu," "add," "imm," and the *skeleton* "diagram_a" models one add-immediate instruction. Other combinations such as "alu," "sub," and "reg" share the same *skeleton* in the description. The decoding statement ensures that at run time, only one *expansion* is bound to the *skeleton*.

To summarize, the dynamic OSM model enables us to apply the AND-OR graph model to ease the description. Like the static OSM model, the dynamic OSM model is also executable. However, its execution is less efficient due to the overhead incurred by dynamic binding. Therefore the MADL compiler will perform a simple transformation to convert the dynamic model to the static version. For each *expansion* in the AND-OR graph, the compiler will duplicate a portion of the state diagram that is reachable from the decoding edge without going through State *I*. It will then statically bind all *actions* of the *expansion* onto the duplicated sub-diagram. The result of the transformation is an expanded state machine with an overall topology similar to that shown in Fig. 9.6. For more details of the core layer, we refer interested readers to the reference manual [8].

9.3.3 The Annotation Layer

Annotations appear as supplemental *paragraphs* in an MADL description. Fig. 9.8 shows the syntax in Backus-Naur Form. A paragraph contains an optional namespace label and a list of declarations and statements. The label specifies the tool-scope of the paragraph and can be used to filter irrelevant annotations. Paragraphs without a label belong to the global namespace.

One feature of the annotation syntax is that it supports the declaration of variables and the relationships between them. The scope of the variables is

```
annot_paragraph : : = claus_list
          | : id : claus_list  // with namespace

claus  : : = decl | stmt

decl   : : = var id : type      // variable
          | define id value  // macro

stmt  : : = id (arg_list)     // command
          | val op val        // relationship

arg   : : = id = value

val   : : = id | number | string
          | (val_list)     // tuple
          | {val_list}     // set

type  : : = int<width> | uint<width> | string
          | (type_list)  // tuple type
          | {type}       // set type
```

FIGURE 9.8

Annotation syntax in BNF.

```
MANAGER
  CLASS
    unit_resource : void -> void;
    $:COMP: SCHED_RESOURCE_TYPE (size=1);
    $$:SIM : USE_CLASS (name="untyped_resource",
            param="1");$$

    ......
```

FIGURE 9.9

Two-layer description example.

determined by the location of the annotation paragraph in the description and its namespace. The relation operators currently supported include normal comparison operators and the set containment operator ("<:" in syntax). Section 9.4 shows how these operators can be effectively used to express irregular architecture constraints.

The annotation paragraphs are associated with syntax elements of the core description. After the MADL description is parsed, the tools can access an annotation paragraph via the pointer to the intermediate representation object of the syntax element that it is associated with. A set of APIs are defined for such access.

An annotation paragraph can either be in a single-line format or in a block format. The former is preceded by a "$" and runs through the end of the line while the latter is enclosed within a pair of "$$"s. Fig. 9.9 gives an example of an MADL description. It shows part of a MANAGER section. The third line defines a token manager class "unit_resource" and its type property. The two subsequent annotation paragraphs are attached to this manager class and provide additional information to different tools. The first paragraph informs the tool in the "COMP" namespace (in this case the compiler) that this manager class can be treated as a resource unit by a scheduler. Such information is not essential to the processor architecture, but a user-supplied hint for the compiler. The second paragraph notifies the "SIM" tool (in this case the cycle accurate simulator) that this manager uses the template class "untyped_resource" with a template argument of "1" as its C++ implementation in the simulator. Note that the interpretation of the annotation commands is solely up to the tools. Any change in the tool implementation may affect the annotation description but not its syntax or the core layer. This feature insulates the MADL syntax from the frequent modifications and extensions of the software tools, thereby lengthening the lifetime of MADL.

9.4 SUPPORT FOR TOOLS

In this section, we describe how MADL can be effectively used to describe information necessary for various software tools. In particular, we show that the core layer is powerful enough to model a wide range of architectures, and that the annotation layer is flexible enough to supply information needed by a variety of tools, specifically, the cycle accurate simulator, the instruction set simulator, the register allocator, and the reservation table-based scheduler.

9.4.1 Cycle Accurate Simulator (CAS)

Cycle accurate simulators are important for the verification of programmable platform designs, as well as the application softwares that run on the platform. Since the OSM model is inherently an executable model, it is straightforward to synthesize cycle accurate simulators from MADL descriptions. The MADL framework is capable of generating two types of cycle accurate simulators: one based on the dynamic-OSM model in MADL, and the other an optimized one based on the translated static-OSM model. The latter is significantly faster than the former. This section describes the optimized simulator.

Fig. 9.10 shows the structure of the cycle accurate simulator synthesis framework based on MADL. The inputs to the framework are the MADL description, the token manager implementations in C++, and the peripheral unit implementations in C++. The peripheral units are those hardware components that do not directly interact with the state machines in our model. They communicate with the state machines indirectly through token managers.

As discussed in Section 9.3, the core layer is used to specify the operation layer information. Thereby, the MADL compiler can extract all state machine details from the *skeleton* and *syntax operations* and translate them into state machine classes

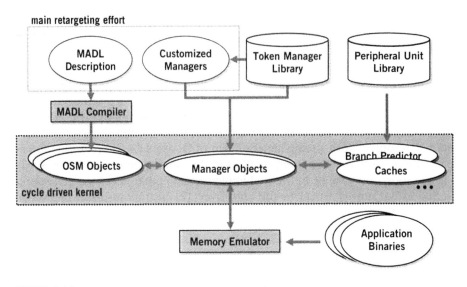

FIGURE 9.10

Cycle accurate simulation framework.

in C++. As significant reuse is expected for the token managers and the peripheral units, they are organized as C++ template class libraries. New token managers or function units can be customized by extending the library classes. As shown in the example of Fig. 9.9, we can use annotations to specify pointers to the actual token manager implementation. The synthesis framework will follow these pointers to statically instantiate the token manager objects. The MADL description and the customized token managers represent the major development effort for targeting the framework to a new processor.

The state machine classes and the token manager classes are instantiated in the cycle accurate simulator. The simulator also contains instances of the relevant peripheral units such as the cache models, the cycle-driven simulation kernel, as well as the memory emulator.

It is worth noting that the use of MADL improves the productivity of simulator development since MADL descriptions are very compact. Moreover, the synthesized simulators most often have better simulation performance than hand-coded OSM simulators. The reason is that when coding by hand, programmers commonly have to trade code speed for code simplicity. For instance, in our hand-coded OSM simulators, all OSM action classes are derived from a common base class. Such use of the C++ polymorphism feature greatly simplifies the implementation as the actions can be treated in a homogeneous fashion, for example, their pointers can be stored in an array. However, this introduces many virtual function calls at run time that slow down the simulation speed notably. While in a synthesized simulator, code complexity concerns are irrelevant and the fastest option is always favored.

9.4.2 Instruction Set Simulator

Instruction set simulators (also called functional simulators) are another important class of software tools that help verify the functionality of application binaries. With two minor modifications, our simulation framework can also be used to generate instruction set simulators.

First, since the instruction set simulator has no notion of time or operation concurrency and executes the operations sequentially, only one state machine instance is needed in the simulation kernel to model the operation stream. Therefore the simulation kernel becomes a simple loop that iteratively activates the state machine instance. In each iteration, the state machine instance will execute all the actions of one operation from its fetch to its retirement without interruption. Second, the token managers used for the instruction set simulator are simplified by removing all timing-related control semantics. Therefore, by supplying a set of simplified token managers, the same MADL description written for the cycle accurate simulator can be used to generate the instruction set simulator.

9.4.3 Disassembler

As mentioned in Section 9.3, assembly syntax chunks can be specified in the "SYNTAX" subsections of the syntax operations. The chunks from an expansion of the AND-OR graph can be assembled together to form an assembly template for the type of operations matching the expansion. The template is the concatenation of a set of string literals and variables. The string literals represent the operation name, the punctuations, and spaces in the assembly while the variables represent the operand fields. For the example description in Fig. 9.7, the assembly template for the expansion involving "alu," "add," and "imm" is the concatenation of "add R," $rs1$, v_imm, "R," and rd.

Given an instruction word, its matching expansion and the assembly template can be located as a result of decoding. The variables corresponding to operand fields can be extracted from the instruction word and filled into the template. The assembly output of the operation is then obtained.

The implementation of the disassembler is straightforward as most of its components are available in the CAS and the ISS synthesis framework. Therefore, it is simply integrated in the simulation framework as a utility. It can be used to print out execution traces for the purpose of debugging the simulated programs.

In principle, the inverse of the disassembler—the assembler—can also be synthesized based on the assembly templates. It is not yet implemented in the MADL framework due to limited manpower.

9.4.4 Register Allocator

Register allocation is considered to be one of the most important optimizations in a compiler. An allocator that produces excessive spill code or too many false dependencies can potentially eliminate the benefits of other optimization phases. Hence it is very critical to have a good register allocator and to provide accurate information to the register allocator from the processor specification. One of the most widely used register allocators is the graph coloring-based allocator. It is elegant, retargetable, and effective for RISC-style ISAs with a large number of homogeneous registers. However, for CISC-style architectures and many embedded processors such as DSPs, graph coloring-based allocators do not work well. This is due to their highly irregular instruction sets, their limited numbers of registers, and sometimes the use of heterogeneous register banks. While several attempts have been made to develop efficient register allocators for such processors, very few are retargetable, efficient, and easy to use. In our work, apart from a graph coloring-based allocator, we also built an Integer Linear Programming (ILinP)-based allocator developed by Appel and George [9]. The AMPL [10] based ILinP model is simple to use, with an expressive specification format for the constraints. It is also retargetable and guarantees an optimal solution. Although we may use different allocators for different domains, the information needed by the allocators is very similar. In this section, we describe the various inputs to a register allocator and how they can be specified in MADL.

The basic input to the register allocator is the different types of register banks, both physical and logical. Properties of a register bank include data type of the bank, number of registers, and overlap information (such as that of single and double precision registers commonly seen in modern microprocessors). Fig. 9.11(a) gives a general overview of the nature of information that we need to capture. The bank information, enclosed within the dotted box, includes the eight registers "R0" to "R7," the register banks "bank1" and "bank2," and register overlap information as indicated by the double arrow between registers "R2" and "R4." Note that a register may be in multiple logical banks like "R3" or may not be a part of any bank like "R7."

Apart from all the register and register bank-related information, the allocators also need to know how the operands of various operations are constrained. This includes the set of register candidates that an operand may be assigned from, operands that have been pre-allocated, and sets of operands that may have to be allocated together. In Fig. 9.11(a), this information is represented by the thick dotted arrows. For example, operands "o1" and "o2" of "oper1" are constrained to be assigned from register banks "bank1" and "bank2" respectively, operand "o3" of "oper2" needs to be pre-assigned to "R7," and the operand pair ("o3," "o4") of "oper1" must be assigned the register pair ("R5," "R6"). In general, for the last constraint there may be more than one option for assignment. All the constraints are collectively termed *allocation constraints*.

Now we describe how all the register information and operand constraint information can be specified in MADL using annotations. In the example shown in Fig. 9.11, Fig. 9.11(b) shows how the register information corresponding to the dotted box of Fig. 9.11(a) can be specified in the MANAGER section of MADL. The *DEFINE* keyword is used to define a macro as the set of register symbols in physical or logical banks. The annotation command "REG_BANK" takes a symbol set and

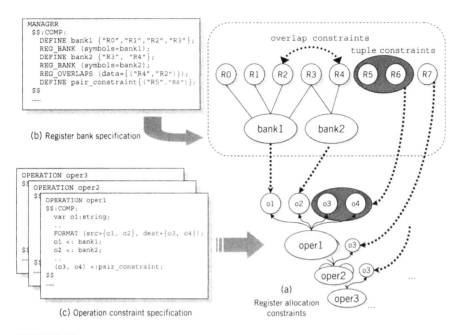

(b) Register bank specification

(c) Operation constraint specification

FIGURE 9.11

Register constraint specification in MADL.

instantiates a register bank. In order to specify all pairs of registers that overlap, the command "REG_OVERLAPS" is used. Fig. 9.11(c) shows how the OPERATION sections of MADL can be annotated to specify the various operand constraints corresponding to the lower half of Fig. 9.11(a). "FORMAT" is used to define the source and destination operands. The set containment operator is then used to constrain each one of the operands specified in "FORMAT," for example, $o1 < : bank1$ constrains operand "o1" to be assigned a register from "bank1." *Tuple constraints* can also be specified using the set containment operator as shown in Fig. 9.11.

The irregular register allocation constraints are largely introduced by non-orthogonal instruction encodings designed to achieve high code density. However, it is not always possible for tools to infer automatically such constraints from the encodings. Therefore, a proper annotation scheme is necessary.

As stated earlier, to demonstrate the capabilities of MADL, we extracted the processor specific ILinP specification for the Hiperion DSP [11]. We chose the Hiperion architecture since it is a highly irregular DSP with all the aforementioned register constraints. The practical experiences for this are discussed in Section 9.5.

9.4.5 Instruction Scheduler

The instruction scheduler is another important component of a compiler as it helps reduce the execution time of a program by effectively hiding instruction latencies.

This is particularly true for embedded processors as many of them are statically scheduled and fully depend on the compiler for best performance. Reservation table-based schedulers not only offer good performance, but are also highly retargetable. Hence, they are among the most widely used types of schedulers. In this section, we describe how we can extract information for a reservation table-based scheduler from an MADL description.

In order to build a reservation table, the scheduler needs to obtain information including the physical or virtual resources that are exported to the scheduler such as issue slots and functional units, all possible execution paths for each operation, resource usages of each operation describing the time slots when different resources will be used, the latencies of source, and destination operands of each operation. It also requires register overlap information to detect data hazards.

Fig. 9.12 gives an overview of how reservation tables can be extracted from an MADL description for an example operation. Fig. 9.12(a) shows a part of a MANAGER section in which a token manager type, "unit_resource," containing a single resource is defined. "IF" is defined as an instance of such a type. As a scheduling resource instance, it is assigned an alias "f_slot." All other physical resources are similarly defined by annotations on their respective manager classes and instances. The MADL compiler can extract these physical resources by searching

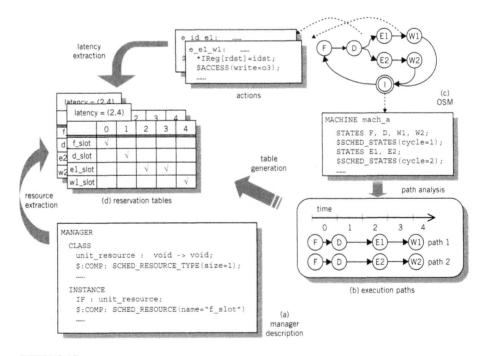

FIGURE 9.12

Overview of reservation table extraction from MADL.

for annotation commands "SCHED_RESOURCE" and "SCHED_RESOURCE_TYPE" in the MANAGER sections.

Fig. 9.12(c) shows the state machine for the example operation. The state diagram and the actions associated with its edges are all specified in the MACHINE and OPERATION sections in an MADL description. By analyzing the connectivity of the state diagram and its associated "SCHED_STATES" annotation commands, we can extract the relevant execution paths and nominal latencies. Moreover, by analyzing the actions on the edges, we can determine the resources used at each state. Take the OSM in Fig. 9.1(b), for example. When the state machine proceeds from state *I* to *F*, the *allocate* request to the fetch manager is fired. Therefore we can immediately deduce that the resource "f_slot" is used by the state machine at state *F*. The resource is given up when the state machine proceeds to state *D* and fires the *release* action. Hence, by examining all the *allocate*, *release*, and *discard* actions along all execution paths, we can obtain the resource usage information of the operation.

The last few pieces of information needed to build the reservation table are the number and the types (source/destination) of the operands, and their read or write latencies. The "FORMAT" annotation statements in OPERATION sections are used to describe names and types of operands, as shown in the example of Fig. 9.11(c). In Fig. 9.12, operand read and write activities are annotated by the "ACCESS" commands on the edges where the corresponding *read* and *write* actions occur. We can combine operand access information with the path latency information to get the operand latencies.

Finally, we now have all the information that we need to build the reservation tables shown in Fig. 9.12(d).

Many embedded processors including DSPs have ISAs that are optimized for code density as well as performance. Their narrow instruction width introduces irregular ILP constraints, where the encoding allows only a certain combination of operations to be issued in parallel. In order to use a reservation table-based scheduler, we adopt a preprocessing tool, the Artificial Resource Allocator (ARA) [12], to translate such encoding constraints to resource based constraints. The ARA takes the set of all combinations of operations that may be issued in parallel from MADL, translates the constraints to artificial resources, and augments the reservation table appropriately.

9.5 RESULTS

In order to demonstrate the effectiveness of the MADL-based software tool development approach, we modeled three architectures for software tool generation. They are the StrongARM core, the Fujitsu Hiperion DSP, and a 128-bit subword parallel media processor named PLX [13]. Each of these represents a class of processors commonly used in embedded systems. The StrongARM core is a 5-stage pipelined implementation of the ARM V4 architecture. Though generally viewed as a RISC processor, the ARM ISA contains some CISC features such as the load-multiple instruction. Hiperion is a 16-bit DSP with an irregular ISA. Its instruction encoding is optimized mainly for code size and therefore is not fully orthogonal. PLX is a research architecture developed by a group of colleagues. It contains more than 200 integer and floating-point instructions. Its MADL model was developed by its designers.

We wrote a specification for the StrongARM in MADL to generate the cycle accurate simulator, the instruction set simulator, and the reservation table for the scheduler. The Hiperion description is also used to generate both simulators—the register allocator and the scheduler. The PLX description only supports generating both simulators. Table 9.1 shows some statistics of the processor models. This contains the line counts of the MADL descriptions, the number of annotation paragraphs, the line counts of the token manager implementations in C++ for both types of simulators, and line counts of the synthesized C++ code from MADL for both types of simulators. The MADL compiler optimizes the synthesized cycle accurate simulators to minimize their code sizes without negatively affecting their performance. This results in smaller source code size of the cycle accurate simulator than that of the instruction set simulator in the StrongARM and PLX cases.

We performed all our experiments on a P4 2.8 GHz Linux system with 1 GB memory. The compilation of the MADL description in each case takes less than ten seconds. The building of the synthesized simulators as done by g++3.3.2. The building time is less than two minutes for each simulator with g++ optimization switches "-O3 -finline-limit=1500-fomit-frame-pointer."

Table 9.2 shows the average simulation speed of all simulators. We used a mix of SPECINT benchmarks and MediaBench [14] to evaluate the StrongARM simulators. The Hiperion model was tested with the DSP-stone benchmarks and the PLX model was tested with several hand-coded assembly kernel loops.

Table 9.1 Description statistics.

	StrongARM	Hiperion	PLX
MADL lines	2,001	3,346	3,389
annotation paragraphs	135	368	10
token managers (cycle accurate)	494	417	349
token managers (functional)	319	417	281
synthesized code (cycle accurate)	45,707	16,804	20,764
synthesized code (functional)	62,816	12,243	27,110

Table 9.2 Simulation speed.

	StrongARM	Hiperion	PLX
cycle accurate (MHz)	2.60	3.82	2.01
ISS (million inst/sec)	8.26	11.5	2.40

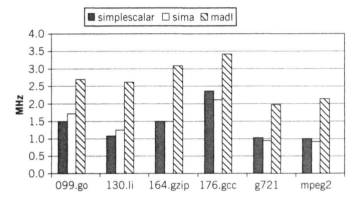

FIGURE 9.13

StrongARM simulation speed comparison.

In general, the synthesized cycle accurate simulators are faster than their hand-coded counterparts. Fig. 9.13 compares the simulation speed of several StrongARM simulators, namely, the SimpleScalar target for StrongARM [15] (configuration file *sa1core.cfg*), the hand-coded OSM simulator (sima) for StrongARM [1], and the simulator synthesized from MADL. The results show that the synthesized simulator has obvious simulation speed advantage, beating the best hand-coded simulator by 80% on average. The synthesized PLX cycle accurate simulator is also faster than the hand-coded simulator written by its designers, which runs at an average speed of 0.76 MHz.

Currently the speed of the instruction set simulators is slower than that of hand-coded ones. In the StrongARM case, the synthesized simulator runs at about one-third the speed of our hand-coded simulator. There are two main reasons for the difference. First, the synthesized simulator utilizes bit-true data types for state variables and for computation, while the hand-coded simulator uses native data types. The overhead of using the bit-true data types slows down the simulation speed. Such overhead also exists in synthesized cycle accurate simulators, but they are less sensitive to the overhead as data copying and computation constitute only a tiny fraction of their execution time. Second, since we use the same MADL description to generate the instruction set simulator as well as the cycle accurate simulator, the synthesized instruction set simulator executes the operations following the state diagrams designed for cycle accurate simulation. So for each operation, it will undergo the normal stages of fetching, decoding, evaluation, and result write back. In contrast, the hand-coded simulator does not have to follow these stages and can thereby benefit from some processor specific optimizations. For example, in the StrongARM case, the hand-coded simulator evaluates the predicate operand of an instruction prior to fully decoding it. If the predicate operand is false, the instruction will be skipped and its decoding/evaluation saved. Such an early evaluation of the predicate operand is not possible in the cycle accurate simulator due to timing constraints. Since the synthesized ISS shares the same description as the cycle accurate simulator, it does not benefit from early predicate evaluation.

The speed of the instruction set simulators can be improved by using different MADL descriptions from those used to generate cycle accurate simulators. However, we believe that it is valuable to synthesize both types of simulators from the same description as it improves productivity and ensures consistency. We expect to improve the instruction set simulation speed by implementing more aggressive optimizations in the MADL compiler.

The MADL specifications for Hiperion and StrongARM include the annotations needed for compiler optimizations, as shown in Table 9.1. From the core and the annotation layers of both descriptions, we extracted all the relevant reservation table information such as scheduling resources, resource usages, and operand latencies in a low-level format. For the StrongARM model, 6 physical resources and 548 reservation tables were generated. For the Hiperion model, 5 physical resources, 7 virtual resources, and 164 reservation tables were generated. The virtual resources for Hiperion were generated by the ARA tool for irregular ILP constraint modeling.

The bulk of the annotation paragraphs for the Hiperion, however, is for the register allocator tool due to the irregular nature of the architecture. From the Hiperion MADL description, we automatically extracted the register allocation constraints for an *AMPL*-based ILinP model. The total number of processor specific *AMPL* constraints generated is 660. In the presence of such a large number of constraints, automating the extraction of constraints from a machine description is of great value.

9.6 RELATED WORK

Different ADLs emphasize different aspects of microprocessors and adopt different architecture models. Their design is commonly influenced by the tools that they intend to support. This section compares MADL with several other ADLs designed for supporting both simulators and compilers.

The MIMOLA [16] and UDL/I [17] were mainly designed to support hardware synthesis. They are based on the discrete event model of computation and describe microarchitectures at the register transfer level. They can be used to model arbitrary digital hardware and therefore are among the most flexible ADLs. However, due to such low level of abstraction, it is inefficient to model complex microprocessors with these languages. Moreover, processor specifications in these languages can only be analyzed and utilized by compilers under certain assumptions that constrain the scope of target architectures to only simple pipelines.

The ISDL [5] and the early version of nML [2] were mainly designed for retargetable compiler generation. nML was later extended to include architecture-template-based pipeline information [18]. Both languages utilize register-transfer-lists to abstract the behavior of individual instructions. These languages can support the generation of both simulators and compilers for a small range of architectures such as DSPs.

The EXPRESSION [19] is a research language developed for the synthesis of both compilers and simulators. It uses a coarse-grained netlist to model the pipeline structure. The language was later extended with functional abstraction models for hardware control specification [20].

The LISA [6] is a pipeline description language. It utilizes the L-chart to model operation flow in the pipeline. It adopts a C-like syntax and several pipeline control APIs to support the specification of pipeline control logic. With manual assistance for instruction latency specification, instruction schedulers can be generated from a LISA description for several DSP and ASIP architectures [21].

In Table 9.3, we compare the existing ADLs and MADL through metrics such as the *range* of supported architectures, the modeling *productivity*, the *efficiency* of synthesized simulators, and the *ease* to extract architecture properties for compilers' use. MADL is less flexible than MIMOLA or UDL/I since it models only processor cores. However, on the whole MADL is well balanced in all four aspects. Therefore, we believe that it is a practical and promising language for use in developing programmable platforms.

Table 9.3 ADL comparisons.

ADL	Range	Productivity	Efficiency	Ease
MIMOLA, UDL/I	++	−	−	−
ISDL, nML	−	+	+	+
EXPRESSION	−	+	+	+
LISA	+	+	+	−
MADL	+	+	+	+

9.7 CONCLUSION

In this chapter, we introduced the OSM model, which naturally captured the concurrency at both the operation and the microarchitecture level in microprocessors. Based on the OSM model, we developed the MADL architecture description language that contained a core layer and an annotation layer. The two-layer structure separated essential architecture properties from volatile implementation-dependent information, thereby lengthening the lifetime of MADL. The language can be used to model a broad range of architectures. We demonstrated that with proper annotation schemes, MADL descriptions can be used to assist the generation of efficient cycle accurate simulators, instruction set simulators, disassemblers, register allocators, and instruction schedulers. Our ongoing work involves generalizing the OSM model so as to cover both the operation and the hardware levels. If such a goal can be achieved, the resulting description language can be used for synthesizing the microarchitecture of microprocessors.

REFERENCES

[1] W. Qin and S. Malik. Flexible and formal modeling of microprocessors with application to retargetable simulation. In *Proc. of Conference on Design Automation and Test in Europe*, pages 556–561, 2003.

[2] A. Fauth, J. V. Praet, and M. Freericks. Describing instructions set processors using nML. In *Proc. of Conference on Design Automation and Test in Europe*, pages 503–507, Paris, France, 1995.

[3] K. Cheng and A. S. Krishnakumar. Automatic generation of functional vectors using the extended finite state machine model. 1(1):57–79, 1996.

[4] W. Qin. *Modeling and Description of Embedded Processors for the Development of Software Tools*. PhD thesis, Princeton University, 2004.

[5] G. Hadjiyiannis, S. Hanono, and S. Devadas. ISDL: An instruction set description language for retargetability. In *Proc. of Design Automation Conference*, pages 299–302, June 1997.

[6] S. Pees, A. Hoffmann, V. Zivojnovic, and H. Meyr. LISA—machine description language for cycle-accurate models of programmable DSP architectures. In *Proc. of Design Automation Conference*, pages 933–938, 1999.

[7] Y. Mahajan, C. Chan, A. Bayazit, S. Malik, and W. Qin. Verification driven formal architecture and microarchitecture modeling. In *Fifth ACM-IEEE International Conference on Formal Methods and Models for Codesign (MEMOCODE '2007)*, April 2007.

[8] W. Qin. *http://www.ee.princeton.edu/MESCAL/madl.html*, 2004.

[9] A. Appel and L. George. Optimal spilling for CISC machines with few registers. In *Proc. of the Conference on Programming Language Design and Implementation*, 2000.

[10] R. Fourer, D. Gay, and B. Kernighan. *AMPL: A Modeling Language for Mathematical Programming*. Duxbury Press, 2002.

[11] Fujitsu Limited. *Hiperion II - Digital Signal Processor User's Manual*, 1998.

[12] S. Rajagopalan, M. Vacharajani, and S. Malik. Handling irregular ILP within conventional VLIW schedulers using artificial resource constraints. In *International Conference on Compilers, Architectures and Synthesis for Embedded Systems*, 2000.

[13] R. B. Lee and A. M. Fiskiran. PLX: A fully subword-parallel instruction set architecture for fast scalable multimedia processing. In *Proc. of the 2002 IEEE International Conference on Multimedia and Expo (ICME 2002)*, pages 117–120, August 2002.

[14] C. Lee, M. Potkonjak, and W. H. Mangione-Smith. MediaBench: A tool for evaluating and synthesizing multimedia and communicatons systems. In *International Symposium on Microarchitecture*, pages 330-335, 1997.

[15] T. Austin, E. Larson, and D. Ernst. Simplescalar: An infrastructure for computer system modeling. *IEEE Computer*, pages 59-67, February 2002.

[16] G. Zimmerman. The MIMOLA design system: A computer-aided processor design method. In *Proc. of Design Automation Conference*, pages 53-58, June 1979.

[17] H. Akaboshi. *A Study on Design Support for Computer Architecture Design*. PhD thesis, Department of Information Systems, Kyushu University, Japan, 1996.

[18] Target Compiler Technologies N.V. *http://www.retraget.com*, 2004.

[19] A. Halambi, P. Grun, V. Ganesh, A. Khare, N. Dutt, and A. Nicolau. EXPRESSION: A language for architecture exploration through compiler/simulator retargetability. In *Proc. of Conference on Design Automation and Test in Europe*, pages 485-490, 1999.

[20] P. Mishra, N. Dutt, and A. Nicolau. Functional abstraction driven design space exploration of heterogeneous programmable architectures. In *Proc. of the International Symposium on System Synthesis*, October 2001.

[21] O. Wahlen, M. Hohenauer, and R. Leupers. Instruction scheduler generation for retargetable compilation. *IEEE Design & Test of Computers*, 20(1):34-41, January 2003.

ADL++: Object-Oriented Specification of Complicated Instruction Sets and Microarchitectures

10

Soner Önder

The ADL++ architecture description language allows automatic synthesis of cycle-accurate simulators, assemblers, and disassemblers from a specification written in the language. The language's compiler adl++ has been implemented as part of the Flexible Architecture Simulation Toolset, or FAST in short. The ADL++ language supports an execution model that is suitable for expressing a broad class of processor architectures. This execution model supports specification of the microarchitecture including pipelines, control, and the memory hierarchy including instruction and data caches as well as the ISA, the assembly language syntax, and the binary representation. ADL++ is a verbose language; the specifications of the architectures written in the language vary from, few thousand lines of ADL++ code to six thousand lines of ADL++ code. On the other hand, the sizes of automatically generated software varies from 20,000 to 30,000 lines of C++ code, which compares well with the size of hand-coded simulators for similar architectures.

The foundation of the language rests on a simple model; the simulators generated from the language explicitly model the instruction flow through the architectural components under programmer control, yielding cycle-accurate, yet architecture-level specifications. This simple model naturally embodies an object-oriented approach. ADL++ permits higher degrees of reusability and also provides an ability to describe complex instruction set processor architectures. Simply stated, ADL++ was born out of a need to describe complicated architectures. Variable length and complex instruction sets such as Intel's IA-32 [1], pose significant challenges for architecture specification, and this chapter along with an overview of the basic underlying model for the language discuss these challenges and illustrates how an object-oriented approach can help the language designer to overcome these difficulties.

This chapter is organized as follows. Section 10.1 provides an overview of the FAST/ADL++ system and discusses its main features. Section 10.2 introduces the **247** object-oriented approach behind the ADL++ language. The section motivates the separation of instruction set architecture specification and microarchitecture specification and illustrates how the instruction and microarchitecture semantics are fused together in an object-oriented setting. In Section 10.3 we discuss the challenges posed by complex instruction set architectures by presenting somewhat unique properties of Intel's IA-32 instruction set architecture. Among these, we discuss the issues brought up by variable length instructions, many addressing modes, overlapping registers as well as mixed arguments in the instruction set, and how these in turn affect the microarchitecture specification. Section 10.4 discusses how set theory can alleviate some of these difficulties. We show that regular expressions are capable of describing addressing modes in a manner suitable for automatic generation of assemblers and simulators for these instruction set architectures. In Section 10.5, we first introduce the concept of instruction templates from which we derive other instructions by using *multiple conditional inheritance*. Multiple conditional inheritance allows instructions to inherit from multiple parents in such a way that different parts of an instruction are inherited from different parents using a *best fit policy*. In Section 10.6 we discuss microarchitecture related issues. Finally, in Section 10.7 we present a summary and in Section 10.8 we give a history of the development of the language and acknowledge people who contributed to its development.

10.1 FLEXIBLE ARCHITECTURE SIMULATION TOOLSET (FAST)

10.1.1 Overview of the Toolset

The toolset provides an implementation of a compiler for ADL++ through which cycle-level simulators as well as support software tools including the assembler, disassembler, and the loader/linker for the architecture are generated. The compiler also generates a cycle-level assembly language debugger that assists in tracing the program behavior and provides support for displaying statistics and monitored information.

The ADL++ compiler uses prototype modules called *templates* to generate the desired software. A template is a prototype module of software that consists of only architecture-independent components. For example, the assembler template contains a complete assembler with the exception of instruction set-specific portions such as the mnemonics tables and specific rules to parse individual instructions and code that converts symbolic addresses to machine addresses. All these portions of an assembler are instruction set architecture (ISA) specific and they are compiled in from the ADL++ program and filled in by the compiler. Similarly, built-in artifacts have been implemented in another template file. For each instance of artifact declaration, the ADL++ compiler obtains the corresponding artifact declaration from the template file and generates the desired artifact implementation. The generation of the simulator system is accomplished by copying a template until a descriptor

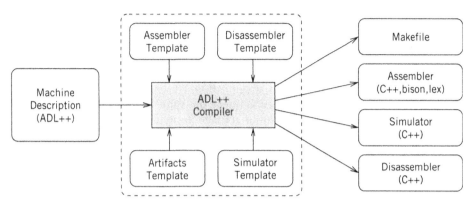

FIGURE 10.1

An overview of flexible architecture simulation toolset (FAST).

marker, indicating the position at which a component should be generated and placed, is encountered. The compiler generates a table, a procedure, or a C++ class described by the marker. Once the required software element is generated, the scanning continues until another marker is found, or the end of file is reached. An overview of the toolset is provided in Fig. 10.1.

10.2 THE FAST/ADL MODEL

One of the distinguishing properties of the ADL++ language is its underlying model. ADL++ follows an object-oriented approach to both the specification of the microarchitecture as well as the instruction set architecture. In this respect, one can view an ADL++ program consisting of two objects, namely, the *ISA artifact* and the *microarchitecture artifact*. The ISA artifact is assembled from a collection of *instruction templates* and *instructions* whereas the microarchitecture artifact represents an implementation of the processor and it is constructed from user-defined artifacts. User-defined artifacts are built from the building blocks of microarchitecture specification, namely, *built-in artifacts*, *pipelines*, and *events*. We discuss each of these building blocks later in more detail. For now, we concentrate on the global view of the model.

The ISA specification is separated from the microarchitecture specification to facilitate the development of different microarchitecture implementations for the same ISA or extend an ISA by adding new instructions without altering the microarchitecture. This separation involves introduction of a new layer of abstraction between the microarchitecture definition and the instruction set definition. Instead of making the microarchitecture specification aware of a particular instruction set, it is made aware of a set of user-defined attributes termed *instruction attributes* and each instruction (or a template) definition declares the values of these attributes. In order to understand the concept better, let us consider the problem of making microarchitecture aware of the values of certain instruction fields such as the destination register number. By making the destination register an instruction attribute and attaching the corresponding instruction field to this attribute as part of the instruction specification, we can provide the microarchitecture specification with the ability to query the value of the destination register field of the current instruction without actually knowing the specifics of the current instruction set, such as the location of the field, or whether it spans multiple fields. In other words, the attribute-based approach gives us the freedom to substitute a different instruction set and use the same microarchitecture specification, as long as the two instruction set definitions share the same attribute value set. Obviously, instruction attributes can be specified as part of instruction templates and instructions may inherit these attributes conveniently.

Fig. 10.2 gives a global view of an ADL++ program. As it can be seen, one can construct an ADL++ program by incorporating an ISA artifact and a microarchitecture artifact. The ISA artifact is built from a set of instruction templates from which instructions are derived and the microarchitecture artifact is built from a set of user-defined artifacts. It is clear that user-defined artifacts can use other user-defined artifacts, or, at the bottom of the hierarchy, incorporate built-in artifacts to implement the desired function.

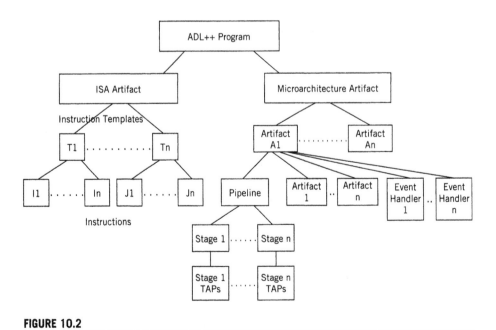

FIGURE 10.2

ADL++ program.

Built-in artifacts are objects that have well-defined architectural semantics such as caches, memory units, and register files. Built-in artifacts are directly supported by ADL++ as built-in *types*. The programmer can use them by simply declaring objects of these types in an ADL++ program. Access to built-in artifacts takes the form of assignments to and from the artifact variables. Different implementations of these components can be used by specifying different attribute values for the artifacts.

As seen, having a clear object-oriented model can greatly simplify the implementation of a cycle-accurate simulator for a particular architecture. Unfortunately, implementing the semantics of a processor architecture is much more involved than what we have covered so far. Although in principle one could create an object for each hardware artifact and glue them together, such an approach appears to be more feasible for hardware description languages [2–5]. ADL++ aims to be an architecture description language; its goal is to provide cycle-accurate simulators without tying the specification to a particular hardware implementation. Therefore, contrary to hardware description languages, implementation of microarchitecture semantics relies on a different approach. Instead of modeling hardware components, ADL++ explicitly models the instruction flow through some placeholders called *pipeline stages*, which make up the microarchitecture. This flow is tightly coupled with a *clock* built into the language and the programmer explicitly specifies when an instruction can move from one stage to the next. In other words, in ADL++, the flow of instructions through the architectural components is explicit. A *pipeline* consists of one or more pipeline stages or simply *stages*. Typically, each instruction is allocated a private data area roughly corresponding to a hardware pipeline register. Contrary to a hardware implementation, an instruction carries this private data along with itself as it moves through various components. This private data is called the *instruction context*. The context is allocated when an instruction enters a *pipeline*. The context is passed from one component to the next and is operated upon by the components till the execution of the instruction is complete. Once the instruction retires, the context is deallocated.

10.2.1 Timing of Events

Modeling of instruction flow through the pipelines alone cannot adequately describe the operation of an architecture unless we can describe the timing of events. In order to provide a cycle-accurate simulation of a processor architecture, ADL++ directly supports the notion of a symbolic machine clock and the operation of the architectural components is described with respect to this clock. The machine clock is viewed as a series of *pulses*. Each discrete pulse is called a *minor cycle*, and a number of minor cycles are grouped together to form a *machine cycle*. The minor cycles in ADL are represented by a series of labels. The first and the last minor cycles of a machine cycle are labeled as the *prologue* and the *epilogue*, and those in between are labeled as *intermissions*. The actions of each component in the system during a machine cycle are divided into the operations that it performs in

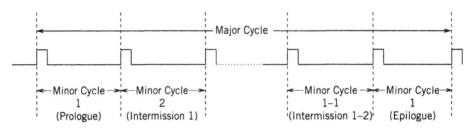

FIGURE 10.3

ADL clock labeling.

each of the minor cycles. During the prologue a component receives an instruction context from another component for processing, during the intermissions it operates upon the instruction context, and during the epilogue it sends the modified context to another component. Fig. 10.3 shows the clock of a machine in which the major cycle is composed of λ minor cycles.

Since pipeline stages are architectural components that exhibit a significant functional variety and their operation is dependent on the microarchitecture as well as the current instruction being processed, ADL++ follows an imperative approach to specify their semantics. This semantics is specified at a *Relaxed Register Transfer Level* (RRTL). The difference between the relaxed and traditional RTL statements is that ADL++ supports concurrency of these statements at a procedure level whereas RTL supports the concurrency at a statement level. Let us see through a simple example of how RRTL can achieve the same effect, and at the same time provide the flexibility of traditonal languages. Consider the expressions:

```
a = b
b = a
```

In traditional languages, after the execution of the two statements the variables *a* and *b* have the same value. In conventional RTL, these two statements will swap the values of *a* and *b*. In RRTL, these two statements appearing in a procedure one after another behave in the same manner as traditional languages; however if the statement $a = b$ and the statement $b = a$ appear in two procedures which are scheduled to execute in the same clock label, the result is identical to that of RTL.

10.2.2 Time Annotated Actions and Parallelism in the Microarchitecture

In ADL++ the procedures which encapsulate RRTL expressions are called *Time Annotated Procedures* (TAPs). A time annotated procedure executes when the system clock has the same label as the procedure itself. Obviously, all TAPs that have the same label execute at the same time, providing the desired RTL effect. Referring back to Fig. 10.2, we see that each stage is attached with a set of

TAPs that implement the semantics of that particular stage. TAPs not only implement what a pipeline stage does in a real piece of hardware, but they also implement this semantics in an instruction-dependent way. As a consequence of the separation of ISA and microarchitecture specifications, ADL++ makes the distinction of a *general component* of the semantics of a stage and an *ISA component* of a stage. The *general component* is considered to be common to all instructions and is specified as part of the TAP that defines the stage semantics for that clock label. The *ISA component*, which depends upon the specific instruction being processed, is included as part of the ISA specification. The procedures that implement the *general component* of actions associated with a processing stage carry the name of the stage and the label of the minor clock cycle during which they are to be executed. Since there are λ minor cycles, there may be up to λ TAPs for a given stage. The *ISA component* associated with an instruction is labeled with the name of the processing stage and optionally with the label of the minor cycle during which it must be executed. These statements are referred to as *labeled register transfer level* (LRTL) segments.

Parallelism at the architectural level is achieved by executing in each machine cycle the actions associated with each component during that cycle, as well as actions associated with an instruction that are annotated with the current cycle. The machine execution is realized by invoking each TAP corresponding to a minor cycle as the clock generates the corresponding label and the parallel operation of individual components is modeled by concurrently executing all TAPs that have the same annotation. During this process, LRTL segments corresponding to the currently processed instruction are *fused* together with the corresponding TAP. Note that all TAPs in all artifacts execute in parallel during the corresponding minor cycle. Therefore, the operation of the machine can be described as follows:

```
do forever
   for clock.label := prologue, intermission 1, ...
          ... intermission (λ − 2), epilogue do
      ∀ Artifact Aᵢ which embody a pipeline do
      ∀ TAP ∈ Aᵢ, TAP.annotation = clock.label do
          { process {TAP; TAP.instruction.LRTL } }
   end
```

10.2.3 Microarchitecture Specification

Let us now observe through a simple example as to how time annotated procedures and explicit instruction flow management actually implement the microarchitecture semantics in ADL++, and how various aspects we have discussed so far fit together. Consider the simple pipeline shown in Fig. 10.4. In this architecture, the Instruction Fetch stage (IF) fetches instructions from the instruction cache and ships them to the Instruction Decode (ID) stage. ID stage decodes the instructions it receives,

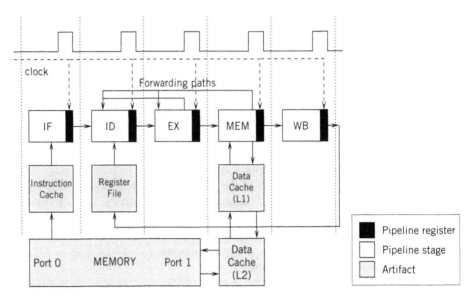

FIGURE 10.4

A simple pipelined processor.

fetches their operands from the register file, and sends them to the Execution Unit (EX). The Memory Access (MEM) stage performs a data memory access for the load and the store instructions, but other instructions pass through this stage unchanged. Finally, the Write Back (WB) stage writes the results back to the register file. In order to eliminate pipeline stalls that would otherwise result, data values are forwarded through forwarding paths to the earlier stages.

The microarchitecture specification involves declaration of built-in artifacts, the instruction context, the stages that make up the pipeline, and the TAPs that implement their semantics.

Built-in artifacts

A sequence of artifact declarations for the example pipelined architecture of Fig. 10.4 is shown in Fig. 10.5. The first declaration declares a register file *gpr* and individual registers in the file are assigned aliases. The names $0, $fp, etc., are ISA visible since the register file itself is ISA visible. RRTL statements may use either form of access (i.e., gpr [31] or $ra). The declaration specifies two memory ports with 12 cycles of access latency and 64-bit data paths. The memory port *mport0* hosts a direct mapped instruction cache of 64 kilobytes with 4 words per cache line. Memory port *mport1* hosts a direct mapped cache of similar attributes and this direct mapped cache in turn services a four-way set associative cache of size 8 kilobytes. Thus, the Cache L1 is at the highest level in the hierarchy and the memory ports are at the lowest level.

```
register file gpr [32,32]  # 32 registers,32 bits each.
   $zero    0,              # $zero is another alias for gpr[0]
   $at      1,              # $at is another alias for gpr[1]
   .....
   $fp      30,
   $ra      31;
memory mport0 latency 12 width 64, # 64 bit path to memory.
       mport1 latency 12 width 64; # 64 bit path to memory.
instruction cache icache of mport0 directmapped 64 kb 4 wpl;
data cache  l2 of mport1 directmapped 64 kb 4 wpl,
            l1 of l2     4 way 8 kb 4 wpl;
```

FIGURE 10.5

Example artifact declarations.

Once declared, artifacts are accessed just like variables by the RRTL statements in the specification. For complicated structures, such as data caches, passing of additional parameters may be required. When an artifact is accessed, the status of the result is queried using the *access-complete* statement. This statement returns a true value if the operation has been completed successfully, and a false value otherwise. A false value may be returned because the artifact is slow, such as in the case of memory ports, or because there is a structural hazard. In these cases the request must be repeated. Further, details of why the operation was not successful may be queried using additional statements.

Stages and pipelines

The primary means of declaring stages of the microarchitecture is the *pipeline declaration*. A pipeline declaration specifies an ordering among pipeline stages such that each stage receives an instruction context from the preceding stage and sends the processed context to a later stage. There may be more than one pipeline declaration in an ADL++ program but the stage names must be unique. Pipelines can be declared as part of a user-defined artifact, or globally. We will further investigate object-oriented specification of microarchitectures in Section 10.6. Once a stage is declared using a pipeline construct, TAPs may be specified for each of the stages and semantic sections of instruction declarations may utilize the stage names as LRTL labels. The following declaration defines the pipeline for the example architecture:

```
pipeline ipipe (s_IF, s_ID, s_EX, s_MEM, s_WB);
```

The instruction context

In ADL, the set of data values carried along with pipeline stages are grouped together in a structure called *controldata*. There is only one such declaration, which means all stages have the same type of context, and the *instruction context* is the *union* of the data required by all the pipeline stages in the system. While in a hardware implementation, pipeline stages may carry different types of contexts, definition of instruction context in this way simplifies the transfer and handling of instruction contexts in the simulator. Since there is a uniform single instruction context for all pipeline stages, each pipeline stage name is also an object of type *controldata*. The following is a simple *controldata* declaration for a pipelined machine:

```
controldata register
  my_pc 32, # Instruction pointer for the instruction.
  simm  32, # Sign extended immediate.
  ....
  dest  32, # dest holds the value to be written.
  lop   32, # lop holds the left operand value.
  rop   32; # rop holds the right operand value.
```

Elements of the *controldata* structure may be accessed from TAPs and by the semantic parts of instruction declarations (i.e., LRTLs). Access to the elements of the structure may be qualified or unqualified. When they are not qualified, the pipeline stage is the stage of the TAP that performs the reference or the label associated with the LRTL segment that performs the reference. In its qualified form, the syntax controldata-element[stage-name] is used to access the instruction context of another stage. This form is primarily used to implement internal data forwarding by either the source stage writing into the context of the sink stage, or the sink stage reading the data from the context of the source stage.

Specifying control and TAPs

The machine control is responsible for checking the conditions for moving the pipeline forward, forwarding the instruction context from one stage to the next, controlling the flow of data to and from the artifacts, and introducing stalls for resolving data, control, and structural hazards. In ADL, the semantics of the control part of the architecture is specified in a distributed fashion as parts of TAPs by indicating how and when instruction contexts are transferred from one stage to another.

The movement of an instruction context through the pipeline, from one stage to the next, is accomplished through the *send* statement. The send is successful if the destination stage is in the idle state or it is also executing a send statement in the same cycle. All pipeline stages execute the send statement during the epilogue minor cycle. In the normal pipeline operation, an instruction context is allocated by the first pipeline stage using the ADL statement *new-context*. This context is then filled in with an instruction loaded to the instruction register. When this stage finishes its processing, it executes the *send* statement to send the context to the downstream pipeline stage. When a context reaches the last pipeline stage, it is

```
procedure s_ID prologue    procedure s_ID epilogue    procedure s_WB prologue    procedure s_WB epilogue
begin                      begin                      begin                      begin
    decode;                    send EX;                   gpr[dest_r]=dest;          retire;
    lop = gpr[lop_r];      end s_ID;                  end s_WB;                  end s_WB;
    rop = gpr[rop_r];
end s_ID;
```

FIGURE 10.6

Example TAPs for the pipelined processor.

deallocated using the ADL statement *retire*. If any of the pipeline stages does not execute a send, send operations of the preceding stages fail. In this case, they repeat their send operations at the end of the next cycle. For decoding the instructions, ADL provides a *decode* statement. The *decode* statement does not take any arguments and establishes a mapping from the current context to an instruction name. This mapping is fully computable from the instruction declarations. Once decoded, all the attributes of the instruction become read-only *controldata* variables and are accessed accordingly. Some TAPs for the microarchitecture example are shown in Fig. 10.6.

The conditions for internal data forwarding can be easily checked by the stage that needs the data. For example, the TAP for the ID stage in the example pipelined machine may check if any of the stages EX and MEM has computed a value that is needed by the current instruction by comparing their destination registers with the source registers of the instruction currently in the ID stage. If that is the case, the stage reads the data from the respective stages instead of the register file.

For the handling of artifact data-flow and the handling of various hazards, ADL provides the *stall* statement through which a stage may stall itself. For the purposes of statistics collection, the stall statement can be followed by a *stall category*. The generated simulator then provides statistics about stalls in the architecture. The *stall* statement terminates the processing of the current TAP and the remaining TAPs that handle the rest of the machine cycle. The net effect of the *stall* statement is that no *send* statement is executed by the stage executing the stall in that machine cycle. In addition to the *stall* statement, ADL also provides statements to *reserve* a stage, *release* a stage, and *freeze/unfreeze* the whole pipeline. When a stage is reserved, only the *instruction* that reserved it may perform a send operation to that stage, and only this instruction can release it, regardless of where in the pipeline the instruction is at. When a stage executes a *freeze*, all stages except the stage that executed the *freeze* statement will stall and only the stage that executed the *freeze* statement may later execute an *unfreeze* statement.

10.2.4 ISA Specification

The ISA is specified by means of instruction declarations which describe the syntax and semantics of both the machine instructions and the macro instructions using a uniform syntax. Let us see through a simple instruction definition how the information needed to generate an assembler, a disassembler, and the cycle-accurate simulator is specified. Such a simplified specification fragment is shown in Fig. 10.7.

The top line of the instruction definition describes instruction mnemonics to be used b the assembler and the disassembler and the *emit* section in an instruction specifi tion describes how the assembler should generate the corresponding binary presentation. Now let us examine the semantic part of the lw instruction declaration shown in this figure. This instruction has the *i_type* attribute *load_type*, which is enumeration valued and all load instructions (e.g., lh, lb, lwl, etc.) have the same value. In MIPS instruction set, the destination register field is found in two locations, *rt* and *rd*. By attaching the instruction attribute *dest_r* to an instruction field, we can hide the irregularities of the instruction set from the microarchitecture specification.

LRTL segments *s_ID*, *s_EX*, and *s_MEM* define the operations for each of the corresponding stages. The LRTL segment s_ID performs a sign extension using powerful ADL++ bit operations. The sign extension is achieved by repeating the bit 15 of the immediate field (|< operator) for 16 bits and then concatenating (|| operator) it with the field itself. The result is then stored into the variable *simm*. Note that the variable *simm* was defined as a *controldata variable* in the earlier sections. When the instruction is moved by the s_ID TAP into s_EX, this data value will be available to be used by that stage. The LRTL segment s_EX performs an address computation by adding the contents of the variable *lop* with the sign extended value computed by the ID stage. Similarly, the LRTL segment s_MEM performs a data cache access using the value computed in the s_EX stage and stores the returned value into the variable *dest*. Since writing back the results of instructions into the register file is common for all instructions, this task is handled by TAPs.

```
declare rt register field 5 bits, rs register field 5 bits,
        rd register field 5 bits, immediate signed field 16 bits,
        address label variable;
instruction
        lw rt address
           emit opcode=35 rs=<address.base> rt immediate=<address.offset>
           attributes (i_type:load_type,dest_r:rt,lop_r:rs)
        begin
           case s_ID: simm=immediate.[15:1] |< 16 || immediate; end;
           case s_EX: lmar=lop+simm; end;
           case s_MEM: dest=dcache[lmar]; end;
        end;
```

FIGURE 10.7

MIPS load word instruction.

10.2.5 Putting It Together

The address space of a TAP consists of the global address space implemented by the artifacts and the local address space defined by the instruction being currently processed. Let us see through an example as to how the address space is exercised by TAPs to implement hazard detection and handling in a pipelined processor. Fig. 10.8 illustrates much simplified TAPs defining the s_ID epilogue and s_IF prologue.

Examples of *hazard handling* using these statements are shown in Fig. 10.8. Fig. 10.8(a) indicates the case where the result of a load instruction may be used immediately by the next instruction. Such data hazards cannot be overcome by forwarding alone, and therefore require insertion of pipeline bubbles. The stage in this case checks for the condition by examining the context of the s_EX stage and its destination register and stalls appropriately. Because of the stall, the s_ID stage does not execute a *send* in this cycle. Since the send operations of following stages are not effected by the *stall* of prior stages, the s_EX stage enters the next cycle in an idle state, which is equivalent to introducing a pipeline bubble. An instruction cache miss in a pipelined architecture is usually handled by freezing the machine state. In Fig. 10.8(b), the instruction fetch stage executes a *freeze* statement whenever there is a cache miss. A *stall* is also executed so that the epilogue will not attempt to execute the *send* statement. Note that an *unfreeze* is always executed whenever the cache access is successful. Executing an *unfreeze* on a pipeline which is not frozen is a null operation. In this way, the stage code does not have to be history sensitive.

We have seen how the instruction set description and the microarchitecture description seamlessly work under the ADL++ model. In the following sections we discuss how this model helps with the description of complex instruction sets and microarchitectures. For this purpose, we first review specific challenges posed by complex instruction sets.

```
instruction register ir;
stall category mem_ic,ld_d_dep,pool_full;
```

```
(a) procedure s_ID epilogue
      begin if i_type[s_EX]== load_type &
            (dest_r[s_EX]==lop_r |
            dest_r[s_EX]==rop_r) then
            stall ld_d_dep;
      end s_ID;
```

```
(b) procedure s_IF prologue
      begin ir=icache[pc];
            if access_complete then
                  begin unfreeze; pc=pc+4 end
            else
                  begin freeze; stall mem_icl end;
      end s_IF;
```

FIGURE 10.8

Handling of hazards.

10.3 REVIEW OF COMPLEX INSTRUCTION SET ARCHITECTURES

The problem of instruction code density has always manifested itself in the form of additional complexity on part of the instruction set and the microarchitecture. Although the art of instruction set design has made significant progress since the early days of computing, one earlier complex instruction set remains to be the dominant instruction set on desktop computing and there are many other newly developed sets, which are commonly found in DSPs, [6] where instruction code density is still a significant issue. In this chapter, we discuss Intel's IA-32 instruction set as it represents many of the features commonly found in complex instruction sets. This instruction set as seen in the latest Intel processors has its roots in the 8086 and 8088 processors from 1978. The ISA embodies a variable length instruction set encoding and the processor supports many memory models including segmented memory. The architecture also includes overlapping registers. There are very few, if any, wasted bits in a typical x86 instruction. All these properties make the Intel IA-32 architecture quite challenging for an ADL specification. Yet, these complications are only the tip of the iceberg. Complex instructions significantly overload individual instructions not only in terms of syntax, but also in terms of the associated semantics. As a result, they become very difficult to specify in a clean manner.

In this section, we take a closer look at these challenges starting with syntax-related challenges and proceed to the involved semantics, and show how an object-oriented approach can help.

10.3.1 Variable Length Instructions

It is well known that variable length instructions pose significant difficulty in the design of microarchitectures, particularly the fetch units. The difficulty for architecture specification languages lie partly in the specification of instruction fields which make up the binary representation of instructions. With variable length instructions the common approach of declaring instruction fields and specifying instruction formats using these fields as building blocks, such as the approach taken by UPFAST, [7] does not work well. This is because, variable length instruction sets unavoidably define many instruction formats that extend in length depending on the number and size of the parameters of a given instruction.

In case of IA-32, an instruction can have up to four prefixes that modify the semantics (e.g., to use 16-bit or 32-bit registers), one or two bytes of opcode (a few reserved opcodes indicate to use the second opcode byte), a ModR/M byte for memory or register arguments, an SIB byte to help the ModR/M if necessary, as well as displacement and immediate fields. In other words, the fields making up the instruction format appear depending on the particular instruction format.

Recall that ADL++ model clearly separates the instruction set specification from the microarchitecture specification. In this respect, in order to hide the details of the instruction set from the microarchitecture specification, the language has to map an individual case of a semantic specification to an unambiguous instance that can be understood by the microarchitecture. This mapping is provided by the *decode* statement, which can uniquely identify a semantically unambiguous operation. In order to generate the necessary code for the decode statement, the ADL++ compiler has to automatically distinguish one instruction from another. For this purpose, the compiler needs to look through all of the defined instructions and try to identify a unique constant valued field for each instruction. If a series of instructions have a constant valued field but share the same value for all the instructions, they can be grouped together and a second field can be searched for (e.g., extended opcodes). The compiler keeps looking until it finds a unique constant valued field (or set of fields) for each instruction. This field (or set of fields) becomes the *opcode*. In case of variable length instructions, this process is particularly difficult. Although in principle one could enumerate all possible instruction formats, this is cumbersome and error prone. Added to this complexity is the usage of many addressing modes, which again overload the same instruction to many instances.

10.3.2 Many Memory Addressing Modes

There are many addressing modes used in IA-32, which are, for the most part, independent of the instruction since they are encoded using the ModR/M byte (and an SIB byte, if necessary).

IA-32 addressing modes are shown together with examples in Fig. 10.9. As we have pointed out, the challenging aspect of the many addressing modes in IA-32 is trying to define them succinctly in ADL, since the fields are mostly independent of the opcode. That is, the opcode alone does not indicate all the fields that follow the opcode. For example, the mov instruction shown in Fig. 10.9 has the opcode of 0x89, which indicates that the opcode byte will be followed by a ModR/M byte, with the Reg/Opcode field of the ModR/M byte selecting a general purpose register.

Displacement (or Absolute)	mov %ecx, 0xDEADBEEF
Base	mov %ecx, [%esp]
Base+Offset	mov %ecx, [%esp-4]
Base+Index+Offset	mov %ecx, [%esp+%edi-4]
Base+(Index*Scale)+Offset	mov %ecx, [%esp+%edi*2-4]
Base+Index	mov %ecx, [%esp+%edi*1]
Base+(Index*Scale)	mov %ecx, [%esp+%edi*2]
Index+Offset	mov %ecx, [%edi*1-4]
(Index*Scale)+Offset	mov %ecx, [%edi*2-4]

FIGURE 10.9

Intel IA-32 addressing modes.

What follows the ModR/M byte, if anything, is indicated by the ModR/M byte. Thus, each byte in the instruction provides a hint as to what comes next.

Again, the simplest way to approach this problem is to enumerate every possible variation of an instruction as if it is a separate instruction, and then use simple overloading to distinguish different instructions, just like functions in C++. For example, the mov instruction overloaded for Displacement and Base addressing modes:

```
mov reg32 disp emit opcode=0x89 mod='00'b regop=reg32 rm='100'b disp ...
mov reg32 base emit opcode=0x89 mod='00'b regop=reg32 rm=base ...
```

However, this leads to the problem of having to overload the same instruction many times due to the many addressing modes. There are nine addressing modes listed earlier, however, three modes (Base + Displacement, Base + Index + Displacement, and Base + (Index*Scale) + Displacement) can use either an 8-bit or a 32-bit displacement, giving us 12 effective modes. Furthermore, there are restrictions on when %esp and %ebp can be used for base or index registers. Treating these restrictions as special addressing modes gives us 6 additional special case modes, for a total of 18 addressing modes! Creating separate ADL++ instruction definitions for every combination of x86 opcode with addressing mode would generate thousands of ADL++ instructions. This is clearly unacceptable.

10.3.3 Overlapping Registers

The IA-32 architecture is backward compatible with the original x86 processor to a large extent. One of the features of this backward compatibility is a set of registers that have overlapping parts. IA-32 includes eight 32-bit general purpose registers: EAX, ECX, EDX, EBX, ESP, EBP, ESI, and EDI. However, in order to maintain backward compatibility, there are aliases for 8-bit and 16-bit parts of the registers. For example, AL, AH, and AX all describe different parts of the EAX register. AL and AH are two 8-bit registers that represent the lower and upper 8 bits of the 16-bit register AX, and AX is the low 16 bits of the 32-bit register EAX (Fig. 10.10(a)).

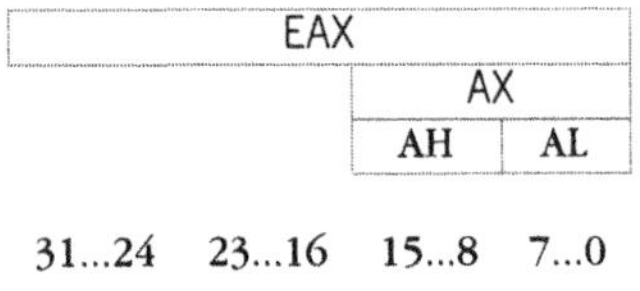

Register	Special Purpose
%EAX	accumulator
%ECX	counter for string and loop operations
%EDX	I/O pointer
%EBX	pointer to data in the DS segment
%ESP	stack pointer
%EBP	pointer to data on the stack (base pointer)
%ESI	source pointer for string operations
%EDI	destination pointer for string operations

```
|            EAX            |
|              |     AX     |
|              | AH  | AL   |
  31...24  23...16  15...8  7...0
```

(a) Overlapping Registers

(b) Special Purpose Registers

FIGURE 10.10

IA-32 registers.

Registers ECX, EDX, and EBX are broken down in a similar fashion. Registers ESP, EBP, ESI, and EDI have aliases for the lower 16 bits SP, BP, SI, and DI, respectively. As a side note, while the earlier stated eight registers are noted as general purpose, the registers occasionally have special uses, hence the seemingly obtuse names of the registers. The special uses are shown in Fig. 10.10(b).

10.3.4 Mixed Arguments

Unlike RISC architectures, instructions in CISC architectures can operate directly on memory. That is, RISC machines, or load-store architectures, have only two instructions that can operate on memory, load and store, while all other instructions must operate on registers. The CISC instructions have no such restrictions. This leads to instructions that can have a variety of arguments. The mov instruction takes two arguments, a source and a destination, one of which must be a register, the other can be either a register or a memory address.

The address can come in any one of the 18 modes listed earlier. Specifying these instructions naively would create a plethora of instructions (also described above), since from a semantic perspective, these are different instructions that happen to be named with the same instruction name.

Although some of these issues have been addressed within the context of *instruction set specification* with the SLED approach [8], as it can be seen, the approach taken by SLED is inadequate for automatic generation of simulators. Although one can describe x86 ISA in less than 500 lines of code in SLED, the language was only designed for encoding and decoding instructions (as the name implies). Many instructions in x86 are encoded/decoded the same way, with the only difference being the opcode, so patterns are used to define many instructions in one line. On the other hand, in order to tie in the microarchitecture specification, one needs to be able to specify the semantics of each instruction. Since the semantics of each instruction are very different, attaching semantics to many opcodes cannot be done with one line, and an alternative technique must be sought.

10.3.5 Assembly Language Syntax

Assembly language syntax significantly affects the complexity of an instruction specification. In case of IA-32, there are two styles of assembly language syntax which are widely used. The AT&T style [9] assembly language is used most often with *nix operating systems (Solaris, Linux, *BSD, etc.) due to AT&T's history with Unix. There are significant differences between the two representations. Among these, the most important one is the usage of l, w, and b suffixes by the AT&T syntax (e.g., movl) to indicate if the instruction is operating on a 32-bit long word, a 16-bit word, or a byte. Intel syntax either derives the operand size by the register name (e.g., %eax implies 32-bit vs. %ax implies 16-bit) or by explicitly using DWORD, WORD, or BYTE before the memory operand. In terms of memory operands, Intel uses a verbose syntax that indicates how the addresses are calculated. Finally, the two syntaxes have opposite order of destination, and source order.

In general, AT&T syntax exacerbates the problem of overloading in an ADL specification because of the suffixes appended to instruction mnemonics. While it is quite feasible to define an instruction template for a particular instruction and then derive variations of it for different operand sizes, AT&T syntax would require a different template for each of the sizes since deriving instruction mnemonics using inheritance does not appear to be practical. As a result, in the rest of this chapter we use the Intel syntax and give the examples accordingly.

10.4 SETS AND REGULAR EXPRESSIONS AS LANGUAGE CONSTRUCTS

Sets are powerful constructs that help in defining both the syntax and the semantics of complex instruction sets. ADL++ sets were in part inspired by the work of Bailey and Davidson [10]. In ADL++, sets are used to describe the registers used by the architecture, as well as registers that are part of the addressing mode specifications. The addressing mode specifications use regular expressions where elements of the regular expressions may be sets. Let us see through an example how sets help in defining the registers. Recall that IA-32 have various *views* when it comes to the registers supported by the architecture.

10.4.1 Registers and Their Specification Using Sets

In order to define the registers used by IA-32, we begin by constructing plain sets with register names and declare that these sets form the machine's registers:

```
# create groups of registers.
# Registers may exist in multiple sets.
set
  reg32 {"%eax","%ecx","%edx","%ebx","%esp","%ebp","%esi","%edi"},
  reg16 {"%ax","%cx","%dx","%bx","%sp","%bp","%si","%di"},
  reg8 {"%al","%cl","%dl","%bl","%ah","%ch","%dh","%bh"},
  reg32_noESP {reg32 - "%esp"},
  reg32_noESBP {reg32_noESP - "%ebp"},
  reg32_noEBP {reg32 - "%ebp"};

meta
  isa_registers reg8 | reg16 | reg32;
```

In this specification the sets such as *reg32_noESP* handle specific cases where a certain register cannot be used due to architectural and instruction set-related idiosyncrasies. For example, the %esp and %ebp registers cannot be used as a base register in base addressing mode, so a special group is created such that those registers are excluded. Note how the set approach makes it possible to define them cleanly. The *meta* construct declares the meta symbols used by the assembler and in this case, the set union defines any member of the sets *reg8*, *reg16*, and *reg32* to be an assembler-recognized register. For the microarchitecture references, we can define the physical registers as generic 32-bit registers using a built-in artifact:

```
register file
    gpr [8, 32]   # 8 registers, 32 bits each
    "%reg0"  0,  "%reg1"  1,  "%reg2"  2,  "%reg3"  3,
    "%reg4"  4,  "%reg5"  5,  "%reg6"  6,  "%reg7"  7;
```

We can now use a union to define the symbolic registers and the parts of the physical register used. In the notation [start_bit:length] bits are numbered from *high bit* down to *zero*.

```
union
    "%eax"  %reg0[31:32],  "%ecx"  %reg1[31:32],  .... # 32 bit registers
    "%ax"   %reg0[15:16],  "%cx"   %reg1[15:16],  .... # 16 bit registers
    "%al"   %reg0[ 7: 8],  "%cl"   %reg1[ 7: 8],  .... # 8 bit low registers
    "%ah"   %reg0[15: 8],  "%ch"   %reg1[15: 8],
    "%dh"   %reg2[15: 8],  "%bh"   %reg3[15: 8];        # 8 bit high registers
```

Sets can be used to group more than just the registers. The opcode prefixes as well as valid scale values used in addressing modes are also handled conveniently using sets.

10.4.2 Regular Expressions and Addressing Modes

Now that we have defined the registers used by the architecture, we define how these registers are referenced by the addressing modes. ADL++ uses an extended regular expression syntax to define them. Unlike real regular expressions, in this form of regular expressions the strings are artifacts (such as sets) and not characters. A short summary of the regular expression meta symbols is given in Table 10.1.

Table 10.1   ADL++ regular expression meta symbols.

Symbol	Description
?	The previous item is optional
*	0 or more of the previous item
+	1 or more of the previous item
n	match previous item exactly n times
n,	match previous item at least n times
n, m	match previous item at least n, at most m times
( )	groups items
|	logical or

The specification of the addressing modes using regular expressions is given here. In these expressions, the instruction fields, sets of registers, and other sets of values appear as artifacts:

```
#- Instruction fields
#
type
    .....
    s8    signed integer variable  8 bit,
    s16   signed integer variable 16 bit,
    s32   signed integer variable 32 bit;
  set
    scaleVals { 1, 2, 4, 8 };

  addressing modes
    disp32            s32,
    base              "[" reg32_noESBP "]",
    base_index        "[" reg32_noEBP [-+] reg32_noESP "]",
    base_index_scale  "[" reg32_noEBP [-+] reg32_noESP "*" scaleVals "]",
    index_scale_disp32 "[" reg32_noESP "*" scaleVals [-+] s32 "]",
    base_disp8        "[" reg32_noESP [-+] s8 "]",
    base_index_disp8  "[" reg32       [-+] reg32_noESP  [-+] s8 "]"
    ... ;
```

The regular expressions in the addressing mode define the syntax of the addressing mode, but they do not define the semantics. For example, in base_index mode stated earlier, the assembler does not know if reg32_noEBP is the base or the index; it only knows to expect a 32-bit register excluding %ebp. Mapping the two registers to some meaningful value like base or index is handled by a user-defined procedure. The procedures are called using the < address_mode.procedure > syntax, which return the register number corresponding to the register in question.

10.5 INSTRUCTION TEMPLATES AND MULTIPLE CONDITIONAL INHERITANCE

One of the most powerful features of ADL++ is the use of the concept of inheritance. Inheritance allows definition of complex instruction sets and architectures in a well-organized manner. In order to succinctly describe IA-32, we define an *encoding pattern* for each of the addressing modes and, we let each instruction inherit the right pattern. This is the key idea for the instruction handling in ADL++: treat instructions as objects and use multiple inheritance with a twist! The encoding patterns are defined by a series of templates, and the real instructions inherit the properties from these patterns. The objects in ADL++ are the instruction templates and the instructions themselves.

10.5.1 Inheritance with a Twist: Multiple Conditional Inheritance

Conditional inheritance borrows from the idea of well-established concept of multiple inheritance. Unlike multiple inheritance, however, instead of inheriting all of the features from the parents, in conditional inheritance a derived object can inherit from the one parent with the *best fit*. This best fit is determined by the arguments to the instruction. Of course for this approach to work, the inherited arguments must be unambiguous. Using conditional inheritance, we can create one template for each addressing mode. Each such template has the emit fields defined as well as any other common properties. The child that inherits from the template then overrides the emit fields as necessary and defines the LRTL segments necessary, implementing the semantics of that particular derivation.

The common addressing modes are then combined using the conditional inheritance into one template that the real instructions inherit from. To reinforce the idea that this is not traditional multiple inheritance, the logical or operator (| operator) is used to split the parents. In Section 10.5.2 we discuss the properties of instruction templates and how they differ from real instructions.

10.5.2 Instruction Templates

In ADL++, instruction templates differ from normal instructions in two ways. Fields in instruction templates can be grouped or made optional with the use of regular expression like syntax. Parentheses group fields together, (field1 field2...), to indicate that all of the fields in the group must appear together. That is, *field1* cannot exist without *field2*, and vice versa. This is quite useful for larger fields like the ModR/M byte in IA-32, which consists of three smaller fields: the 2 bit mod field, the 3 bit reg/op field, and the 3 bit r/m field, used to describe how memory and/or registers will be addressed. A '?' following a field indicates the field is optional. For example, the SIB byte is optional depending on the ModR/M byte, thus, it appears as (scale index base)?. A ? is really just shorthand for {n,m} syntax (where n = 0 and m = 1), which says the previous item must appear at least n times but no more than m times. Finally, a | indicates logical-or, useful for fields that vary in size. Some instructions have 8-bit immediates, others 16-bit, and others 32-bit, and others none at all, so, putting it all together: (imm8 | imm16 | imm32)?.

Templates do not exist in the actual instruction set. That is, when the generated assembler is assembling code, it will never try to match a template with a real assembly instruction. Templates can inherit properties from other instruction templates and override fields or sections from the parent. This allows the creation of a master template. A master template is really just another template (i.e., it is not a special type of template), but it helps the programmer to avoid syntactical errors. In IA-32, there is one general instruction format:

| Prefixes | Opcode | ModR/M | SIB | Disp | Immediate |

In this format there may be up to 4 prefixes, where each prefix is 1 byte long. The opcode field can be 1 or 2 bytes long and is followed by the optional ModR/M and SIB bytes. Displacement and the Immediate fields can be anywhere from 0 to 4 bytes. Templates and instructions inherit the properties illustrated in the following template:

```
instruction template
begin
    intel           # no arguments given
      emit prefix1? prefix2? prefix3? prefix4?
          opcode{1,2}
          (mod reg_op rm)?
          (scale index base)?
          (disp8 | disp16 | disp32)?
          (imm8  | imm16  | imm32)?
      attributes
          (i_class : intel_class,op_type : intel_ops )
      begin
        exact s_MEM_LD ...
        end;
        exact s_EX ...
        end;
        exact s_MEM_ST ...
        end;
      end, ....


  intel_r8_bis rd8 base_index_scale inherits intel
      emit opcode=0xF1 mod=00 reg_op=rd8 rm=100
      scale=<base_index_scale.scale>
      index=<base_index_scale.index>
      base=<base_index_scale.base>
begin
   exact s_MEM ...
         # calculate address = base + (index * scale)
   end;
end,
```

The first item to notice is on the third line, no arguments are given to the generic instruction name intel. The nonexistent arguments will be overridden by the following templates. The emit line, on the other hand, defines every possible field that might be emitted by a descendant and uses the ? and {n,m} modifiers to indicate optional fields. Only two example attributes are defined; depending on the microarchitecture specification it is clear that many more are needed.

An instruction that inherits from this intel master template is free to override the arguments, any of the emit fields, any of the attributes, or any pipeline stage. (Note that if a pipeline stage is overridden, the entire stage must be overridden, even if only one line is changed.) Inheritance is indicated by the inherits keyword following the instruction's arguments.

The template intel_r8_bis has two arguments, an 8-bit destination register, rd8, and a memory location addressed by base_index_scale mode. It inherits from the intel master template and then defines exactly which fields will be emitted for this type of instruction. The scale, index, and base functions are user-defined functions and, with the help of the regular expressions for the addressing modes, return the respective values for scale, index, and base. Finally, the s_MEM pipeline stage is used to load a byte into a temporary pipeline register that will be used by the s_EX stage in instructions that inherit from this template.

Now we are ready to use conditional inheritance to combine everything together. Instructions that have a 32-bit register for a source and a 32-bit word in memory for a destination would inherit from the intel_r32_rm32 template. A real instruction that inherits from this template is shown here:

```
intel_r32_rm32 inherits (
intel_r32_d32      | intel_r32_b      | intel_r32_bi      |
intel_r32_bis      | intel_r32_id32   | intel_r32_isd32   |
intel_r32_bd8      | intel_r32_bid8   | intel_r32_bisd8   |
intel_r32_bd32     | intel_r32_bid32  | intel_r32_bisd32  |
intel_r32_b_esp    | intel_r32_b_ebp  | intel_r32_b_bd8_esp |
intel_r32_bd32_esp | intel_r32_bi_ebp | intel_r32_bis_ebp )

mov inherits intel_r32_rm32  emit opcode=0x8B
begin
   exact s_EX
        rd32 = ....;
   end;
end,
```

Each of the 18 templates which intel_r32_rm32 inherits from defines an addressing mode (there are 12 modes plus 6 special modes for using %esp or %ebp as a base register). To see this in action, consider the following x86 instructions:

```
mov %eax, DWORD PTR [%esp - 4]
mov %eax, DWORD PTR [%esp + %ebp*4 - 4]
mov %ax, WORD PTR [%esp - 4]
```

The instruction is the same in all cases, mov, but the arguments differ. However, they differ in a unique and unambiguous way allowing the compiler generated code to match it against only one parent. The first instruction matches intel_r32_rm32 and its parent intel_r32_bd8 (base + 8-bit-displacement) (technically it also matches intel_r32_bd32, but the compiler can be made smart enough to choose an 8-bit-displacement via a *pragma*, if optimizations in this sense are desired). Likewise, the second instruction matches intel_r32_rm32 but with

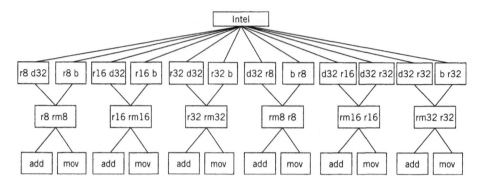

FIGURE 10.11

A portion of the inheritance tree for IA-32.

a different parent, intel_r32_bisd8. The third instruction matches none of the parents in intel_r32_rm32 but matches mov inherits intel_r16_rm16 and its parent intel_r16_bd8).

Fig. 10.11 shows the conditional inheritance tree with only two addressing mode templates in the second level. There are many more templates in the final tree.

We have discussed how the ADL++ conditional inheritance technique can help in defining complex instruction sets. In Section 10.6, we see how objects help in succinctly describing the micro-architechture and at the same time promote code reusability.

10.6 OBJECT-ORIENTED MICROARCHITECTURE SPECIFICATION

We have already covered the basic building blocks of microarchitecture specification. In order to facilitate the ease of use and reuse of code, users can create ADL++ objects which are called *artifacts*. The ADL++ artifacts are similar to C++ objects in terms of use of inheritance and inclusion of data objects and member functions. The ADL++ objects differ in the use of *events* that provide a simultaneous entry into all instances of objects that define them, as well as inclusion of pipelines and TAPs which together also provide simultaneous entry into all instances of objects that have the same clock label and the stage name. Event handlers provide a means of broadcasting across all objects, whereas embodied pipelines and TAPs provide synchronous parallel operation of the objects in the system.

10.6.1 Artifacts as Objects

As with the rest of the chapter, we begin by introducing an example artifact:

```
artifact functional_unit
begin
    integer transition_count;
```

```
    statistics
    begin
       printf("%s:transition count %d\n",myself.name,
           transition_count);
    end statistics;

    initialization
    begin
       transition_count  = 0;
    end initialization;

    restart
    begin ...
    end restart;
end functional_unit;
```

In this declaration, *statistics*, *initializaton*, and *restart* are event handlers. An ADL++ statement, namely *cause*, executed by any other artifact will activate all event handlers that have the same name as the cause statement parameter. The generated simulator *causes* initialization at the beginning of simulation and causes the event *statistics* at the end of the simulation. Like other object-oriented languages, before the artifact *functional_unit* can be used, an instance of it should be created. Similarly, artifacts can be derived from other artifacts. The following will create a new artifact called *simple_stage* from the earlier mentioned artifact:

```
    artifact simple_stage inherits functional_unit
    begin
       pipeline IPIPE (s_All);

       procedure s_ALL prologue
       begin ....
       end s_ALL:

       procedure s_ALL epilogue
       begin ....
          retire;
          newcontext;
       end s_ALL:
    end simple_stage;
```

Note that in this declaration there is only a single pipeline and a single pipeline stage. The epilogue TAP shown retires the current instruction and allocates a new one for the next cycle. Suppose we would like to use the MIPS instruction set definition discussed earlier to create a simple functional simulator using the object description given earlier. In order to create this simulator we can declare our machine:

```
simple_stage  MACH;
map [if,id,ex,mem,wb] to s_ALL;
```

The map statement completes the abstraction layer provided by instruction attributes by mapping labels of LRTL segments to physical pipeline stages. In this example, it maps all LRTL segments described in the instruction set definition to a single pipeline stage *s_ALL* in the indicated order; in other words, all LRTL segments will execute one after another in the same pipeline stage, essentially providing functional simulation.

10.6.2 Deriving Complex Architectures From Objects

ADL++ artifacts can be used to describe the behavior of any synchronous piece of hardware. More important, it is possible to create arrays of artifacts and use simple artifacts as the building blocks of a more complicated hardware structure. For example, one could create an artifact that embodies the floating pipeline of the processor and use the resulting artifact as a floating point unit of a superscalar. The actual power of the artifacts in ADL++ manifests itself when units of great similarity need to be defined. A derived object is free to replace any of the event handlers, as well as TAPs and other functions embodied in the artifact.

We have successfully built branch predictors, instruction fetch units, and memory pipelines, as well as a complete superscalar processor in an object-oriented way using ADL++.

10.7 EPILOGUE

Like many other languages in this book, ADL++ is a constantly evolving language. It is hoped that not only the language but also its underlying model will be helpful to the designers. The FAST system is available free of charge to educational institutions. To request a copy please send an e-mail to soner@acm.org.

10.8 HISTORY OF FAST AND ADL++

The original FAST system was developed by Dr. Soner Önder in 1998 under the guidance of Prof. Rajiv Gupta and was called UPFAST (University of Pittsburgh Flexible Architecture Simulation Toolset) at the time. The toolset was renamed in 1999 to FAST, dropping the *UP* prefix. The original language and the compiler mostly rehauled during this time and calling convention specifications were removed from the language. Most of the object-oriented structure of the language in terms of microarchitecture aspects has been developed by Dr. Önder as part of the work for DARPA grant PAC/C Program, including parts of the new ADL++ compiler. In terms of handling complex instruction sets, Dr. Robert Pastel [11], under the direction of Dr. Önder, has introduced the regular expressions for addressing modes as part of his M.Sc. thesis. These extensions were the result of investigating how to describe a complex instruction set DSP, namely, StarCore SC-140 [6]. Extending the language for full-fledged object-oriented instruction set specification was achieved by Mr. Jeff Bastian as part of his M.Sc. work under the direction of Dr. Önder. Mr. Bastian has introduced the key idea of conditional inheritance enabling descriptions such as IA-32 to be defined succinctly [12].[1]

REFERENCES

[1] *IA-32 Intel Architecture Software Developer's Manual.* Intel Corporation, Document Numbers 24547004, 24547104, and 24547204, 2001.

[2] D. Perry. VHDL: Programming by Example. McGraw-Hill, New York, USA, 2002.

[3] J. Armstrong and F. Gray. *Structured Logic Design with VHDL.* New Jersey: Prentice Hall, 1993.

[4] J. Morison and A. Clarke. *ELLA2000 A language for Electronic System Design.* McGraw-Hill, New York, USA, 1993.

[5] D. Thomas and P. Moorby. *The Verilog Hardware Description Language 4th ed.* Kluwer Academic Publishers, Norwell, MA, USA, 1998.

[6] *SC140 DSP Core Reference Manual.* Second Revision, Motorola Corporation, 2001.

[7] S. Önder and R. Gupta. Automatic generation of microarchitecture simulators. In *IEEE International Conference on Computer Languages,* pages 80–89, Chicago, May 1998.

[8] N. Ramsey and M. F. Fernandez. Specifying representations of machine instructions. *ACM Transactions on Programming Languages and Systems,* 19(3):492–524, May 1997.

[9] J. Molenda. *GNU Assembler Documentation.* GNU http://http://www.gnu.org/manual/manual.html, February 2008.

[10] M. W. Bailey and J. W. Davidson. A formal model and specification language for procedure calling conventions. In *The 22nd Annual ACM SIGPLAN-SIGACT Symposium on Principles of Programming Languages,* pages 298–310, San Francisco, California, 1995.

[11] R. Pastel. Describing vliw architectures using a domain specific language. Master's thesis, Michigan Technological University, 2001.

[12] J. Bastian and S. Önder. Specification of intel ia-32 using an architecture description language. In *Workshop on Architecture Description Languages WADL-04,* pages 33–42, Toulouse, France, August 2004.

[1]Special Thanks: This work is supported in part by a grant from DARPA, PACC Award no. F29601-00-1-0183 to the Michigan Technological University and a CAREER award (CCR-0347592) from the National Science Foundation to Soner Önder.

Processor Design with ArchC

11

Guido Araujo, Sandro Rigo, and Rodolfo Azevedo

11.1 OVERVIEW

With increasing complexity of electronic systems, System on Chip (SoC) design, based on heterogeneous prevalidated modules, has become a major methodology in the hardware design of modern electronic products. On the other hand, as the demand for flexible programmable devices continues to grow, and software development time has become the largest share of the design time, accounting for a critical portion of product development.

Software design starts after the basic hardware–software architecture interface is defined. In traditional design flows, this occurs only after the hardware is prototyped, and its basic firmware is in place, considerably delaying product development.

In recent years, Electronic System Level (ESL) design methodologies [1] have been proposed to enable the construction of virtual prototypes for system architecture. This brings two advantages to designers: (a) an early definition of the hardware–software interface, thus allowing a fast start of software development (even before the hardware platform in finished); and (b) an approach to enable platform trade-off, analysis, and debugging.

In any SoC design, microprocessors are central modules to platform coordination and control. Efficient microprocessor ESL models are thus relevant pieces in the design of modern virtual platforms. Such models should be very fast, and accurate enough only to the level where they allow a well-defined interface to the software boundary.

ArchC [2] is an open-source Architecture Description Language (ADL) based on SystemC [3]. It was designed to enable the fast generation of microprocessor simulators. ArchC takes as input a processor description and automatically generates a SystemC program containing: an object code decoder, a Linux syscall interface, and a simulator for the given microprocessor. The processor simulator allows TLM communication capabilities: external communication and interrupt modeling, compliant with the OSCI TLM standard. Multiple instances of the simulator can coexist in the same platform, thus enabling the easy construction

275

of multiprocessors and complex heterogeneous systems. ArchC also allows the automatic retargetability of GNU assemblers and linkers.

SystemC [3–5] is included in a new trend of design languages and extensions being proposed to raise the abstraction level for hardware design and verification. SystemC is an open source and entirely based on C++, being composed by a set of C++ class libraries that extends the language to allow hardware and system-level modeling. Although it is possible to model in low levels of abstraction using SystemC, its main goal is not to replace traditional HDLs (VHDL, Verilog) in RTL design, but to allow system-level design.

Though SystemC supports a wide range of computation models and abstraction levels, it is not possible to extract from a generic SystemC processor description all necessary information in order to automatically generate tools to experiment and evaluate a new *Instruction Set Architecture* (ISA). That is what first motivated the ArchC design.

Functional models for the MIPS-I, SPARC, PowerPC, i8051 and PIC architectures, and cycle-accurate models of R3000 and i8051 were designed using ArchC and are freely available [2]. These models have been thoroughly tested using programs from the MediaBench [6] and MiBench [7] benchmarks. Program traces larger than 100 Bi. (dynamic) instruction count have been tested. Models follow a strict version roadmap that certifies their maturity level.

ArchC is based on two simulators: an interpreted-based simulator and a compiled-simulator. As an example, the performance range achieved by running MediaBench and Mibench programs on a SPARC compiled-simulator was 110–225 MIPS, with an average of 134 MIPS (the host is a 2.8 GHz Pentium 4). Recently, some new optimizations have pushed model performance of the interpreted-based simulator to the range of 40 MIPS for the MIPS-I, SPARC, and the PowerPC processor.

The goal of this chapter is to describe ArchC and to explain its application in microprocessor modeling. The chapter is divided into four sections as follows. Section 11.2 describes the ArchC constructs and semantics and how they can be used to design a basic processor model. Section 11.3 shows how ArchC generated processor models can be interconnected to other SystemC modules through its OSCI TLM interface, which is the most common use of ArchC simulators nowadays, enabling users to develop multicore and SoC platform models. It also describes an example of how to use the processor TLM port to design an interrupt handling mechanism. Finally, Section 11.4 puts everything together by describing a multicore platform constructed using ArchC processor models.

11.2 SYNTAX AND SEMANTICS

ArchC [2, 8, 9] is an ADL that relies both on structural and instruction-level information in order to automatically generate processor simulators. Its main goal is to provide enough information, at the right level of abstraction, in order to allow users to explore and verify a new architecture by automatically generating software tools like assemblers, simulators, linkers, and debuggers.

ArchC was designed to be mainly adopted by SystemC users, and so its syntax is based on C++ and SystemC. The main goal was to make it pretty simple for SystemC designers to start writing ArchC processor models, since they would find language constructions that are very similar to those they are used to, with the exception of a small set of keywords.

Basically, a processor description in ArchC is divided in two parts: (1) the **Instruction Set Architecture** (AC_ISA) description and (2) the **Architecture Resources** (AC_ARCH) description. In the AC_ISA description, the designer provides details about instruction formats, size, and opcodes combined with all the necessary information to decode it, as well as the behavior of each instruction. The AC_ARCH description declares storage devices, pipeline structure, etc. Based on these two descriptions, ArchC can generate interpreted simulators (using SystemC) and compiled simulators (using C++), among other software tools. The following sections cover these two descriptions in more detail. For a complete specification of the ArchC language, readers should refer to the *ArchC Language Reference Manual* [10].

Simulation performance is certainly one of the main goals that motivates designers to move to higher levels of abstraction when designing processor models. That is why multicore or SoC platform models targeting to architecture exploration are usually built relying on a functional model of the processor. A functional-level description is composed by an instruction behavior description and a few structural information of the architecture. Fig. 11.1 illustrates the structural information needed in order to build an ARMv5 functional model in ArchC. The syntax is pretty similar to a SystemC module declaration, and the most important section here is the storage resources specification, in lines 4–5, for a local memory and register bank.

```
1   AC_ARCH(armv5e){
2
3     ac_mem     MEM:256M;
4     ac_regbank RB:31;
5     ac_wordsize 32;
6
7     ARCH_CTOR(armv5e) {
8
9       ac_isa("armv5e_isa.ac");
10      set_endian("little");
11
12    };
13  };
```

FIGURE 11.1

Functional architecture resource description.

This is called an AC_ARCH description. In Section 11.3 we show how to use external memory devices modeled in SystemC.

The main effort when describing a functional model in ArchC is in the instruction set specification, which is enclosed within the AC_ISA description. The designer must provide details on instruction formats, decoding, and behavior. Fig. 11.2 is an excerpt from our ARMv5 instruction set description that illustrates several ArchC constructions. The keyword ac_format is used to specify instruction formats (lines 4–10), providing field names and bit width. Afterward, those formats are associated to instructions in their declarations through the keyword ac_instr, as showed in lines 13–16, using a syntax similar to C++ templates. Besides format

```
1   AC_ISA(armv5e) {
2
3       /* ALU */
4       ac_format Type_DPI1 = "%cond:4 %op:3 %func1:4 %s:1 %rn:4 %rd:4
5                               %shiftamount:5 %shift:2 %subop1:1 %rm:4";
6       /* Branch */
7       ac_format Type_BBL = "%cond:4 %op:3 %h:1 %offset:24";
8       /* Load/store */
9       ac_format Type_LSI = "%cond:4 %op:3 %p:1 %u:1 %b:1 %w:1 %l:1
10                              %rn:4 %rd:4 %imm12:12";
11      ...
12      /* Instruction declaration - ALU */
13      ac_instr<Type_DPI1> and, eor, sub, rsb, add, adc, sbc, rsc,
14                          tst, teq, cmp, cmn, orr, mov, bic, mvn;
15      /* Instruction declaration - load/store */
16      ac_instr<Type_LSI> ldrt, ldrbt, ldr, ldrb, strt, strbt, strb;
17      ...
18      ISA_CTOR(armv5e) {
19
20      /* Decoding sequence */
21      and1.set_decoder(op=0x00, subop1=0x00, func1=0x00);
22      and1.set_asm("and%[cond]s %reg, %reg, %reg, RRX", cond, rd,
23                  rn, rm, shift=3, shiftamount=0, s=1);
24      b.set_asm("b%[cond] %imm(bimm)", cond, offset, h=0);
25      b.set_asm("bl%[cond] %imm(bimm)", cond, offset, h=1);
26      b.set_decoder(op=0x05);
27      ...
28      };
29  };
```

FIGURE 11.2

Instruction set description.

and instruction declaration, some important data is initialized inside the ISA_CTOR constructor. Based on this AC_ISA description, ArchC automatically generates an instruction decoder for the simulator. The key information for this purpose is the decoding sequence provided through the set_decoder method. Fig. 11.2 (line 21) shows the decoding sequence of instruction and1, where fields op, subop1, and func1 must all have zero value in order to have a bitstream fetched from the memory be considered an and1 instruction. It is important to notice that the order here is very important, since the decoder will look into the fields in exactly the same order they are passed to the set_decoder method.

This chapter focuses on using ArchC processor models to automatically generate simulators and to integrate them to other SystemC modules. But ArchC models can also be used for assembler generation, through an automatic retargeting of the GNU Binutils suite [11]. For this purpose, it is important to include a set_asm declaration in your processor description. Lines 24–25 in Fig. 11.2 show an example for a branch instruction. The assembly syntax specification follows a syntax similar to the printf function in C. A string is used to specify how assembly instructions must be printed, and fields declared in the format associated to the instruction are used to fill up the string. If the model is not intended for assembler generation, the designer can just omit this initialization, or provide an empty string. More details on GNU Binutils retargeting using ArchC and on how to embed assembly information on processor models can be found in [12-14].

Finally, we have the most important part on a processor description in ArchC: instruction behavior. A behavior description in ArchC is a set of SystemC C++ methods. An ArchC description is instruction-centric, which means that the designer must fill up one behavior method for each instruction declared in the architecture. In order to make this task easier and more efficient, ArchC introduced a concept called behavior hierarchy.

If we consider a reasonably complex instruction set architecture (ISA), it is common to have a set of operations being shared among all instructions in a given class, like arithmetic instructions, or even among all the instructions, like program counter (PC) increment. For example, architectures like ARM have a conditional field that must be checked to determine whether the instruction must be executed or cancelled. All instructions in the ARM ISA must perform this task before executing its specific behavior. That is why ArchC has three levels in its behavior hierarchy: generic, format, and instruction. Every operation defined in the generic behavior method will be executed for the whole ISA, no matter which instruction has been fetched from memory. The format behavior method is used to describe operations that are executed by all instructions associated to a given format. Finally, the instruction behavior is where the designer specifies individual operations for each instruction. The behaviors are always executed in the same order, which is generic, format, and individual. Let us see some examples to clarify how each of these behavior types can be used.

Fig. 11.3 presents an example of a generic instruction behavior method extracted from an ARMv5 description. It includes the conditional field checking, named

```
1  //!Generic instruction behavior method.
2  void ac_behavior( instruction ) {
3
4    // Handling conditional (COND) field
5    execute = false;
6
7    switch(cond) {
8      case  0: if (flags.Z == true) execute = true; break;
9      case  1: if (flags.Z == false) execute = true; break;
10     case  2: if (flags.C == true) execute = true; break;
11     case  3: if (flags.C == false) execute = true; break;
12     case  4: if (flags.N == true) execute = true; break;
13     case  5: if (flags.N == false) execute = true; break;
14     case  6: if (flags.V == true) execute = true; break;
15     case  7: if (flags.V == false) execute = true; break;
16     case  8: if ((flags.C == true)&&(flags.Z == false))
17                  execute = true; break;
18     case  9: if ((flags.C == false)||(flags.Z == true))
19                  execute = true; break;
20     case 10: if (flags.N == flags.V) execute = true; break;
21     case 11: if (flags.N != flags.V) execute = true; break;
22     case 12: if ((flags.Z == false)&&(flags.N == flags.V))
23                  execute = true; break;
24     case 13: if ((flags.Z == true)||(flags.N != flags.V))
25                  execute = true;  break;
26     case 14: execute = true; break;
27     default: execute = false;
28   }
29
30   // PC Increment
31   ac_pc += 4;
32   RB.write(PC, ac_pc);
33
34   if (execute)
35     dprintf("Executing\n");
36   else {
37     dprintf("Instruction annulled\n");
38     ac_annul();
39   }
40 }
```

FIGURE 11.3

Generic instruction behavior.

cond in the formats declared in Fig. 11.2, that must be performed by all ARM instructions. It is important to notice that for a behavior description, ArchC accepts pure C++ methods that may even use any SystemC-specific data type or method. Notice that storage devices, like those declared in Fig. 11.1, may be directly accessed from behavior methods, like those illustrated in line 32 of Fig. 11.3.

Fig. 11.4 illustrates the kind of operation that can be described in the format behavior methods. It is another situation that could appear in an ARMv5 model, where register 15 is the PC, and some instructions must be cancelled if this value appears in their register operands. In ArchC, to annul the current executing instruction, the ac_annul() method (line 8) must be called inside any of the behavior methods the instruction executes. This method in Fig. 11.3 would be executed by any instruction of the Type_DPI2 format fetched from the program memory.

Finally, Fig. 11.5 includes an example of operations that typically appear in specific instruction behaviors. Again, it could be part of an add instruction behavior for an ARMv5 description, where add instructions may be used for updating the PC register (ac_pc). Notice that users can define their own types, local and global variables, and even methods to be used inside ArchC behavior descriptions, just like in any regular C++ code.

Verification of processor models is no easy task. ArchC developers adopted a strict roadmap that uses version numbers to state the maturity level of each model. Table 11.1 presents this roadmap. Four benchmarks suites are involved in the process. The first to be applied is called AC_STONE, which is composed of dozens of small C programs conceived to stress different classes of instructions each: adds, subtractions, multiplies, logical, etc. These are small programs created by ArchC developers that do not need external libraries or operating system calls to run, making them ideal to check models on early development stages.

The next three benchmark suites, Mediabench, Mibench, and SPEC 2000, are composed of real-world programs from several different application areas like multimedia, networking, image processing, automotive, telecom, etc. They are used to ensure a good stability level to each model by running a huge amount of instructions with real data samples as input. For each model available on the ArchC website

```
1  void ac_behavior( Type_DPI2 ) {
2
3    int rs40;
4    reg_t RS2, RM2;
5
6    // IF PC is used as operand  ... instruction is invalid
7    if ((rd == 15)||(rm == 15)||(rn == 15)||(rs == 15)) {
8      ac_annul();
9    }
```

FIGURE 11.4

Format behavior.

```
1   void ac_behavior( add ) {
2
3     reg_t RD2, RN2;
4     r64bit_t result;
5
6     RN2.entire = RB.read(rn);
7     if(rn == PC) RN2.entire += 4;
8     result.hilo = (RN2.entire + dpi_shiftop.entire);
9     RD2.entire = result.reg[0];
10    RB.write(rd, RD2.entire);
11    if (rd == PC)
12        ac_pc = RB.read(PC);
13  }
```

FIGURE 11.5

Instruction behavior.

Table 11.1 ArchC model roadmap.

Version	Development Stage	Benchmark	Certifies That ...
0.0.x	Writing AC_ISA and AC_ARCH description	—	—
0.1.0	AC_ARCH and AC_ISA declarations finished	—	All instructions are correctly decoded
0.2.0	Instruction behavior description finished	—	Individual behaviors are working properly
0.3.0	AC_ARCH and AC_ISA completed	AC_STONE	All programs run successfully
0.4.0	ABI design finished	—	Individual system calls are working properly
0.5.0	Model description completed	MediaBench	All selected programs passed successfully
0.6.0	Testing	MiBench (small)	All selected programs passed successfully
0.7.0	Testing	MiBench (large)	All selected programs passed successfully
1.0.0	Final test	SPEC 2000	All selected programs passed successfully

(www.archc.org), outputs were compared to native executions of each program, with no the failure detected.

11.3 INTEGRATION THROUGH A TLM INTERFACE

This section presents the integration of SystemC components and ArchC simulators. This is accomplished by means of a Transaction Level Modeling (TLM) interface that may be included in an ArchC processor description since the ArchC 2.0 release.

SystemC allows hardware description into several abstraction levels, from the algorithmic to the Register Transfer Level (RTL). It is quite common for a platform model to be composed of modules described in different abstraction levels. In this case, the connection between two modules must be accomplished through a *wrapper*, which is a specialized module capable of converting the communication between two different abstraction levels. As stated earlier, ArchC adopts a TLM interface for external communication, but this does not mean that ArchC processor simulators may only be connected to other SystemC modules that implement the same interface, using exactly the same protocol for communication as described here.

Often, it is very simple to design wrappers to serve as adaptors between two TLM modules that do not share the same TLM interface and protocol. The TLM communication is based on function calls and packet transmissions. Interfaces define the functions available for transmission and the protocol specifies how packets must be built. Suppose that we have to connect two modules, one implementing an interface called intf_A, and the other intf_B. We could design a wrapper, which is yet another SystemC module that implements, through inheritance, both intf_A and intf_B. Moreover, the wrapper can have a process to perform the translation between packets used in both interfaces, which usually is a simple copy of several data fields.

It is also possible to design wrappers to connect TLM modules to RTL modules, although it is necessary to build a more sophisticated wrapper, capable of converting RTL signals into TLM packets and, probably, of communicating both through blocking and nonblocking interfaces. More details on TLM modeling and implementation can be found at [15, 16].

Fig. 11.6 shows how to include a TLM communication port in the ArchC architecture resource description presented in Section 11.2. Notice that where we had a memory declaration before, we now have an `ac_tlm_port` declaration. In this case, this port will be used for connecting the processor to an external memory model, and the address space accessible through this port is from address 0x0 to 256 megabytes. From inside behavior methods, nothing changes, and the designer will use the TLM port through `read` and `write` methods, just like it is done with storage devices.

When ArchC generates a simulator for the description in Fig. 11.6, it automatically includes an external port named `MEM_port`, which is bound to an external

```
1  AC_ARCH(armv5e){
2
3    ac_tlm_port   MEM:256M;   <== TLM port declaration
4    ac_regbank RB:31;
5    ac_wordsize 32;
6
7    ARCH_CTOR(armv5e) {
8
9      ac_isa("armv5e_isa.ac");
10     set_endian("little");
11
12   };
13  };
```

FIGURE 11.6

ARM functional architecture resource description.

module using the regular SystemC constructions for port binding. This is usually done just after modules instantiation in the main file of a system model description. The processor simulator will use this MEM_port to communicate with the external modules, and the read and write methods mentioned earlier will actually trigger calls to the methods in the TLM interface implemented by the communication port.

11.3.1 ArchC TLM Interfaces and Protocol

ArchC TLM was developed so that designers can integrate ArchC -generated simulator modules to their PV (Programmer's View) system-level SystemC models. The PV models are suitable for the function call communication style imposed by the TLM standard. In particular, the interface chosen for ArchC TLM was the tlm_transport_if. It is part of the SystemC standard since its 1.0 release, and provides useful features for PV system models, like its bidirectional nature and the tightly coupled requests and responses, exactly like in a method call, thus requiring a small design effort from the users.

The ArchC TLM protocol consists of request and response packet definitions, which can be easily understood even from the source files in the library. In summary, they are structures whose fields are as follows:

The request packet:

type : Type of the transaction. It must contain one of five values: READ, WRITE, LOCK (for locking a device or bus), UNLOCK, and REQUEST_COUNT (for debugging reasons, requests a count of the transaction packets).

dev_id : Device ID. Every ArchC TLM initiator port gets automatically assigned a device ID, which provides useful information for multiport devices like buses. More on initiator ports ahead.

addr : Address.
data : Data sent.

The response packet:

status : Transaction status: ERROR or SUCCESS.
req_type : Type of the transaction.
data : Data received.

11.3.2 TLM Interrupt Port

ArchC 2.0 also enables communication initiated from an external module, with the processor module working as a slave, via its interrupt mechanism. To enable an interrupt mechanism in your model, you have first to declare an ac_tlm_intr_port in the *project_name*.ac file. As an example, you can add an interrupt port to the mips1 model below by just doing the following:

```
AC_ARCH(mips1){

  ac_cache    DM:5M;
  ac_regbank RB:32;

  ac_tlm_intr_port inta; // Add this line
//...
```

Now we have an interrupt port on our mips1 model, called inta. By running the acsim tool on it, we will see that it generates some extra files related to interrupts. The one we modify is *project_name*_intr_handlers.cpp. First, we copy *project_name*_intr_handlers.cpp.tmpl to *project_name*_intr_handlers.cpp. Then we edit the file.

When you open the file, you will see that it contains an ac_behavior definition, pretty much as in the *project_name*_isa.cpp file:

```
// Interrupt handler behavior for interrupt port inta.
void ac_behavior(inta, value) {
}
```

Inside this behavior method, you can write any standard behavior code, altering register values, the program counter, writing or reading the memory, etc. Basically, this behavior code represents whatever the processor hardware does when it encounters an interrupt. Note that it receives a value from the interrupt, which can be used in the interrupt handling, and it is your choice whether to use it or not. Another possibility is to declare more than one ac_tlm_intr_ports in the *project_name*.ac file, each of them having its separate interrupt handler method defined in the *project_name*_intr_handlers.cpp file. There are lots of options to implement interrupt behavior in ArchC models. The ArchC interrupt system is very flexible, as it does not hinder the developers' options.

Now that we know how interrupts work inside the processor, what we have to understand is how an external module interrupts the processor. First, whenever you declare an ac_tlm_intr_port in your *project_name*.ac file you will have an ac_tlm_intr_port as a member of your processor module.

This `ac_tlm_intr_port` has the exact same name you declared in project_name.ac, which in this case is inta. The ac_tlm_intr_port *is-a* (inherits from) sc_export<ac_tlm_transport_if>, and this export is bound to itself, because ac_tlm_intr_port also implements the transport() method mentioned earlier, and this method has a fixed implementation, which calls an interrupt handler that is bound to a port (those are the ones you implemented on the *project_name*_intr_handlers.cpp, one for each port).

Finally, to provide an external module with the ability to interrupt a processor, all you have to do is binding an sc_port<ac_tlm_transport_if> of this module to the ac_tlm_intr_port of the processor, as follows:

```
ext_module.out_port(proc_module.inta);
```

and then make the external module call the transport() method via this port, passing an ac_tlm_req transaction request packet. Any value available on the data field of the packet will be passed as the "value" parameter of the interrupt handler. Ideally, direct usage of the transport() function should be delegated to a user–layer method (with an appropriate name like interrupt()). This user–layer method is the one invoked by the module's behavior or process methods.

11.3.3 A Word on ArchC Simulators

ArchC may generate simulators using two different techniques: interpreted or compiled. The interpreted simulators use SystemC as the simulations engine. So, if you plan to use those kinds of simulations it is mandatory to have a working SystemC installation in your machine. On the other hand, the compiled simulators are pure C++ programs, and do not demand a working SystemC installation to run.

Interpreted simulators provide a slower performance when compared to compiled simulators. That is because interpreted simulators have to fetch, decode, and execute each instruction in the program. They normally use a cache of decoded instructions to avoid repeating the process to an instruction that was already decoded in the same execution of the simulation. Decoding instructions on the fly is still a complex work to be done.

Compiled simulators statically compute operations that are normally performed dynamically by the interpreted technique. Mainly, they move instruction decoding to compile time. This is accomplished by providing not only a processor architecture description, but also the target application, to the simulator generator. In short, we get a simulator of a given processor that runs a specific given application. In order to change the application, the simulator must be regenerated. By mixing the code of the simulator with the application, it is possible to apply several optimizations and achieve a much faster simulator.

As depicted in Fig. 11.7, the simulator generation flow is pretty straightforward. The designer provides both the AC_ISA and the AC_ARCH description files to the simulator generation tool and gets a source code for the simulator. This source code can be compiled with standard C++ compilers, like GCC, and an executable specification of the processor will be ready for experimentation. The difference between interpreted and compiled simulators is that the second one also requires the target application code, represented by the dashed box in Fig. 11.7. ArchC interpreted simulator generator is called acsim and the compiled simulator generator is called accsim.

Let us take a look at an example to illustrate the kind of optimizations ArchC applies to its compiled simulators. One of the tasks a compiled simulator does is to check the modification in the PC the application will perform in order to execute the next step in the simulation. A possible optimization requires the identification of instructions that can change the execution flow. This can be easily provided by the designer in the ISA specification on the ArchC processor model. Using this information, ArchC creates a simulator that tests for changes in the PC, only after those instructions that have the ability to change it. As the simulation works at instruction-level, the delay slot feature of some architectures has to be treated carefully. The program flow can change not after the control instruction, but only after the instruction in the delay slot. For this reason, the size of the delay slot must also be given in the ISA description. Fig. 11.8 shows how the extra information should be included on the ac_isa description. We have two instructions, call and be, that may change the execution flow. The designer tells that through the jump() (line 4) and branch() (line 10) methods. Delay slots are informed through the delay() method (lines 5 and 11). More optimizations and details on compiled simulation in ArchC are available in [8, 17].

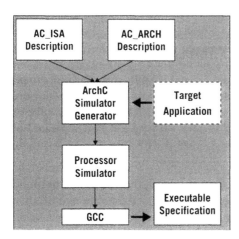

FIGURE 11.7

ArchC simulator generation flow.

```
1   . .
2   call.set asm("call %disp30");
3   call.set decoder(op=0x01);
4   call.jump();
5   call.delay(1);
6
7   be.set asm("be %an, label");
8   be.set decoder(op=0x00, cond=0x01, op2=0x02);
9   be.branch();
10  be.delay(1);
11  ...
```

FIGURE 11.8

Optimization for compiled simulation.

Nowadays, ArchC interpreted simulators are capable of running up to 40 MIPS (million instructions/sec) while compiled simulators may reach up to 200 MIPS. Which simulator to pick is a choice that depends on your goals. If integration into a SystemC platform model is the final goal for the simulator, the interpreted simulator is most likely your best choice. The main reason is the capability of automatically instrumenting the simulator with the TLM communication port. Till now, only interpreted simulators provided this feature. It should be possible to integrate the compiled simulator to other SystemC components, since they are written in C++, but this would require some coding effort from the designer. Will the difference in performance pay for this effort? Probably not, because the simulator performance will be limited by the SystemC simulation kernel, which will bring it down to the interpreted performance level. It will get worse if the SystemC TLM library is also applied.

11.4 A MULTICORE PLATFORM EXAMPLE

This section shows in more detail how to build a SystemC multicore platform model with interrupt support. ArchC simulators are used as components, following the methodology described in Section 11.3. Interrupts are used to make it possible to boot the processors from the same code, do some startup stuff, and separate the execution flow to two different functions. Fig. 11.9 shows the diagram of our sample platform, which is composed of:

Processor: Two MIPS processors that will start executing the same code from the same address. These processors are freely available from the ArchC site [2].

Interrupt Controller: In this example, the interrupt controller is used to wake up the processors after they execute their basic boot sequence.

System Bus: Designed to make a fast connection between the processors and the peripherals.

Memory: A simple 5 Mb memory.

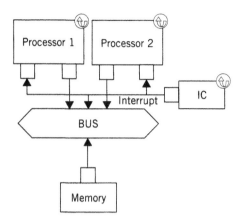

FIGURE 11.9

The multicore platform example with interrupt support.

All the components described here were described in SystemC with TLM support.

The original ArchC MIPS model was modified to include the two TLM ports: the output TLM port (called DM in this example) and the interrupt TLM port (called intp). These are two of the three required changes in the MIPS model (the third one is described later and in Fig. 11.12). Moreover, switching to another processor architecture is just a matter of downloading the model from the ArchC site and including the corresponding lines, as in the following example:

```
1   ac_tlm_port DM:5M;
2   ...
3   ac_tlm_intr_port intp;
```

The platform is described in plain SystemC using the ArchC generated processor model as any other SystemC module. Lines 9 and 10 of Fig. 11.10 show how to declare the two processors used in this platform. This way, building a heterogeneous multicore platform is as simple as declaring two different SystemC modules. Following the example, the memory, bus, and interrupt controller are declared. Line 16 binds the first processor to the bus. The interrupt controller is bound to the processor in line 22. These two lines show the two binding sequences between processors and the outside world.

Each processor module must be initialized for proper execution. Line 25 shows how to initialize Processor 1. It is also easy to load a binary code to the memory by using the ArchC loader, as in line 26. The ArchC loader is capable of recognizing an ELF binary file, which is generated directly from the gcc compiler. The ELF format specifies the memory position to load the file and also the execution start address. The ArchC loader relies on the start address to automatically set the first instruction to be executed (the start PC). The second load statement 30 is only used to correctly

```
1   #include <systemc.h>
2   #include "mips1.H"
3   #include "fastBus.h"
4   #include "memory.h"
5   #include "intr_controller.h"
6
7   int sc_main(int ac, char *av[])
8   {
9     mips1 mips1Proc1("mips1_1");
10    mips1 mips1Proc2("mips1_2");
11
12    memory mem("memory", 0x0500000);
13    fastBus bus("bus");
14    intr_controller intController("Interrupt_Controller");
15
16    mips1Proc1.DM_port(bus.target_export);
17    mips1Proc2.DM_port(bus.target_export);
18
19    bus.Memory_port(mem);
20    bus.IntController_port(intController);
21
22    intController.proc1_port(mips1Proc1.intp);
23    intController.proc2_port(mips1Proc2.intp);
24
25    mips1Proc1.init();
26    mips1Proc1.load("boot.x");
27    mips1Proc1.set_instr_batch_size(1);
28
29    mips1Proc2.init();
30    mips1Proc2.load("boot.x");
31    mips1Proc2.set_instr_batch_size(1);
32
33    sc_start();
34
35    mips1Proc1.PrintStat();
36    mips1Proc2.PrintStat();
37
38    return mips1Proc1.ac_exit_status;
39  }
```

FIGURE 11.10

Multicore platform main SystemC file.

set the PC of the second processor to the same address of the first, since the binary programs are the same.

When dealing with platform simulation, the user can specify how many instructions the ArchC simulator will execute before giving the control back to the SystemC kernel. The ArchC set_instr_batch_size(n) processors method is used to specify that the simulator must call the SystemC wait function once every (n) instructions. Line 27 shows the example requesting the processor to give the control back to the SystemC simulation kernel after every instruction. The default ArchC value is 200; that is a good compromise between speed and platform responsiveness for a single-core architecture. If you are designing a multicore architecture with rigorous communication steps, you should consider reducing this value, as in the example here.

One can easily notice that both processors were configured exactly in the same way. So, to boot our platform, we will need an external component to split the execution flow. This is to be done by the interrupt controller (IC). Both processors start from the same address in the same code and enter in loop, waiting for the interrupt.

In our example shown in Fig. 11.11, the interrupt routine is programmed in the IC by writing the interrupt handler address in its base address (0x0800000, as shown in line 13). Both processors print the greetings message from line 12 and follow to an infinite loop in line 14. After some cycles, the interrupt controller sends the first interrupt to the first processor, waking it up and moving its execution flow to the boot function. The first processor prints a message (line 20) and continues its execution. The interrupt controller is designed to include a small delay between the generated interrupts, so that the first processor will be able to increment the processor_number variable without incurring any concurrence problem. By using this variable as a control flag, each processor can easily start executing a different code as expected (line 21).

Inside ArchC , after the interrupt port is declared, it is also necessary to declare the interrupt behavior, using the keyword ac_behavior with the first parameter being the interrupt port (intp in our example) and the second parameter the value passed by the interrupt. This value can be used to pass more information from the interrupt controller to the processor. In this example, in Fig. 11.12, the value passed is the program counter (PC) of the instruction to be executed. This method complements the source code presented before (Fig. 11.11) where the interrupt handler routine was located. Line 8 shows the received parameter being assigned to the PC register.

Going further with this example, the user can easily change the number of processors by instantiate them in Fig. 11.10, including different processors if needed. Line 21 of Fig. 11.11 also provides a placeholder for the specific code for each processor, so no different binary is necessary for each program. If the platform designer wants to use several different pieces of code without mixing them together, or use heterogeneous processors, line 26 of Fig. 11.10 provides the placeholder for each processor does load its own binary code. It is still possible to externally load the binary code to memory and only ask each processor to boot to different

```
1  #include <stdio.h>
2  #include <stdlib.h>
3
4  #define INT_CONTROLLER 0x0800000;
5  volatile int processor_number = 0;
6  int boot();
7
8  int main(int argc, char *argv)
9  {
10   volatile int *int_controller_cfg = (int *) INT_CONTROLLER;
11
12   puts("Basic booting sequence for both processor.");
13   *int_controller_cfg = (int) boot;
14   while(1) {};
15   return 0;
16  }
17
18  int boot()
19  {
20   printf("Continuing to boot processor %d ...\n", processor_number++);
21   switch (processor_number) {
22    ...;
23   }
24  }
```

FIGURE 11.11

Boot code for the dual MIPS example.

```
1  #include "ac_intr_handler.H"
2  #include "mips1_intr_handlers.H"
3
4  #include "mips1_ih_bhv_macros.H"
5
6  // Interrupt handler behavior for interrupt port intp.
7  void ac_behavior(intp, value) {
8    PC_register = value;
9  }
```

FIGURE 11.12

ArchC TLM interrupt handling function.

addresses by using the set_ac_pc method of the processor module. In this way, ArchC modules can be used to easily match the requirements of your platform.

11.5 CONCLUSIONS

In this chapter, we discussed the main aspects of the designing processor models, including multicore platform prototypes, with a SystemC-based ADL called ArchC. The simulator generation tools and several processor models which already reached a good stability level are available on the ArchC website (*www.archc.org*) for free download. Readers will also find some platform examples, that are called the ArchC Reference Platforms, and software tools like binary utilities retargeting, compilers, and benchmarks.

Besides their well-known suitability in the industry, as an important ESL tool for designing and experimenting with new architectures, ADLs can also be very useful as a research and education tool. On one hand, models of well-known architectures are appropriate to a first contact with complex ISAs used in modern general purpose and embedded processors. Those models may become an important asset on computer architecture courses, specially in those ones involving laboratory experiments. On the other hand, researchers can use ADLs to model modern architectures and experiment with their ISA and structure with all the flexibility demanded in research projects. Moreover, processor simulators are key components for prototypes in projects related to SoC design, network-on-chip (NoC), hardware/software co-design, functional verification, The TLM methodologies, and almost any ESL related research topic. With the increasing popularity of SystemC as the standard language for system-level design and TLM development, we have already seen ArchC processor models being successfully applied in research projects around the world in many of these areas. In particular, platform prototype models for game consoles, cell phones, embedded IPs like GSM, JPEG, and MPEG decoding, as well as for research on transactional memories [18], verification methodologies [19], and mixed language co-simulation frameworks [20] have already been demonstrated.

REFERENCES

[1] B. Bailey, G. Martin, and A. Piziali. *ESL Design and Verification: a prescription for electronic system-level methodology*. Morgan Kaufmann series on systems on silicon.san Francisco, USA, 2007.

[2] http://www.archc.org. The ArchC Resource Center, 2007.

[3] http://www.systemc.org. SystemC homepage, 2007.

[4] T. Grotker, S. Liao, G. Martin, and S. Swan. *System Design with SystemC*. Kluwer Academic Publishers, Boston, May 2002.

[5] D. C. Black, J. Donovan, B. Bunton, and A. Keist. *SystemC: From the Ground Up*. Kluwer Academic Publishers, Boston, 244 pages, 2004.

[6] C. Lee, M. Potkonjak, and W. H. Mangione-Smith. MediaBench: A Tool for Evaluating and Synthesizing Multimedia and Communications Systems. In *Proc. of the 30th Annual International Symposium on Microarchitecture (Micro-30)*, pages 330-335, December 1997.

[7] Guthaus, M., Ringenberg, M., Ernst, D., Austin, T., Mudge, T. and Brown, R. MiBench: A Free, Commercially Representative Embedded Benchmark Suite. In *Proc. of the* IEEE 4th Annual Workshop on workload characterization, Pages 3-14, December 2001.

[8] R. Azevedo, S. Rigo, M. Bartholomeu, G. Araujo, C. Araujo, and E. Barros. The ArchC Architecture Description Language. *International Journal of Parallel Programming*, 33(5): 453-484, October 2005.

[9] S. Rigo, G. Araujo, M. Bartholomeu, and R. Azevedo. ArchC: A SystemC-Based Architecture Description Language. In *Proc. of the 16th Symposium on Computer Architecture and High Performance Computing*, pages 66-73, October 2004.

[10] T. A. Team. *The ArchC Architecture Description Language Reference Manual*. Computer Systems Laboratory (LSC)—Institute of Computing, University of Campinas, *http://www.archc.org*, 2007.

[11] R. H. Pesch and J. M. Osier. *The GNU binary utilities*. Free Software Foundation, Inc., May 1993. Version 2.15.

[12] A. Baldassin, P. Centoducatte, and S. Rigo. Extending the ArchC Language for Automatic Generation of Assemblers. In *Proc. of the 17th International Symposium on Computer Architecture and High Performance Computing*, pages 60-67, October 2005.

[13] T. A. Team. *The ArchC Assembler Generator Manual*. Computer Systems Laboratory (LSC)— Institute of Computing, University of Campinas, *http://www.archc.org*, 2005.

[14] A. Baldassin, P. Centoducatte, S. Rigo, D. Casarotto, L. Santos, M. Schultz, and O. Furtado. Automatic Retargeting of Binary Utilities for Embedded Code Generation. In *Proc. of the IEEE Computer Society Annual Symposium on VLSI (ISVLSI '07)*, pages 253-258, May 2007.

[15] Frank Ghenassia, editor. *Transaction-Level Modeling with SystemC*. Springer, 2005.

[16] SystemC TLM Working Group. *SystemC Transaction-Level Modeling Standard Version 1.0*. *http://www.systemc.org*, 2005.

[17] M. Bartholomeu, R. Azevedo, S. Rigo, and G. Araujo. Optimizations for Compiled Simulation Using Instruction Type Information. In *Proc. of the 16th Symposium on Computer Architecture and High Performance Computing (SBAC '04)*, pages 74-81, October 2004.

[18] F. Kronbauer, A. Baldassin, B. Albertini, P. Centoducatte, S. Rigo, G. Araujo, and R. Azevedo. A Flexible Platform for Rapid Transactional Memory Systems Prototyping and Evaluation. In *Proc. of the 18th IEEE/IFIP International Workshop on Rapid System Prototyping (RSP '07)*, 2007.

[19] B. Albertini, S. Rigo, G. Araujo, C. Araujo, E. Barros, and W. Azevedo. A Computational Reflection Mechanism to Support Platform Debugging in SystemC. In *Proc. of the International Symposium on System Synthesis (CODES+ISSS)*, pages 81-86, October 2007.

[20] R. Maciel, B. Albertini, S. Rigo, G. Araujo, and R. Azevedo. A Methodology and Toolset to Enable SystemC and VHDL Co-Simulation. In *Proc. of the IEEE Annual Symposium on VLSI Design (ISVLSI)*, pages 351-356, May 2007.

MAML: An ADL for Designing Single and Multiprocessor Architectures

12

Alexey Kupriyanov, Frank Hannig, Dmitrij Kissler, and Jürgen Teich

MAchine **M**arkup **L**anguage (MAML) is an architecture description language for modeling, simulation, and architecture/compiler co-exploration of domain-specific processor architectures that are designed for special purpose applications. Since design time and cost are critical aspects during the design of processor architectures, it is important to provide designers with efficient modeling, simulation, and compiler techniques in order to evaluate architecture prototypes before actually building them.

The MAML is based on XML, which allows one to characterize the resources of a complex processor architecture at both structural and behavioral levels in a convenient manner. The MAML has its roots in designing application-specific instruction set processors (ASIPs)[1–3]. Fig. 12.1(a) gives an example of such a single-processor architecture.

Today, the technological progress allows implementations of hundreds of microprocessors and more on a single die. Therefore, massively parallel data processing has become possible in portable and other embedded systems where high power consumption is often a luxury. An example of a generic parallel multiprocessor architecture is shown in Fig. 12.1(b). Here, the array contains 48 interconnected *processing elements* (PE). The input and processed data can be transmitted through the input/output ports placed at the borders of the array. The MAML has recently been extended for the characterization of such multiprocessor architectures [4, 5]. The parameters extracted from MAML can be used on the one hand to generate fast interactive cycle-accurate *simulators*, and on the other for *compiler retargeting*. Moreover, these parameters can also be used for *architecture/compiler co-exploration* of a given architecture. Finally, a processor architecture described in MAML and accompanied by hardware libraries can be automatically synthesized

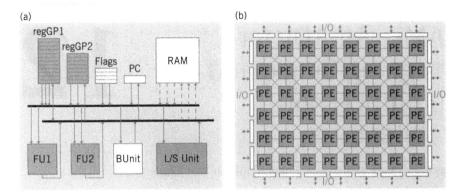

FIGURE 12.1

In (a), Example of a single processor architecture [3], in (b), Example of a multiprocessor architecture.

for the purpose of *rapid prototyping*. The benefits of MAML can be summarized as follows:

(a) Capability to characterize parallel processor architectures,
(b) Efficiency of MAML (code size) due to clustered domain-based specification of processors and interconnect,
(c) XML format of MAML provides automatic validation of the syntax and is easily extendable,
(d) The MAML is suitable for cycle-accurate simulator generation, library-based synthesis, and automated compiler generation.

This chapter is organized as follows. Section 12.1 describes the history of MAML. Sections 12.2 and 12.3 describe the various features (syntax and semantics) of the ADL for single and multiprocessor architectures using illustrative examples, and a case study. Section 12.4 gives an overview of various approaches driven by MAML. Finally, Section 12.5 concludes this chapter and offers an outlook on future work.

12.1 HISTORY OF MAML

The MAML was developed within the scope of the BUILDABONG framework (building special computer architectures based on architecture and compiler co-generation) [1–3] in 2002 at the University of Paderborn and developed further at the University of Erlangen–Nuremberg. The framework for the design of ASIPs aims at a semiautomated exploration of optimal architecture/compiler co-designs. Fig. 12.2 from [2] shows the overall software architecture of the BUILDABONG framework.

In the first step, a processor's control- and data path is entered graphically using a library of customizable building blocks such as register files, memories, arithmetic

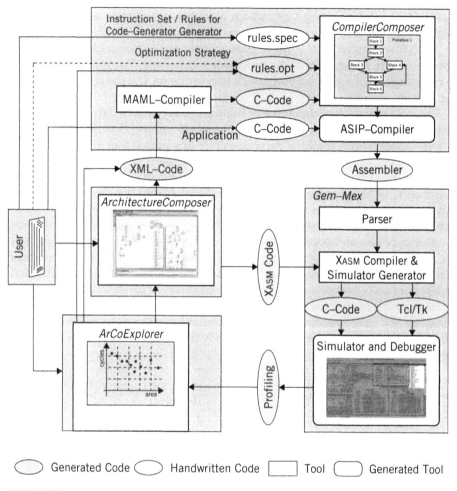

FIGURE 12.2

Overview of the BUILDABONG project [2].

and logic units, buses, and others. This editor called *ArchitectureComposer* is able to generate synthesizable VHDL- and Verilog design models for purpose of rapid prototyping. Further, it is able to generate a cycle-accurate model of the register-transfer architecture using the formalism of Gurevich's Abstract State Machines (ASMs) [6], a mathematical model of computation based on first-order logic formulas.

Based on such an ASM model, a debugging and simulation environment is automatically generated. In order to achieve a high-speed cycle-accurate simulation of large processor architectures at register transfer level (RTL), *RASIM* [7, 8], an efficient compiled simulator generator can be used.

The input of the simulator consists of the application program which has been translated by the ASIP-compiler into target-specific assembler code. The compiler is driven by a script-file (rules.opt), which describes the number and frequency of preferred compiler optimizations and transformations [1].

The ASIP-compiler itself is automatically generated by the tool *CompilerComposer* using the parameters of the processor architecture. Therefore, the user has to enter a specification of the processor's instruction set and the grammar rules for the code-generator generate.

The machine model of the compiler is automatically extracted from the graphical architecture description by the tool *ArchitectureComposer*. Here, MAML is used as an interchange format between the architecture description tool *ArchitectureComposer* and compiler generation tool *CompilerComposer* [3].

Finally, the simulator generates a profiling protocol containing activity and timing information of the simulation run. The exploration tool *ArCoExplorer* analyzes this protocol and modifies the architecture/compiler parameters with the goal to explore a set of optimal (in terms of hardware cost, execution time, code size, etc.) designs. The architecture parameters extracted from the MAML description are required by the compiler back-end modules such as functional unit assignment, scheduling, and register assignment. The reason for this interface between the RTL architecture and *CompilerComposer* is the complexity of the description: the back-end is compiled (retargeted) automatically for each architectural change. Before that, a number of C code header files describing the architecture and a number of C-files enabling the access of the back-end modules to the data structures described in the header files are generated.

The BUILDABONG framework performs a semiautomated architecture/compiler co-exploration of single processor architectures as, for example, the *SimpleProc* architecture [3] shown in Fig. 12.1(a). The main constructs of MAML capable to describe such architectures are given in Section 12.2.

12.2 DESCRIPTION OF SINGLE PROCESSOR ARCHITECTURES

An MAML description contains a clearly arranged list of the architecture's *resources* (functional units, pipeline stages, register files, etc.), *operation sets* (binding possibilities of operations to functional units, operand directions, etc.), *communication structures* (buses, ports), and *timing behavior* (latency of operations, behavior of multicycle operations). In the following, the basic MAML elements for single processor architectures are described.

12.2.1 Syntax

The main constructs used within MAML are listed in Fig. 12.3. An MAML notation consists of two types of XML—elements distinguished by their number of appearance (unique or multiple) and attributes optionally associated to the elements. Moreover, the elements can be semantically distinguished by defining the *structure* and the *behavior* of a processor architecture. The structure is defined by

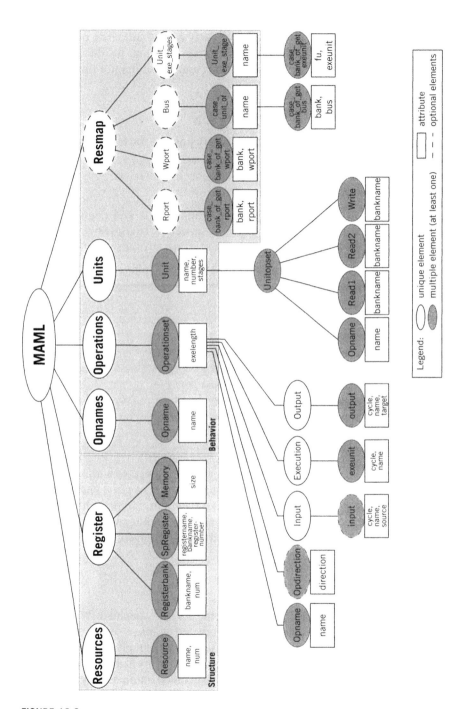

FIGURE 12.3

Structure of MAML for single processor architectures.

the specification of communication resources, such as read and write ports, buses, register files, and functional units. The behavior is described by the instruction set, operations, their latency, multicycle behavior, and its binding of operations to functional units.

Communication resources and storage elements

The structure of a single processor architecture is defined by the elements Resources, and Register within MAML.

Resources

The element provides a definition of communication ports and communication units, for instance, buses. Each resource is parameterized by two attributes: its name and number of instances. For bus resources, the number attribute specifies the bus width.

Example 12.1

Definition of internal communication resources (buses).

```
<Resources>
    <Resource name = "wPort1" num = "1">
    <Resource name = "wPort2" num = "1">
    <Resource name = "rPort1" num = "3">
    <Resource name = "rPort2" num = "3">
    <Resource name = "bus1" num = "4">
</Resources>
```

Example 12.1 shows a specification of a processor architecture with one bus and two assumed register files. Each of them contains three read ports and one write port. Here, only the available communication resources of the architecture are listed. The specified width of the bus (num = "4") defines an architecture with possible fourfold parallel data transfer. The exact port mapping is given in the behavioral section of MAML.

Register

With this element, both register files and memories are specified. Whereas the memory specification only consists of the annotated number of bytes (memory capacity), registers have to belong to a register bank (a set of registers) according to the registers purpose. A register bank has to be specified by its name and its number. In general, the name and number of registers is specified automatically. In case of special registers the user has to name the respective register and to assign it to a dedicated position inside the register bank. In the following example, the special register sp is located at position 0 in register bank regB.

Example 12.2

Specification of storage elements.

```
<Register>
    <Registerbank bankname="regA" num="8"/>
    <Registerbank bankname="regB" num="4"/>
    <SpRegister    registername="sp"
                   bankname="regB"
                   registernumber="0"/>
    <Memory        size="4096"/>
</Register>
```

Example 12.2 shows a section of MAML which describes the storage elements of a particular architecture. This architecture contains two register banks regA, regB, and a memory of size of 4096 bytes. Register bank regA contains eight registers whereas register bank regB contains only four registers. Moreover, the first register sp in register bank regB is a special register (i.e., stack pointer). However, at this step the behavioral characteristics of it are still undefined. This is generally done in the corresponding section of MAML which is responsible for the specification of the architecture's instruction set and is described in the next example.

Functional units and operations

The behavior of a single processor architecture is basically defined by the behavior of its functional units and the corresponding operations executed by them. That is, the description of resource usage, pipeline stages, latencies, delay slots, operand directions, and other properties of multicycle operations are provided. The elements Opnames, Resmap, Operations, and Units describe the behavior of the architecture within MAML. The auxiliary element Resmap is used to define the particular assignment rules of read and write ports to the register banks.

An illustrative example of a simple very long instruction word (VLIW) processor architecture is shown in Fig. 12.6. The corresponding MAML specification is listed in Fig. 12.5.

Opnames

This element provides a listing of the architecture's instructions.

Resmap

The Resmap element maps read and write ports to register banks, describes connections of functional units to storage elements via buses or dedicated direct links, and defines the pipeline stages. For instance, the assignment of the ports is performed using the subelement Rport or Wport. The appropriate attributes define the name of a register bank and corresponding read or write ports. The port assignments specified here may be used in a behavioral definition of the operations by the entry of a keyword get_rport, get_wport, or get_bus.

Operations

This element provides the description of operation-specific parameters. Here, latency, the operand's communication directions (in, out, inout), and the resource requirements for each operation are specified. Operations with the same resource requirements, latency, and communication directions form an operation set. Elements of such an Operationset are handled identically. Three operation phases are distinguished: input, execution, and output phase. Each element of a phase has to be duplicated according to the operation's number of source operands, target operands, and stages of the execution unit. The flag cycle indicates whether a new cycle begins or the current instruction phase runs in parallel to the previous phase in the same cycle. Each input and output element is either assigned to a fixed communication unit and a read and write port, or the communication path can be constructed dynamically according to respective resources allocated by the functions get_rport, get_wport, or get_bus. The possible results of the function are determined by the element Resmap. The length of the execution phase (number of cycles) is determined by the exelength attribute.

The value of the integer attribute cycle starts with cycle = "0". If the fetching of data requires extra cycles, the first execution cycle in element exeunit is greater than one. Also, the storage of the result at the target can require one or more additional cycles. The following example will explain the notations of resource usage in case of multicycle operations.

Example 12.3

Description of resource usage in case of multicycle operations.

```
<Operationset exelength="3">
    <Opname name="mul" />
    <Opdirection direction="out" />
    <Opdirection direction="in" />
    <Opdirection direction="in" />
    <Input>
        <input cycle="0" name="rPort1" source="1" />
        <input cycle="0" name="rPort2" source="2" />
    </Input>
    <Execution>
        <exeunit cycle="1" name="exe1\_MULT" />
        <exeunit cycle="2" name="exe2\_MULT" />
    </Execution>
    <Output>
```

```
                <output cycle="4" name="wPort1" target="1" />
         </Output>
    </Operationset>
```

Here, an operation set with one multiplication operation mul is described. The complete execution of it is performed in five cycles whereof, the length of the execution phase is three cycles (exelength=3). The operation mul operates with two input operands and assigns the result to one output. Fig. 12.4 shows the diagram of the resource usage in each cycle of this operation. During the operation fetch phase (cycle= "0"), input operands are read from the register bank through the appropriate read ports rPort1 and rPort2. Subsequently, the execution phase starts and takes three cycles. In the first cycle of the execution phase, the pipeline stage exe_1 of functional unit MULT is processed. In the next two cycles the execution of the pipeline stage exe_2_MULT is performed. In the last cycle (cycle = "4"), the result is written to the register bank through the appropriate write port wPort1.

Units: The section Units of MAML contains a set of Unit elements describing the different functional units of a processor. Here, the source and target operands of each operation are specified and the operations are bound to functional units. The attributes number and stages specify the number of functional units of the same type and the number of execution stages, respectively. For each type of functional units, a unique name must be defined. The operations Opname with the same source and target operands (Read1, Read2, and Write attributes) can be grouped together in unit operations sets (Unitopset). In case no register bank is assigned to the operand, the keyword *norb* (no register bank) should be given as attribute value. The name of an execution unit contains the number of the unit instances and its stages. These names are required in the element Operationset for the description of the usage of each pipeline stage during the execution of an operation.

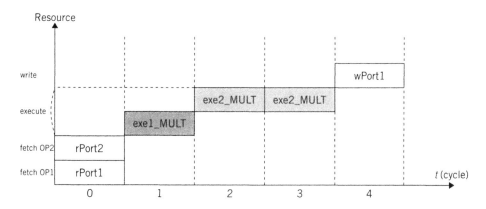

FIGURE 12.4

Resource usage in case of multicycle operations.

```
<maml>                                            |   <Output>
<Resources>                                       |    <output cycle="2" name="wPort2" target="1"/>
 <!-- Example 12.1 -->                            |   </Output>
    ...                                           |  </Operationset>
</Resources>                                      | </Operations>
<Register>                                        | <Units>
 <!-- Example 12.2 -->                            |  <!-- Example 12.4 -->
</Register>                                        | </Units>
<Opnames>                                          | <Resmap>
 <Opname        name="add"/>                       |  <Rport>
 <Opname        name="sub"/>                       |   <case_bank_of_get_rport bank="regA"
 <Opname        name="mul"/>                       |                          rport="rPort1"/>
</Opnames>                                          |   <case_bank_of_get_rport bank="regB"
<Operations>                                        |                          rport="rPort2"/>
 <Operationset   exelength="3">                     |  </Rport>
  <Opname        name="mul"/>                        |  <Bus>
  <!-- Example 12.3 -->                              |   <case_unit_of name="ADDER">
    ...                                              |    <case_bank_of_get_bus bank="regA" bus="bus1"/>
 </Operationset>                                      |    <case_bank_of_get_bus bank="regB" bus="bus1"/>
 <Operationset   exelength="1">                       |   </case_unit_of>
  <Opname        name="add"/>                          |  </Bus>
  <Opname        name="sub"/>                          |  <Unit_exe_stages>
                                                        |   <Unit_exe_stage name="exe1">
  <Opdirection direction="out"/>                        |    <case_unit_of_get_exeunit fu="MULT"
  <Opdirection direction="in"/>                         |                              exeunit="exe1_MULT"/>
  <Opdirection direction="in"/>                         |    <case_unit_of_get_exeunit fu="ADDER"
  <Input>                                               |                              exeunit="exe1_ADDER"/>
   <input cycle="0" name="get_rport" source="1"/>  |   </Unit_exe_stage>
   <input cycle="0" name="bus1" source="1"/>            |   <Unit_exe_stage name="exe2">
   <input cycle="0" name="get_rport" source="2"/>  |    <case_unit_of_get_exeunit fu="MULT"
   <input cycle="0" name="bus1" source="2"/>            |                              exeunit="exe2_MULT"/>
  </Input>                                              |   </Unit_exe_stage>
  <Execution>                                           |  </Unit_exe_stages>
   <exeunit cycle="1" name="exe1_ADDER"/>               | </Resmap>
  </Execution>                                          | </maml>
```

FIGURE 12.5

MAML specification of a simple VLIW processor architecture.

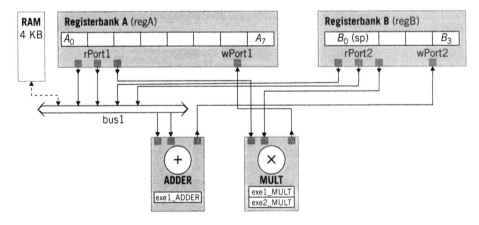

FIGURE 12.6

Structure of a simple VLIW processor architecture described by MAML in Fig. 12.5.

Example 12.4

Definition of functional units.

```
<Unit name="ADDER" number="1" stages="1">
   <Unitopset>
        <Opname name="add" />
        <Opname name="sub" />

        <Read1 bankname="regA" />
        <Read1 bankname="regB" />
        <Read2 bankname="regA" />
        <Read2 bankname="regB" />
        <Write bankname="regB" />
   </Unitopset>
</Unit>
<Unit name="MULT" number="1" stages="2">
   <Unitopset>
        <Opname name="mul" />

        <Read1 bankname="regA" />
        <Read2 bankname="regB" />
        <Write bankname="regA" />
   </Unitopset>
</Unit>
```

In Example 12.4, two functional units are described. The ADDER performs its defined operations in one execution stage, whereas the MULT operates in two stages. The ADDER can access both register banks regA and regB, as defined in Example 12.2, for reading the input operands. RegB is used for storing the result. MULT is constrained to read the first and second operands from register banks regA and regB, respectively, and can only store the result in register bank regA.

12.2.2 Example of a VLIW Processor Architecture

Fig. 12.5 finally shows a MAML description of a simple VLIW processor using the code sections already introduced in Examples 12.1 to 12.4. The appropriate structure of the described architecture is illustrated in Fig. 12.6. This simple VLIW architecture contains two register banks regA, regB, and a small memory, (cf. Example 12.2). The architecture possesses two functional units: the adder/-subtractor ADDER and the multiplier MULT. The complete execution of the listed operations on the adder (including fetching the source operands and writing the result to the target) is performed in three cycles whereas the multiplication operation is executed in five cycles (see Fig. 12.4).

12.3 DESCRIPTION OF MULTIPROCESSORS

Today, the steady technological progress in integration densities and modern nano-technology allows the implementation of hundreds of 32-bit microprocessors and more on a single die. Due to these advances, massively parallel data processing has become possible in portable and other embedded systems (SoC - System-on-a-Chip technology). Such devices can handle increasingly computationally-intensive algorithms like video processing or other digital signal processing tasks, but on the other hand, they are subject to strict limitations in their cost and/or power budget. These kind of applications can only be efficiently realized if design tools are able to identify the inherent parallelism of a given algorithm and if they are able to map it into correctly functional, reliable, and highly optimized systems with respect to cost, performance, and power dissipation. However, technical analysts foresee the dilemma of not being able to fully exploit next generation hardware complexity because of a lack of design tools. Hence, efficient simulators, parallelization techniques, and dedicated compilers will be of utmost importance in order to map computationally-intensive algorithms efficiently to these parallel architectures.

At all times, there was the exigence (demands in terms of speed, die size, cost, power, etc.) to develop dedicated, massively parallel hardware in terms of ASICs (Application-Specific Integrated Circuits). For instance, let us consider the area of image processing, where a cost-benefit analysis is of crucial importance: on a given input image, sequences of millions of similar operations on adjacent picture elements (e.g., filter algorithms, edge detection, Hough transformation) have to be computed within splits of a second. The deployment of general purpose parallel computers like MIMD or SIMD multiprocessor machines is not viable because such systems are too large and very expensive. Such machines are also of no use in the context of mobile environments, where additional criteria such as energy consumption, weight, and geometrical dimensions exclude solutions with (several) general purpose processors or even multicore architectures.

In order to allow the specification of domain-specific multiprocessor architectures, MAML for single processor architectures as described in Section 12.2 is used as a basis for the characterization of multiple processing elements (PE). In order to model an inter-processor network and behavior (e.g., geometry of an array, location of PEs, interconnect topology, I/O ports), appropriate extensions of MAML are provided [4, 5]. The architectural description of an entire multiprocessor system is thereby subdivided into two main abstraction levels: (1) *PE-level* characterizes the internal structure of each *type of processing element*, (2) *array-level* specifies system-level parameters such as the topology of the array, number, and location of processors and I/O-ports.

General constructs and the syntax for characterizing the processing elements are described in Section 12.2. This section describes various features and semantical extensions of MAML required for the characterization of multi-processor architectures. The section concludes with a case study of a tightly coupled processor array specified with MAML. But first, an overview of currently existing approaches and ADLs aiming at description of multiprocessor systems is given.

12.3.1 Related Work

Many architecture description languages have been developed in the field of retargetable compilation. However, only few of them partly provide a capability to characterize multiprocessor architectures. In the following paragraph, we list only some of the most significant ADLs that allow for the description of parallel architectural properties.

For instance, at the ACES laboratory of the University of California, Irvine, the architecture description language EXPRESSION (cf. Chapter 6) has been developed. The architecture can be defined as a collection of programmable SoC components including processor cores, coprocessors, and memories. From an EXPRESSION description of an architecture, the retargetable compiler Express and a cycle-accurate simulator can be automatically generated. The machine description language LISA [9] is the basis for a retargetable compiled simulator approach developed at RWTH Aachen, Germany (cf. Chapter 5). The project focuses on fast simulator generation for already existing architectures to be modeled in LISA. Current work in the domain of multicore system simulation [10, 11] enables a co-simulation of multiple processor cores with buses and peripheral modules described in SystemC. ArchC (cf. Chapter 11) is an open-source ADL based on SystemC. It is used to enable fast generation of microprocessor simulators. Multiple instances of the simulator can coexist in the same platform, thus enabling the evaluation of multiprocessor architectures and complex heterogeneous systems. ArchC also allows the automatic retargeting of GNU assemblers and linkers. The architecture description language MADL (cf. Chapter 9) was developed at Princeton University. The language supports a hardware and software co-design framework for multiprocessor embedded systems. The MADL is based on operation state machine computation model and provides the capability of specifying individual application-specific instruction-set processors. It also supports the generation of instruction-set and cycle-accurate simulators. Another architecture description language is [12] Philips Research Machine Description Language (PRMDL). The target architectures for PRMDL are clustered VLIW architectures. The PRMDL features explicitly separate software and hardware views on the processor architecture. The hardware view describes the processor state, the distribution of functional units among VLIW issue slots, the processor data paths, and the hardware operations. The description of the processor state can include diverse memory structures (e.g., stacks, queues, and random access register files). An explicit specification of the processor data paths allows one to describe clustered VLIW architectures with incomplete resource interconnects (e.g., partial bypass network).

The common property of the earlier listed ADLs is that they have been mainly developed for the design of single processor architectures such as ASIPs that might contain VLIW execution [13–15] and only partly allow for indirect description of multiprocessor architectures. To the best of our knowledge, there exists no ADL that conveniently and efficiently covers the architectural aspects of massively parallel processor arrays. Of course, one could use hardware description languages such as Verilog or VHDL, but the abstraction of these languages is too low, and offers only insufficient possibilities to describe behavioral aspects.

12.3.2 Description of Parallel Processing Elements

The internal structure of a processing element (PE) is described in a PE-level architecture specification section of MAML. Generally, parallel processor architectures may consist of hundreds of processing elements with various kinds of internal structure and behavior. For example, a particular processor array is designed to run a set of image processing algorithms such as filtering or edge detection. It is reasonable to design the processing elements located at the boundaries of such an architecture so that they contain enough memory in order to read the input data or to temporarily store the processed output image data. These PEs could use dedicated functional units that access internal memory, whereas, the PEs located in the middle of the array can exclusively implement arithmetical instructions performing computations.

In this respect, MAML defines the architectural properties of PEs within so called *PE-classes*. The properties of one class can be instantiated either on one PE or on a set of several PEs. MAML also allows the specification of multiple PE-classes to describe nonhomogeneous multiprocessor architectures. One PE-class can extend or implement another earlier defined PE-class. This feature enables the *inheritance* among PE-classes in the architecture specification and thus provides a high code efficiency. PE-classes are defined by the PEClass element. This element specifies the internal architecture of the PE or a set of the processor elements (PE-class) within a massively parallel processor architecture. It covers such architectural issues as: characterization of I/O ports (bitwidth, direction, control path or data path, etc.), internal resources (internal read and write ports, functional units, buses, etc.), storage elements (data or control registers, local memories, instruction memory, FIFOs, register files, etc.), resource mapping (interconnection of the ports with internal elements), instructions (instruction coding, functionality), and functional units (resource usage, pipelining, etc.).

The structure of the PEClass element is shown in Fig. 12.7. The PEClass element contains the attributes name, implements and requires the specification of its subelements Resources, StorageElements, Resmap, Opnames, Operations, and Units. Beside the name of the PE-class, the implements attribute provides the name of another PE-class that inherits all subelements and parameters of the current PE-class. Further description of any subelement in this class will overwrite the appropriate subelement. The implements attribute may contain no value, which would mean that the PE-class is supposed to be composed from scratch.

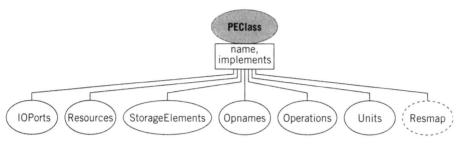

FIGURE 12.7

Structure of the PEClass element.

In the following sections, we discuss the extensions of MAML that are needed in order to describe the internal structure and behavior of processing elements in array architectures.

I/O communication ports

Input and output communication ports are needed to provide the interconnection between several processing elements. Section 12.3.4 describes the interconnect modeling within MAML in detail. The IOPorts element specifies the input and output ports. Ports are distinguished by their bitwidth, direction, and type. The ports of a single PE are represented by the Port element and are characterized by its appropriate attributes within MAML. The type attribute defines whether the port belongs to the data or the control path of the processing element via the corresponding string value *data* or *ctrl*. Moreover, a name and an index value is assigned to each I/O port in the attributes name and index to arrange the ports in a fixed order. The fixed order of the ports is important for the proper modeling of PE interconnection networks (see Section 12.3.4).

Example 12.5

Definition of PE I/O ports.

```
<!--    Data ports    -->
<Port name="ip" ~index="0" bitwidth="32" direction="in" type="data"/>
<Port name="ip" ~index="1" bitwidth="32" direction="in" type="data"/>
<Port name="op" index="0" bitwidth="32" direction="out" type="data"/>
<Port name="op" index="1" bitwidth="32" direction="out" type="data"/>
<!--    Control ports    -->
<Port name="ic" ~index="0" bitwidth="1" direction="in" type="ctrl"/>
<Port name="ic" ~index="1" bitwidth="1" direction="in" type="ctrl"/>
<Port name="oc" index="0" bitwidth="1" direction="out" type="ctrl"/>
<Port name="oc" index="1" bitwidth="1" direction="out" type="ctrl"/>
```

Example 12.5 shows the MAML description of a PE architecture with eight I/O ports. Two 32-bit inputs and two 32-bit outputs of them are data path ports, whereas two 1-bit inputs and two 1-bit outputs belong to the control path of the architecture.

Storage elements

The MAML element StorageElements (similar to a Register element in Section 12.2) is used for characterization of register banks and local memories of a PE. A processing element may contain not only simple control and data registers, but also FIFOs or shift registers suited for data reuse. In order to distinguish between ordinary general purpose register banks and FIFOs, different elements are defined. Each of them has two attributes specifying the name of the register bank/FIFO and the number of registers or FIFOs in it. Additionally, the bank is characterized by its bitwidth, type (data or control path), and namespace. The name attribute assigns a name to the bank whereas the namespace attribute defines a name space (a name without index) for all registers/FIFOs in the bank. The instruction memory is separated from the local memory by introduction of the elements InstructionMemory and LocalMemory, respectively. The size of each of them is defined in bytes.

The PortMapping element declares the direct connections between the registers or FIFOs and I/O ports, defined in the IOPorts element (see Section 12.3.2). Thus, the routing between the internal storage elements of different processor elements is established (see Section 12.3.4).

Arbitrary instruction behavior

In order to provide a complete design flow, starting from the architecture specification and finishing with the compiler generation, the results of compilation must be represented in binary code. This binary code is interpreted as stimuli entry data for architecture simulation. In order to handle this, a MAML description uses an instruction image binary coding. All operations and binary image coding for them are listed by the Opnames element. The operation name is specified by the attribute name and the operation image binary coding is set by the code attribute. The attribute function describes the functionality of the operation, which is given in C-code and is directly embedded into the code of an automatically generated simulator.

12.3.3 Parametric Domains as a Description Paradigm

The array-level properties of multiprocessor architecture are described by the element ProcessorArray within MAML. This element specifies the parameters of the entire processor array in general, that is, for instance the name of the array, the interconnection topology, the number and types of processing elements, and so forth. Parameterizable topologies such as tree, ring, mesh, honeycomb, torus, and others can be selected. For the sake of brevity, only mesh-connected tightly coupled processor arrays are discussed in the rest of the chapter. In this case, the basis of our "drawing-board" is a grid structure where processors can be placed on and a geometry of the processor array can be defined by the number of columns and rows.

As mentioned earlier, processor arrays may contain hundreds of processing elements. Obviously, the characterization of each processors internal architecture and interconnection separately leads to an enormous size of the specification code. Often in such architectures, the groups of several processing elements with the same internal behavior, structure, or common regular interconnection topology, and finally program binary can be located. Therefore, in order to efficiently describe large regular processor architectures, MAML supports the concept of *parametric domains*, meaning that a set of PEs with homogeneous internal architecture or common regular interconnection topology can be described explicitly without recurrences. The MAML distinguishes two types of parametric domains: (1) *Interconnect domain* characterizes a regular interconnection of the sets of selected processing elements, (2) *PE-class domain* assigns one of the available PE-classes (defined by the PEClass element) to the set of selected homogeneous processors. Fig. 12.8 depicts an example of a particular processor array architecture described by interconnect and PE-class domains. The external memory and I/O-elements are not considered. Four interconnect domains $d1, d2, d3, d4$ and three PE-classes $c1, c2, c3$ are shown. For instance, the interconnect domain $d1$ contains the PEs of class $c1$.

Each ProcessorArray element is identified by its name and an optional design version. It also contains a set of subelements: PElements, PEInterconnectWrapper, ICDomain, and ClassDomain. Multiple definition of two last subelements is admissible.

PElements

This defines the number of PEs in the whole processor array, and assigns the referring name for them. The number of elements is specified as a two-dimensional array with a fixed number of rows (rows attribute) and columns (cols attribute). The number of rows multiplied by the number of columns does not necessarily correspond

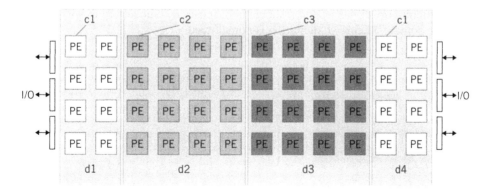

FIGURE 12.8

Characterization of interconnect and PE-class domains.

to the total number of processors within the array, since, as discussed earlier, the grid serves only as a basis to place different types of processors, memories, and I/O-elements. Furthermore, each grid point does not necessarily correspond to one element, because the size the elements could be different. Here, the size denotes not the physical area but rather the logical size in terms of connectors. For example: `<PElements name="pe" rows="4" columns="12">`. The set pe with possibly up to 48 PEs (4 × 12 array) is defined. Each element can be referred to by the name of the set and the two-dimensional indices. In the example of Fig. 12.8, each PE can be referred to as: `pe[1,1]...pe[4,12]`.

ICDomain

Interconnect Domain specifies the set or domain of the processor elements with regular interconnect topology. The PEs here are either subsets of PEs defined within the PElements section of MAML or another domain. Recursive definition is not allowed. The ICDomain element has an attribute name and the set of subelements Interconnect, ElementsPolytopeRange, ElementAt, and ElementsDomain is explained here. The Interconnect subelement specifies the interconnect network topology (see the next example) of the domain. The ElementsPolytopeRange subelement is used to define a subset of PEs that are grouped together in order to organize a domain. The set of PEs is defined by the points of the integer lattice defined as follows:

$$\begin{pmatrix} i \\ j \end{pmatrix} = \left\{ \begin{pmatrix} i \\ j \end{pmatrix} \in \mathbb{Z}^2 \mid \begin{pmatrix} i \\ j \end{pmatrix} = L \cdot \begin{pmatrix} x \\ y \end{pmatrix} + m \wedge A \cdot \begin{pmatrix} x \\ y \end{pmatrix} \leq b \right\}$$

$A \cdot \begin{pmatrix} x \\ y \end{pmatrix} \leq b$ describes a polytope that is affinely transformed by $L \cdot \begin{pmatrix} x \\ y \end{pmatrix} + m$.
An example of this concept is shown in Fig. 12.9. The processor array contains two domains of processor elements: domain *D1* has a triangular shape and domain *D2* is a set of PEs placed in the shape of a square.

The ElementAt subelement is used to select one PE by index of row and column, whereas the ElementsDomain subelement reuses the subset of PEs specified in another domain. Recursive definition is not allowed.

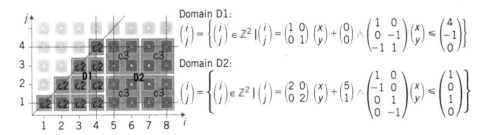

Domain D1:
$$\begin{pmatrix} i \\ j \end{pmatrix} = \left\{ \begin{pmatrix} i \\ j \end{pmatrix} \in \mathbb{Z}^2 \mid \begin{pmatrix} i \\ j \end{pmatrix} = \begin{pmatrix} 1 & 0 \\ 0 & 1 \end{pmatrix} \begin{pmatrix} x \\ y \end{pmatrix} + \begin{pmatrix} 0 \\ 0 \end{pmatrix} \wedge \begin{pmatrix} 1 & 0 \\ 0 & -1 \\ -1 & 1 \end{pmatrix} \begin{pmatrix} x \\ y \end{pmatrix} \leq \begin{pmatrix} 4 \\ -1 \\ 0 \end{pmatrix} \right\}$$

Domain D2:
$$\begin{pmatrix} i \\ j \end{pmatrix} = \left\{ \begin{pmatrix} i \\ j \end{pmatrix} \in \mathbb{Z}^2 \mid \begin{pmatrix} i \\ j \end{pmatrix} = \begin{pmatrix} 2 & 0 \\ 0 & 2 \end{pmatrix} \begin{pmatrix} x \\ y \end{pmatrix} + \begin{pmatrix} 5 \\ 1 \end{pmatrix} \wedge \begin{pmatrix} 1 & 0 \\ -1 & 0 \\ 0 & 1 \\ 0 & -1 \end{pmatrix} \begin{pmatrix} x \\ y \end{pmatrix} \leq \begin{pmatrix} 1 \\ 0 \\ 1 \\ 0 \end{pmatrix} \right\}$$

FIGURE 12.9

Polytope domains representation.

ClassDomain

This specifies the set or domain of processor elements with a common homogeneous architectural structure (PEClass). The PEs here are either subsets of PEs defined by the PElements, or another domain. Recursive definition is not allowed. The name attribute of ClassDomain names the domain and the class attribute specifies the class of the PE architecture for all PEs in this domain. The classes of PE architecture are described in the PE-level section of the MAML description (see Section 12.3.2). The subelements ElementsPolytopeRange and ElementsDomain are the same as in the ICDomain element.

Finally, the subelement PEInterconnectWrapper is used for the description of reconfigurable interconnection networks in processor arrays. It is described in the Section 12.3.4 using illustrative examples.

12.3.4　Description of Adaptive Interconnect Topologies

The interconnect domain is used to specify the interconnect topology for a set of PEs. In order to model and specify a reconfigurable interconnect topology, a PE *interconnect wrapper* (IW) [5] concept is introduced. An interconnect wrapper describes the ingoing and outgoing signal ports of a single processing element. Each interconnect wrapper has a constant number of inputs and outputs on each of its four sides, which are connected to the inputs and outputs of neighbor IW instances.

In Fig. 12.10(a), the IW is represented as a rectangle around the PE and consists of input and output ports on the northern, eastern, southern, and the western side of it. The number of input ports on each side is represented by the integer numbers N^{in}, E^{in}, S^{in}, and W^{in}, respectively. The same holds true for the output ports: N^{out}, E^{out}, S^{out}, and W^{out}. Each *directed interconnect channel* represents a pair of one input and one output port on the opposite sides of the interconnect wrapper with a certain common bitwidth. The direction of the channel is determined by the position of the output port. For example, if we consider a pair of northern input and southern output IW ports, then the direction of corresponding interconnect channel is southward. Therefore, the numbering of the ports is done, as shown in Fig. 12.10(a). The PE is placed inside of the interconnect wrapper. The set of input ports P^{in} and the set of output ports P^{out} are shown on the top and on the bottom of the PE.

In the most complex case, the IW is a configurable full crossbar switched matrix. But in practice it will be less complex, since a compromise between routing flexibility and cost has to be made. By the *configuration of an IW*, we mean the definition of the mapping of the possible connections between the ports of an interconnect wrapper and a processor element. The configuration is specified by the so-called *Interconnect Adjacency Matrix* (IAM) (see Fig. 12.10(b)). The rows of this matrix depict the input ports of the IW, except of the few last rows (dependent on the number of the PEs output ports), which represent the output ports of the PE. The columns represent the output ports of the IW, except of the few last columns (dependent

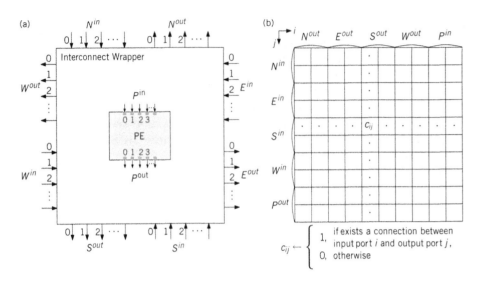

FIGURE 12.10

In (a), interconnect wrapper. In (b), Interconnect Adjacency Matrix.

on the number of the PEs input ports), which represent the input ports of the PE. The matrix contains the values c_{ij}, with $c_{ij} = 1$ if there exists a connection between input port $p_i \in P^{in}$ and output port $p_j \in P^{out}$, and $c_{ij} = 0$ otherwise. The last rows and columns of the IAM represent the port mapping between the PE and IW ports. The positions of input PE ports are interchanged with the positions of the output PE ports in the IAM. This allows the avoidance of the configuration of such incorrect connections as a connection between IW input and PE output or a connection between PE input and IW output. If many input signals are allowed to drive a single output signal, a multiplexer with an appropriate number of input signals is generated. The inputs of the multiplexer are connected to the corresponding IW input signals, and the output is connected to the corresponding IW output signal. The control signals for such generated multiplexers are stored in configuration registers, and can therefore be changed dynamically. By changing the values of these configuration registers inside an interconnect wrapper component, different interconnect topologies can be implemented and changed dynamically at run-time. The interconnect wrapper is characterized within the PEInterconnectWrapper element of MAML, whereas the IAM is a part of the ICDomain (see Section 12.3.3). In the following paragraphs, the syntax of the IAM definition is explained.

The Interconnect subelement of ICDomain specifies the interconnect network topology. Here, the type of the interconnect is defined. The value of the attribute type is one of the following: *static* or *dynamic*. They specify whether the defined interconnect is static or can be configured dynamically during run-time (i.e., configurable during processor array run-time). Depending on the interconnect type, the different parameters in the body of the Interconnect subelement need to be specified.

The AdjacencyMatrix subelement defines the IAM of the interconnect wrapper for the PEs that belong to the current domain. The matrix is described row by row using the tags NInput, EInput, SInput, WInput, and WOutput with the attributes idx and row, specifying the rows of the matrix (rows with all zeros can be skipped). The value of idx defines the index of the matrix row or the port for the appropriate side of the IW.

An example of the interconnect topology of PEs is shown in Fig. 12.11. The interconnect wrapper has four interconnect channels on each of its edges. It contains a PE with four input and four output ports. The corresponding interconnect

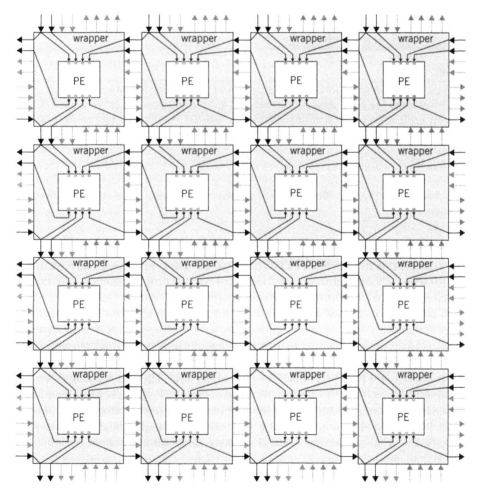

FIGURE 12.11

Exemplary interconnect topology of the PEs belonging to domain *d*2 of the processor array shown in Fig. 12.8.

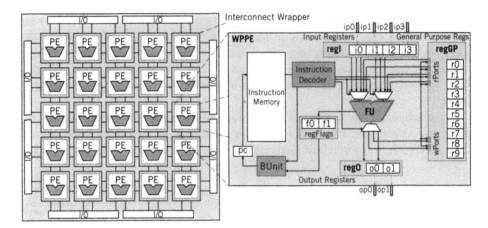

FIGURE 12.12

Example of a WPPA with WPPEs. A WPPE consists of a processing unit which contains a set of functional units. The processing unit is connected to input and output registers. A small data memory exists to temporary store computational results. An instruction sequencer exists as part of the control path that executes a set of control instructions from a "tiny" local program memory.

adjacency matrix that defines the shown ten connections is described in Section 12.3.5.

12.3.5 Case Study of a Tightly Coupled Processor Arrays

Generally, MAML models a wide class of massively parallel processor architectures. In [16], a new class of tightly coupled massively parallel programmable architectures called *weakly programmable processor arrays* (WPPA) is described, see also [17]. Such architectures consist of an array of Weakly Programmable Processing Elements (WPPE) that may contain subword processing units with a small local memory and a regular interconnect structure. In order to efficiently implement a certain algorithm, each PE may implement only a certain constrained functionality. Also, the instruction set is limited and may be configured at compile-time. The PEs are called weakly programmable because the control overhead of each PE is optimized to a minimum. For example, there is no support for interrupts and exceptions. An example of such an architecture is shown in Fig. 12.12. The massive parallelism might be expressed on different levels: (i) several parallel working processing elements, (ii) multiple functional units within one PE, and finally (iii) subword parallelism within the PEs. The WPPAs can be seen as a compromise between programmability and specialty by exploiting architectures realizing the full synergy of programmable processor elements and dedicated processing units.

In Fig. 12.13, the MAML description of the schematically shown processor array from Fig. 12.8 is listed. The processor array "WPPA v1.0" contains 48 processing

```
<ProcessorArray name="WPPA" version="1.0">        |    <MatrixA row="-1 0"/>
<PElements name="pe" rows="4" columns="12"/>      |    <MatrixA row=" 0 1"/>
<PEInterconnectWrapper>                           |    <MatrixA row=" 0 -1"/>
 <Channels>                                       |    <VectorB value=" 2"/>
  <Southward index="0" bitwidth="32"/>            |    <VectorB value="-1"/>
  <Southward index="1" bitwidth="32"/>            |    <VectorB value=" 4"/>
  <Southward index="2" bitwidth="1"/>             |    <VectorB value="-1"/>
  <Southward index="3" bitwidth="1"/>             |   </ElementsPolytopeRange>
  <Northward index="0" bitwidth="32"/>            |  </ICDomain>
  <Northward index="1" bitwidth="32"/>            |  <ICDomain name="d2">
  <Northward index="2" bitwidth="1"/>             |   <Interconnect type="static">
  <Northward index="3" bitwidth="1"/>             |                   <!--NNNNEEEESSSSWWWWPPPP-->
  <Eastward index="0" bitwidth="32"/>             |    <AdjacencyMatrix>  <!--0123012301230123-->
  <Eastward index="1" bitwidth="32"/>             |    <NInput  idx="0" row="00000000000010001000"/>
  <Eastward index="2" bitwidth="1"/>              |    <NInput  idx="1" row="00000000000000000100"/>
  <Eastward index="3" bitwidth="1"/>              |    <EInput  idx="0" row="00000000000000000010"/>
  <Westward index="0" bitwidth="32"/>             |    <EInput  idx="1" row="00000000000000000001"/>
  <Westward index="1" bitwidth="32"/>             |    <WInput  idx="3" row="00000000100000000000"/>
  <Westward index="2" bitwidth="1"/>              |    <POutput idx="0" row="00000000000001000000"/>
  <Westward index="3" bitwidth="1"/>              |    <POutput idx="1" row="00000000100000000000"/>
 </Channels>                                      |    <POutput idx="2" row="00000000010000000000"/>
 <PElementPorts>                                  |    <POutput idx="3" row="00000001000000000000"/>
  <!-- max number of PE inputs-->                 |    </AdjacencyMatrix>
  <Inputs number="4">                             |   </Interconnect>
  <!-- max number of PE outputs-->                |   <ElementsPolytopeRange>
  <Outputs number="4">                            |    <MatrixA row=" 1 0"/>
 </PElementPorts>                                 |    <MatrixA row="-1 0"/>
</PEInterconnectWrapper>                           |    <MatrixA row=" 0 1"/>
<ICDomain name="d1">                              |    <MatrixA row=" 0 -1"/>
 <Interconnect type="static">                     |    <VectorB value=" 6"/>
  <AdjacencyMatrix>  <!--0123012301230123-->      |    <VectorB value="-3"/>
  <WInput  idx="0" row="00000000000000001000"/>  |    <VectorB value=" 4"/>
  <WInput  idx="1" row="00000000000000000100"/>  |    <VectorB value="-1"/>
  <WInput  idx="2" row="00000000000000000010"/>  |   </ElementsPolytopeRange>
  <WInput  idx="3" row="00000000000000000001"/>  |  </ICDomain>
  <POutput idx="0" row="00000001000000000000"/>  |
  <POutput idx="1" row="00000100000000000000"/>  |  ...
  <POutput idx="2" row="00001000000000000000"/>  |  <ClassDomain name="dc1" peclass="c1">
  <POutput idx="3" row="00001000000000000000"/>  |   <ElementsDomain instance="d1"/>
  </AdjacencyMatrix>                              |  </ClassDomain>
 </Interconnect>                                   |  <ClassDomain name="dc2" peclass="c2">
 <ElementsPolytopeRange>                           |   <ElementsDomain instance="d2"/>
  <MatrixA row=" 1 0"/>                            |  </ClassDomain>
  <MatrixA row="-1 0"/>                            |  ...
                                                  | </ProcessorArray>
```

FIGURE 12.13

MAML specification of a tightly coupled processor array shown in Fig. 12.8.

elements that are connected by an interconnect wrapper with four channels on each of its edges. Each PE has four input and four output ports. The array contains four interconnect domains *d1*, *d2*, *d3*, *d4* and three class domains *dc1*, *dc2*, *dc3*, *dc4*. For the sake of brevity, only the interconnect domains *d1*, *d2* and the class domains *dc1*, *dc2* are described.

12.4 APPROACHES AND TOOLS AROUND MAML

The ultimate goal of MAML is to provide an efficient modeling basis for automated architecture/compiler co-design of processor architectures, in particular multiprocessor SoC systems, that are designed for special purpose applications from the domain of embedded systems. In order to best meet application-specific

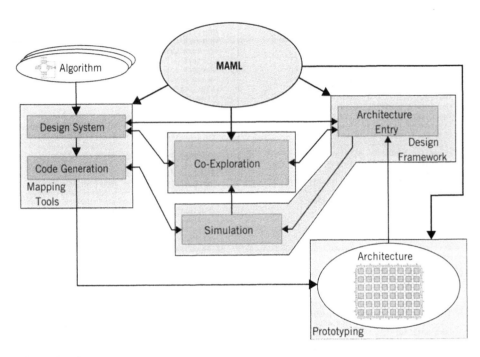

FIGURE 12.14

Architecture/compiler co-design flow for multiprocessor architectures.

requirements (e.g., economic viability, maximum power dissipation, and performance), an exploration of several possible design alternatives has to be conducted. The tools developed in this context centrally use MAML and can be separated into parts shown in Fig. 12.14: (1) Design framework for architecture entry and modeling, (2) Simulation, (3) Generation of architecture prototypes, (4) Mapping tools, and (5) Co-exploration. In the following section, we give a brief overview of the approaches and tools using MAML.

12.4.1 Application Mapping

If done by hand, the mapping of applications could be a crucial task, that is, each processor or groups of processors have to be programmed separately. Especially, the configuration of the interconnection network and the synchronization of the processors could become a time-consuming task. Therefore, efficient compilation techniques are of utmost importance in order to exploit the large degree of architecture's parallelism. Field of application for tightly coupled parallel architectures like WPPAs is the acceleration of data-flow dominant applications. Often these applications are expressed by nested loop programs and have multidimensional array accesses. Loop parallelization in the polytope model [18] has been proven to be very efficient when targeting processor arrays [19]. Thus, the

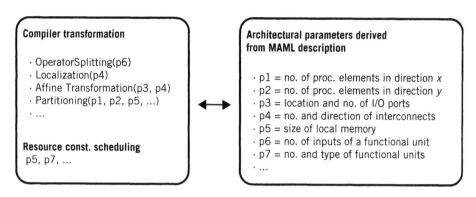

FIGURE 12.15

Relationship between compiler transformation parameters and architectural parameters (MAML).

starting point of our mapping approach is the so-called class of *dynamic piecewise linear algorithms* [20], a functional programming language in the polytope model.

In order to map a given program to a specific architecture, the architectural parameters have to be considered by a number of loop transformations as well as during the allocation, scheduling, and binding phases of the compilation. In Fig. 12.15, a brief overview of the relation between loop transformations and architectural parameters is given.

For instance, to adapt a given algorithm to an architecture, complex statements have to be split into operations with only $p6$ operands according to the number of possible inputs of a functional unit. Affine transformations might be used to fit algorithm dependencies to the array topology or to transform input/output variables to the physical location of I/Os. Partitioning of loop programs is done according to constraints imposed by the physical hardware. Here, the tiling matrices have to be chosen according to the size of the array, the local memory, I/O-bandwidths, and other parameters.

12.4.2 Design Framework

In order to simplify the process of processor array architecture entry and prototyping, a *Design Framework* consisting of a front- and a back-end has been developed [21]. The design of a parallel processor array implementing an edge detection application is shown in Fig. 12.16. The tool supplies standard components necessary for interactive design of a parallel processor array, in particular, a WPPA. It also provides a simulation and debugging environment with interactive visualization of the processor array architecture. The multiprocessor architecture can be either loaded into the *Design Framework* from a MAML-file or can be created from scratch and saved in a MAML-file. At the back-end, a highly parameterizable template in VHDL

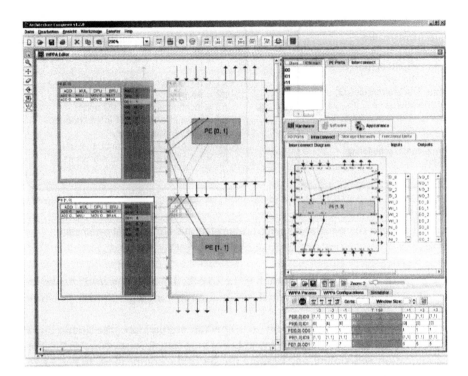

FIGURE 12.16

Designing a parallel processor array implementing an edge detection application using *Design Framework*.

is used to instantiate a weakly programmable processor array implementation with the given architectural parameters.

12.4.3 Interactive Visual Architecture Entry

On the left-hand side of Fig. 12.16, the *architecture design area* is shown where the user can observe the entire processor array with all processing elements surrounded by their interconnect wrappers. The designer can switch the design area to a mode showing the interconnection network between the processor elements. On the right-hand side of Fig. 12.16, the software/hardware configuration window is presented. Here, the basic design elements of PE can be parameterized. The parameters of each PE can be manipulated at any time via a dialog. For example, Fig. 12.17 illustrates editing the VLIW program of the currently selected processor. The hardware specification of the PE is done step by step. First, the hardware parameters are defined. The user specifies the number of inputs/outputs of the PE and bit widths separately for data and control path. Then, the interconnections between the PE ports and the ports of the interconnect wrapper are set. Interconnections between the ports are validated automatically. The user interface enables

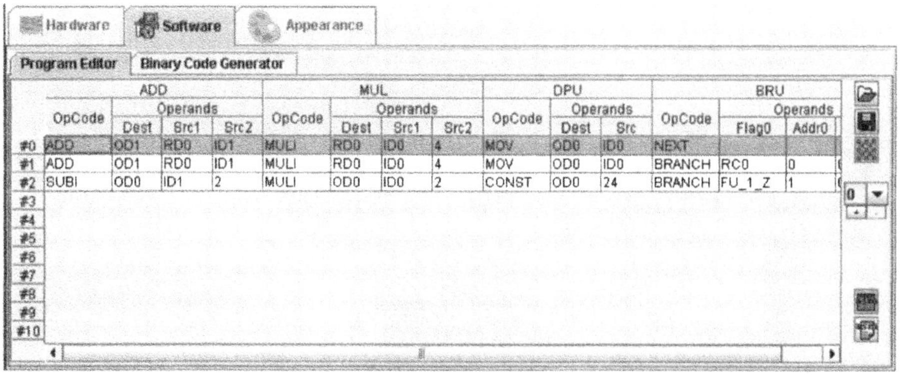

FIGURE 12.17

Interactive entry of processor parameters in *Design Framework*.

the designer to establish only correct interconnections. Furthermore, the storage elements of the PE are defined for the data as well as for the control path. The numbers of input shift registers with arbitrary widths, general purpose registers, and output registers can be defined. The set of different functional units and the size of the instruction memory can be configured in the next dialog window. As soon as the hardware parameters of a particular processor are defined, the user can consider the software programming part by specification of the VLIW program in the appropriate VLIW table (see Fig. 12.17) or in textual form transformed by an appropriate parser. After architecture entry, the C++ code generator creates a corresponding instruction set level cycle-accurate simulation model for further debugging and evaluation.

12.4.4 Simulator Generation

The *Design Framework* is able to generate an external mixed compiled/interpretive *instruction set simulator* (ISS) with a flexible debugging environment for visualization of the architecture simulation run. The flow of conventional compiled simulation, particularly *Levelized Compiled Code* (LCC) [22] simulation, consists of circuit description, code generation, compilation, and execution of the standalone compiled simulator, which reads input vectors and generates the circuit's output values. In Fig. 12.18, the hardware architecture description is extracted from the *Design Framework*, where the architecture is graphically defined or imported from a file during the *architecture description phase*. The simulator is compiled from a C++ program, which is generated automatically by *SimulatorGenerator*, a built-in tool of the *Design Framework*. In order to be able to simulate arbitrary bitwidths accurately, the proposed simulation engine uses the GNU Multiple Precision (GMP) library [23]. However, it is also possible to use the standard integer data type for bitwidths up to 32-bit, in order to improve simulation performance.

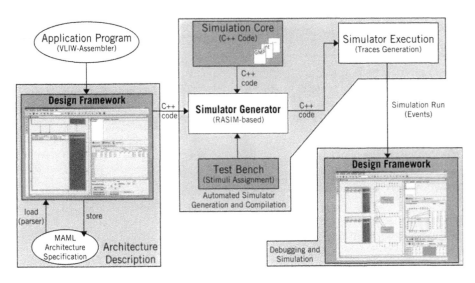

FIGURE 12.18

A workflow of the simulation.

According to this, the simulation core is organized modularly and provides a library of implemented simulation functions parameterizable with different bitwidths. The processing elements are automatically generated as instances of a generic C++ class implementing the PE simulation model described in [21]. The architectural parameters of all PEs as well as the interconnection structure between the processors are stored in a decoupled C++ header file that can be easily integrated into the complete generated simulator. The structure of simulation function calls reflecting the interconnections between the processors is generated automatically, but the assignment of the stimuli to the external inputs of the array should be performed manually in the *test bench* part of the generated simulator. Then, the C++ code is compiled and executed. Before the simulation starts, it is first determined as to which processor elements are active. For this purpose, a Data Dependence Graph (DDG) for each PE is built. Next, it is determined whether the DDG of a PE contains a loop. Further, the DDG for those processors that contain no loops, the *activity-periods* (a minimal number of simulation cycles needed to perform the computation after the simulation event arrives and to store the results) as well as patterns of simulation events are determined for every input according to the methodology presented in [21]. As a result of the compiled simulator execution, a generated textual file is produced with the simulation trace stored in it. However, to enable even higher simulation speed, the simulation run can be stored in memory. The simulation trace is presented by the list of events at appropriate simulation cycles and the corresponding content of the storage elements. Then, the generated simulation results can be directly represented in the debugging and simulation environment.

Since the simulation model is mixed (compiled and interpretive simulation), there is no need to regenerate and to recompile a new simulator for the changed application (binary program code) running on the same architecture. The simulation model presented in [21] also supports reconfigurable parallel architectures, meaning that the interconnection and the local instruction memory contents of the processing elements can be reprogrammed at run-time without time consuming recompilation of the generated simulator. As soon as the new program is loaded into the instruction memory of the PE or the interconnection between certain processors is changed, the appropriate instructions are decoded (fetch-decode-dispatch process) and stored in the internal list of decoded instructions during simulator run-time. This is generally done only once just after the program code or interconnection of certain PEs has been changed. After that, the decoded instructions can be directly reused in a repeated program execution. Such a simulation technique is similar to a *just-in-time cache compiled simulation* (JIT-CCS) [24].

The simulation efficiency was experimentally verified. On a host computer with an Intel Pentium 4 (3 GHz) processor a reconfigurable WPPA architecture with 4 PEs (16 FUs, 20 registers, 8 input shift registers), 810,373 cycles/s or approximately 13 MIPS (FIR, edge detection algorithms) could be simulated. This allowed us to perform a QCIF image processing simulation in real-time.

The change of contents in each register can be recorded and stored during simulator execution. Profiled by the simulator, this information is used for energy consumption estimation for the entire multiprocessor architecture. For each given class of the storage elements (register, FIFO, memory, etc.), an appropriate amount of energy, which is supposed to be consumed after one change of element's contents is annotated. Thus, a coarse estimation of the switching activity of the entire architecture or one individual PE can be obtained. Beside the debugging and simulation of a given architecture, *Design Framework* allows for interactive visualization of the switching activity in parallel processors over time. This may help the designer to increase the efficiency of implemented algorithms running on a given reconfigurable architecture.

Another important aspect is the interfacing between the generated simulators for the processor array and its peripherals, that is, a standard processor or host interface. As mentioned earlier, increasing manufacturing density allows the development of entire SoCs or even *MultiProcessor Systems-on-Chip* (MPSoC). They typically contain one or more cores with considerable on-chip memory and complex communication buses. Therefore, it is very important to develop modeling concepts for MAML in order to generate the simulators being able to interact with other processor cores, which are often legacy or third-party components. To permit a validation of the complete multicore system, a coupling of different simulators responsible for each core in the system is needed.

Instruction set simulators offer an acceptable level of granularity for a core simulation. The ISSs can provide detailed functional information, such as register values, program execution times as well as other timing information. Therefore, the ISS level is sufficient for validation of multiprocessor architectures.

12.4.5 Architecture Synthesis for Rapid Prototyping

As mentioned in Section 12.4.2, a highly parameterizable template written in VHDL is used during the back-end design phase of a weakly programmable processor array. The template parameters on the array level are, for instance, array size, configuration memory size, possible interconnect topologies, etc. On the PE level, the number and size of input and output ports, number of functional units, size of instruction memory, and local register file can be specified for each PE individually. All these parameters are taken directly from the MAML architecture description.

To obtain some initial area, speed, and power consumption results for a WPPA, two dynamically reconfigurable algorithms from the area of digital signal processing were chosen and implemented with the help of our design framework on a single 2×2 WPPA: An FIR filter and an edge detection algorithm. In the data path, 16-bit fixed point arithmetics was used.

Using our framework, the parameters of a WPPA architecture were entered graphically and simulated cycle-accurately. With the help of the parametric domain concept described in Section 12.3.3, it was possible to perform the architecture and program entry within few minutes. The automatic simulator generation allows for fast cycle-accurate application validation. Thereafter, a synthesizable VHDL file containing the top-level entity with current architectural parameters from the MAML description was also generated automatically. In order to obtain a first running prototype, the design was synthesized and tested on a FPGA platform. The resulting maximum clock frequency was 100 MHz. Finally, a semicustom design toolchain was applied to receive a modern standard cell CMOS implementation. There, the clock frequency could further be increased to 200 MHz, allowing for extremely short reconfiguration times of different applications, in this case up to three orders of magnitude shorter than for reconfiguring today's FPGA platforms [25].

12.5 CONCLUSIONS AND FUTURE WORK

In this chapter, we presented the architecture description language MAML (MAchine Markup Language) for modeling and simulation of both single and multiprocessor architectures. It is based on XML, which allows one to characterize the resources of complex processor architectures at both structural and behavioral levels in a convenient manner. The extracted parameters are used, on the one hand for a fast interactive cycle-accurate simulation, and on the other, for compiler retargeting. Finally, the processor architecture described within MAML can be automatically synthesized for rapid prototyping.

Currently, we study the automatic parallelization and compilation of loop nests in order to generate code for WPPA. In the future, we would like to develop integration concepts for efficient co-simulation of array architectures with other cores in the scope of MPSoC at different abstraction levels (e.g., transaction level or bus cycle-accurate).

Further challenges include the development of efficient area and power estimation models for the goal of architecture/compiler co-exploration. Here, MAML serves again as an integral part. For a given set of benchmark algorithms the vast design space of architectural and compiler parameters has to be explored in order to find optimal architectures in terms of performance and energy efficiency.

12.6 ACKNOWLEDGMENTS

Supported in part by the German Science Foundation (DFG) in project under contract TE 163/13-1. We would like to thank Ferdinand Großmann for his support during the review process.

REFERENCES

[1] D. Fischer, J. Teich, M. Thies, and R. Weper. Design Space Characterization for Architecture/Compiler Co-Exploration. In *ACM SIG Proc. International Conference on Compilers, Architectures and Synthesis for Embedded Systems (CASES)*, pages 108-115, Atlanta, GA, USA, November 2001.

[2] D. Fischer, J. Teich, M. Thies, and R. Weper. Efficient Architecture/Compiler Co-Exploration for ASIPs. In *ACM SIG Proc. International Conference on Compilers, Architectures and Synthesis for Embedded Systems (CASES)*, pages 27-34, Grenoble, France, October 2002.

[3] D. Fischer, J. Teich, M. Thies, and R. Weper. BUILDABONG: A Framework for Architecture/Compiler Co-Exploration for ASIPs. *Journal for Circuits, Systems, and Computers, Special Issue: Application Specific Hardware Design*, pages 353-375, 2003.

[4] A. Kupriyanov, F. Hannig, D. Kissler, J. Teich, J. Lallet, O. Sentieys, and S. Pillement. Modeling of Interconnection Networks in Massively Parallel Processor Architectures. In P. Lukowicz, L. Thiele, and G. Tröster, editors, *Proc. of the 20th International Conference on Architecture of Computing Systems (ARCS)*, Lecture Notes in Computer Science (LNCS), pages 268-282, Zurich, Switzerland, March 2007.

[5] A. Kupriyanov, F. Hannig, D. Kissler, J. Teich, R. Schaffer, and R. Merker. An Architecture Description Language for Massively Parallel Processor Architectures. In *GI/ITG/GMM-Workshop - Methoden und Beschreibungssprachen zur Modellierung und Verifikation von Schaltungen und Systemen*, pages 11-20, Dresden, Germany, February 2006.

[6] Y. Gurevich. Evolving algebras 1993: Lipari guide. In E. Börger, editor, *Specification and Validation Methods*, pages 9-36, Oxford University Press, 1995.

[7] A. Kupriyanov, F. Hannig, and J. Teich. Automatic and Optimized Generation of Compiled High-Speed RTL Simulators. In *Proc. of Workshop on Compilers and Tools for Constrained Embedded Systems (CTCES)*, Washington DC, USA, September 2004.

[8] A. Kupriyanov, F. Hannig, and J. Teich. High-Speed Event-Driven RTL Compiled Simulation. In A. Pimentel and S. Vassiliadis, editors, *Computer Systems: Architectures, Modeling, and Simulation, 4th International Samos Workshop (SAMOS), Proc.*, Volume 3133 of *Lecture Notes in Computer Science (LNCS)*, pages 519-529, Island of Samos, Greece, July 2004.

[9] S. Pees, A. Hoffmann, and H. Meyr. Retargeting of Compiled Simulators for Digital Signal Processors Using a Machine Description Language. In *Proc. Design Automation and Test in Europe (DATE)*, pages 669–673, Paris, France, March 2000.

[10] S. Künzli, F. Poletti, L. Benini, and L. Thiele. Combining Simulation and Formal Methods for System-Level Performance Analysis. In *Proc. of the Conference on Design, Automation and Test in Europe (DATE)*, pages 236–241, Munich, Germany, March 2006.

[11] F. Angiolini, J. Ceng, R. Leupers, F. Ferrari, C. Ferri, and L. Benini. An Integrated Open Framework for Heterogeneous MPSoC Design Space Exploration. In *Proc. of the Conference on Design, Automation and Test in Europe (DATE)*, pages 1145–1150, Munich, Germany, March 2006.

[12] A. Terechko, E. Pol, and J. Eijndhoven. PRMDL: A Machine Description Language for Clustered VLIW Architectures. In *Proc. of the Conference on Design, Automation and Test in Europe (DATE)*, page 821, Munich, Germany, March 2001.

[13] W. Qin and S. Malik. Architecture Description Languages for Retargetable Compilation. In Y. N. Srikant and P. Shankar, eds., *The Compiler Design Handbook: Optimizations & Machine Code Generation*, pages 535–564, CRC Press, 2002.

[14] H. Tomiyama, A. Halambi, P. Grun, N. Dutt, and A. Nicolau. Architecture Description Languages for System-on-Chip Design. In *Proc. of Asia Pacific Conference on Chip Design Languages (APCHDL)*, Fukuoka, Japan, October 1999.

[15] P. Mishra and N. Dutt. Architecture Description Languages for Programmable Embedded Systems. In *IEEE Proc. on Computers and Digital Techniques*, pages 285–297, Toronto, Canada, May 2005.

[16] D. Kissler, F. Hannig, A. Kupriyanov, and J. Teich. A Highly Parameterizable Parallel Processor Array Architecture. In *Proc. of the IEEE International Conference on Field Programmable Technology (FPT)*, pages 105–112, Bangkok, Thailand, December 2006.

[17] F. Hannig, H. Dutta, A. Kupriyanov, J. Teich, R. Schaffer, S. Siegel, R. Merker, R. Keryell, B. Pottier, D. Chillet, D. Ménard, and O. Sentieys. Co-Design of Massively Parallel Embedded Processor Architectures. In *Proc. of the first ReCoSoC workshop*, pages 27–34, Montpellier, France, June 2005.

[18] P. Feautrier. Automatic Parallelization in the Polytope Model. Technical Report 8, Laboratoire PRiSM, Université des Versailles St-Quentin en Yvelines, 45, avenue des États-Unis, F-78035 Versailles Cedex, June 1996.

[19] F. Hannig, H. Dutta, and J. Teich. Mapping a Class of Dependence Algorithms to Coarse-grained Reconfigurable Arrays: Architectural Parameters and Methodology. *International Journal of Embedded Systems*, 2(1/2):114–127, January 2006.

[20] F. Hannig and J. Teich. Resource Constrained and Speculative Scheduling of an Algorithm Class with Run-Time Dependent Conditionals. In *Proc. of the 15th IEEE International Conference on Application-specific Systems, Architectures, and Processors (ASAP)*, pages 17–27, Galveston, TX, USA, September 2004.

[21] A. Kupriyanov, D. Kissler, F. Hannig, and J. Teich. Efficient Event-driven Simulation of Parallel Processor Architectures. In *Proc. of the 10th International Workshop on Software and Compilers for Embedded Systems (SCOPES)*, pages 71–80, NY, USA, April 2007.

[22] L. Wang, N. Hoover, E. Porter, and J. Zasio. SSIM: A Software Levelized Compiled-Code Simulator. In *Proc. of the 24th ACM/IEEE Conference on Design Automation*, pages 2–8, Miami Beach, FL, USA, June 1987.

[23] T. Granlund. The GNU multiple precision library, edition 2.0.2. Technical report, TMG Datakonsult, Sodermannagatan 5, 11623 Stockholm, Sweden, 1996.

[24] A. Nohl, G. Braun, O. Schliebusch, R. Leupers, H. Meyr, and A. Hoffmann. A Universal Technique for Fast and Flexible Instruction-Set Architecture Simulation. In *Proc. of the 39th Design Automation Conference (DAC)*, pages 22–27, June 2002.

[25] J. Teich, F. Hannig, H. Ruckdeschel, H. Dutta, D. Kissler, and A. Stravet. A Unified Retargetable Design Methodology for Dedicated and Re-Programmable Multiprocessor Arrays: Case Study and Quantitative Evaluation. In *Proc. of the International Conference on Engineering of Reconfigurable Systems and Algorithms (ERSA), Invited paper*, pages 14–24, Las Vegas, NV, USA, June 2007.

GNR: A Formal Language for Specification, Compilation, and Synthesis of Custom Embedded Processors

Bita Gorjiara, Mehrdad Reshadi, and Daniel Gajski

13.1 INTRODUCTION

The ever-increasing complexity of embedded systems requires design methodologies that focus on improving both designer productivity and design quality. Application Specific Processors (ASPs) can potentially bridge the gap between productivity and quality. However, the adoption of Application Specific Instruction-set Processor (ASIP) Technologies has been relatively slow, mainly because designing efficient instruction-set is challenging and complex in current ASIP solutions. In addition to finding custom instructions for improving application performance, the designer must also consider the compilation and efficient hardware implementation of each instruction. Since all ASIP technologies are based on ADL, a practical ADL must allow the designer to control both the behavior and the hardware implementation of the final custom processor. Clearly, the goal of ADL-based approaches is to automate as much as possible, without sacrificing the quality or flexibility of designs in a certain domain. However, most ADLs in the past have focused on either the compilation or hardware synthesis aspect, but not on both.

Fig. 13.1 illustrates the different ways that ADLs view and model their target processors. Any processor consists of a controller and a datapath. The controller reads a sequence of binary instructions (or configurations) from the program memory (PM) and converts them to proper control words (CWs) that control the behavior of the datapath. In processors, the controller's main responsibility is to decode and schedule the instructions. In *behavioral ADLs*, the combined behavior of controller and datapath is described in terms of instruction behaviors

329

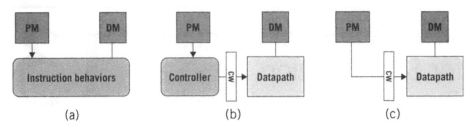

FIGURE 13.1

(a) Behavioral, (b) structural, and (c) NISC ADL views.

(Fig. 13.1(a)). These ADLs may be very lengthy because they have to capture all possible configurations of instructions. Explicit description of instruction behaviors is good for compilation, but prevents the designers from controlling their hardware implementation. Among the few ADL-based approaches that support automatic RTL generation for hardware synthesis, optimization techniques such as resource sharing or forwarding path customization are very challenging or even impossible, because the required information is not captured in the ADL. This problem compromises the productivity gains of behavioral ADLs because at the end, the designer must spend significant amount of time to develop an efficient hardware implementation for the processor. In *structural ADLs*, the details of the controller (i.e., instruction decoder) and datapath netlist are explicitly described (Fig. 13.1(b)). These ADLs are mainly suitable for hardware generation and synthesis since they enable designers to employ their clever ideas for efficient implementation. However, capturing the instruction decoder significantly complicates these ADLs and makes them error-prone. Additionally, extracting the high-level instruction behaviors from these ADLs for compilation is very complex and has been done only for limited architectural features and programming languages.

The challenge is to find a balanced modeling approach that supports both effective synthesis and effective compilation. As we explained earlier, other ADLs have either generated structure (for hardware synthesis) from instruction behaviors, or extracted instruction behaviors (for compilation) from structural descriptions. Now the question is: how can we avoid the complexities and limitations of each approach? To answer this question, we need to understand why we need instructions at all! Traditionally, instruction-set has provided three major benefits: (1) binary compatibility, (2) code-size reduction, and (3) simplification of compiler development. Binary compatibility is not critical in custom embedded processors since the major goal is to compile the application on a customized processor. On the other hand, storing instruction binaries in the program memory is more efficient than storing the control words. The instruction decoder in the processor converts each instruction binary to a sequence of wider control words. In other words, the decoder acts as a decompression stage as well. Instruction-abstraction is also a good way of relieving the compiler developers from dealing with low-level hardware details, but at the same time it prevents the compiler from directly controlling the low-level hardware resources! In ADL-based approaches the compiler and other tools are automatically generated (retargeted) from the description of the custom processor. Now the question is: if we can develop advanced compiler algorithms to deal with hardware details and can use compression to deal with the code-size issue, do we still need instructions? A designer is mainly concerned about how quickly a custom processor can be developed to execute an application efficiently. The intermediate steps, including instruction-set design, are not the major concern of the designer.

To simplify the design of the custom embedded processors, we remove the instruction abstraction in the No-Instruction-Set-Computer (NISC) Technology. We use a formal language called Generic Netlist Representation (GNR) to describe the datapath and then automatically generate the controller for a given application (Fig. 13.1(c)). In NISC, almost all responsibilities of instruction decoder are trans-ferred to a *cycle-accurate compiler* [1]. Using the detailed structural information of the datapath, the compiler generates and schedules the control words to deter-mine the behavior of the datapath in every clock cycle. In NISC, designers only specify the structure of the datapath in GNR. Therefore, the GNR descriptions are very concise and can be used to synthesize high-quality hardware. The controller is automatically generated by the NISC toolset. Since control words (a.k.a. nanocodes) are directly exposed to the compiler, it can statically schedule operations and apply more aggressive optimizations to efficiently utilize the resources in the datapath.

The target architecture in NISC is a variation of microcoded architectures [2]. Instead of using microcode for describing register transfer operations, NISC uses nanocode that determines actual control configuration of every low-level datapath component such as multiplexers, register, functional units, etc. To overcome the increased code size of NISC, the nanocodes are compressed and a low-overhead decompression stage [3] is added during the controller synthesis [4]. This compres-sion scheme can reduce the code size of NISC to within 16% of a RISC's code size. Despite the similarities of NISC and Very Large Instruction Word (VLIW) architec-tures, the controller in NISC is substantially simpler than that of VLIW. The code size of VLIW machines is typically reduced by first packing the instructions and removing the No Operations (NOPs), and then compressing them. Fig. 13.2(a) illus-trates the controller of a VLIW machine that is more complex than that of an NISC

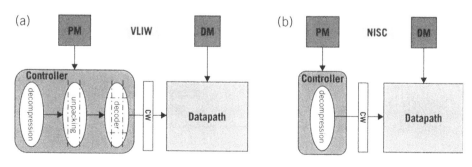

FIGURE 13.2

(a) VLIW, and (b) NISC, after code-size optimization.

(Fig. 13.2(b)) after code-size optimization. For example, in TMS320C6x [5], fetch/unpacking/decode require six pipeline stages, while a similar architecture in NISC only needs one decompression stage.

NISC technology was originally developed in the Center for Embedded Computer Systems (CECS) in University of California Irvine and its toolset is available for public use [6]. The rest of this chapter first presents an overview of NISC design flow and then describes the GNR language for capturing a system consisting of one or more NISC components.

13.2 OVERVIEW OF NISC TECHNOLOGY

A NISC architecture is composed of a datapath and a controller. The datapath of NISC can be simple, or as complex as the datapath of a processor. The controller drives the control signals of the datapath components in each clock cycle. These control values are generated by the NISC compiler. These values are either stored in a memory, or generated via logic in the controller. Both the controller and the datapath can be pipelined. Fig. 13.3 shows a sample NISC architecture with a pipelined datapath that has partial data forwarding, multicycle and pipelined units, as well as data memory and register file. The forwarding paths in the datapath are used for transferring short-lived variables between functional units without going through register file or memory. The NISC compiler generates "0," "1" or "don't care" values for the bits of the control words. A "don't care" value (denoted by 'X') indicates that the corresponding unit is idle at a given cycle and its control signal can be assigned to "0" or "1" without affecting program behavior. Of course, some of the control signals, such as register file write enable (RF_we) are never assigned "X" because a wrong "X"-resolution can affect the program behavior. The "X" can be utilized to reduce

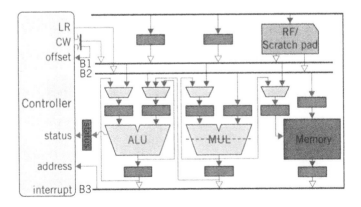

FIGURE 13.3

A sample NISC architecture.

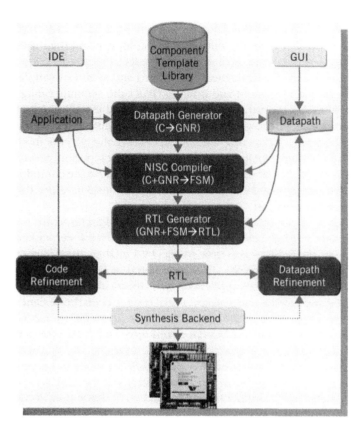

FIGURE 13.4

NISC design flow.

code size, power consumption, or both [7]. Detail of the compilation algorithm is presented in [1, 8].

Fig. 13.4 shows the design flow for designing a custom NISC for a given application. In NISC, the controller is generated after compiling the application on a given datapath. Therefore both the application and the datapath description are considered input to the NISC compiler. The datapath can be generated (allocated) using different techniques. For example, it can be an IP, reused from other designs, or specified by the designer. The datapath can also be generated automatically based on the application behavior as explained in [4]. The datapath is captured in the GNR format (Section 13.3), which describes the datapath as a netlist of components and assigns a set of attributes to each component. A component in datapath can be a register, register-file, bus, multiplexer, functional unit, memory, etc.

The GNR description of the datapath and the high-level description of the application (e.g., C code) are then given as input to the NISC compiler. The NISC compiler maps the application directly on the given datapath and generates a set of control words that determine the behavior of the datapath in each clock cycle. The RTL generator synthesizes a controller from the output of the compiler, and then uses the datapath information to generate the final synthesizable RTL design (described in Verilog). This RTL is then used for simulation (validation) and synthesis (implementation). After synthesis and Placement and Routing (PAR), the accurate timing, power, and area information can be extracted from the layout and used for further datapath *refinement*. For example, the user may add functional units and pipeline registers, or change the bit-width of the components and observe the effect of modifications on precision of the computation, number of cycles, clock period, power, and area. In NISC, there is no need to design the instruction-set because the compiler automatically analyzes the datapath and extracts possible operations. Therefore, the designer can refine the design very fast.

With NISC technology, it is possible to generate a customized architecture for a given application, similar to High-Level-Synthesis (HLS) tools. However, unlike traditional HLS techniques, the NISC design flow in Fig. 13.4 enables the designer to iteratively refine and improve the results. This is depicted in Fig. 13.5. In this flow, the designer can start with an initial application description and use an initial datapath for executing the application and generate initial results. Then the designer can iteratively modify the application or the datapath and use the NISC toolset to generate a new set of results. An important benefit of this approach is that in each iteration the designer can focus on one quality metric. For example, the available parallelism in the application can be improved in one iteration, the clock frequency of the datapath can be improved in another iteration separately, and then the area of the datapath can be improved in yet another separate iteration. In this way, multiple optimizations can be applied to the design without one optimization complicating another. At the end, from several design points, the designer can select the one that best meets the design requirements.

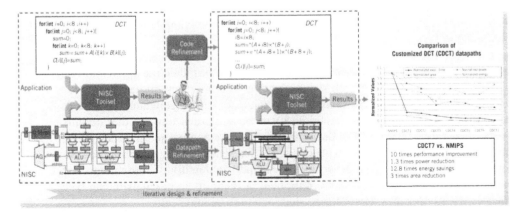

FIGURE 13.5

Iterative design using NISC technology.

The performance and power consumption of a NISC datapath can be improved in several ways: (1) by adjusting the number and type of functional units, number of register file ports, pipeline registers, and forwarding paths to match the application characteristics; (2) by adding custom functional units (ranging from simple bitwise operations to complex multioperand operations); (3) by adding one or more external accelerators, which may be custom NISCs themselves. The difference between (2) and (3) is that in (2) the compiler schedules the communication between the custom unit and the rest of the datapath, while in (3) the software should explicitly define the communication. In Section 13.3, we discuss how a simple NISC architecture as well as a multi-NISC system are specified in GNR.

13.3 MODELING NISC ARCHITECTURES AND SYSTEMS

Today's designs are composed of many cores that communicate with each other through buses, shared memories, and on-chip networks. A successful design paradigm should not only facilitate the efficient design of a single core, but also should simplify its system integration. GNR targets a system composed of several cores. This section explains how a system composed of NSIC cores is modeled for compilation, simulation, and synthesis.

In structural descriptions commonly used in Hardware Description Languages (HDLs), components are instantiated from a library and connected to each other. Each component behavior is captured in HDL as well. Fig. 13.6 shows one such example. To analyze and/or implement the netlist, synthesis tools compile the HDL description of components and map them to one or more known hardware basic blocks such as register, multiplexer, finite state machine, functional units, etc. In this approach, the designer is responsible for correctly connecting the components and controlling them in a meaningful way.

In NISC, the application behavior is captured in a high-level language (i.e., C) and is compiled on a given datapath. Therefore, the structural description of an NISC is used not only for implementation, but also for datapath completion, validation, optimization, and compilation. The goal of datapath completion is to complete a partially described netlist by adding missing connections and components. The goal of validation is to avoid common mistakes, such as connecting wrong ports. The goal of optimization is to automatically change the structure of the netlist so that the overall quality of the design is improved. Finally, the goal of compilation is to map operations and variables in a given program to functional units and storages in a given datapath so that the high-level behavior can be executed on the datapath. All these tools need to process the datapath and its components, and extract the information they need. The HDLs could be extended to capture the information that different tools need. However, this requires extending an HDL compiler and an RTL/logic synthesis tool.

To address this problem, the GNR captures structural details augmented with *types* and tool-specific information called *aspects*. The types enable the tools to know *what* each structural component is, without analyzing *how* that component's behavior is described. To understand this concept, consider Fig. 13.7, which is the same as Fig. 13.6, but it uses shapes and colors for different types of structural elements. In this figure, control lines are thinner and have a lighter color than the input/output data lines. The shapes of register file, functional units, multiplexers, etc. are also different. Fig. 13.7 is also visually more descriptive than Fig. 13.6.

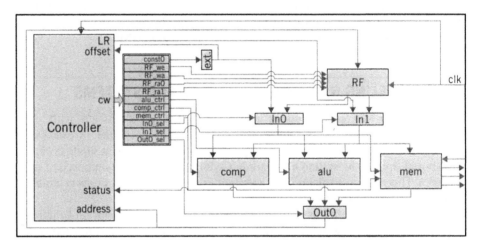

FIGURE 13.6

Block diagram of an NISC captured in structural level.

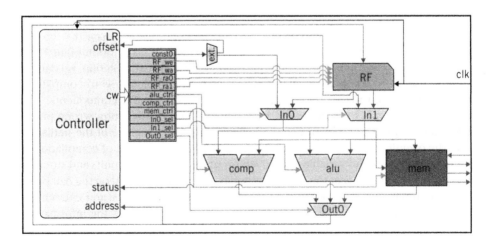

FIGURE 13.7

Type information added to the NISC architecture of Fig. 13.6.

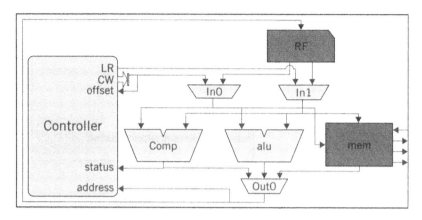

FIGURE 13.8

Simplified block diagram of NISC architecture of Fig. 13.7.

Adding types not only simplifies analysis & processing of the architecture description, it also increases the designer productivity by enabling the automatic completion of partially described datapaths, or by enabling verification of basic properties of the netlist early on in the design process. For example, we know control wires can only connect the CW port of the controller to the control ports of the components. Therefore, we do not need to explicitly describe them and the GNR parser can automatically generate them. In this way, Fig. 13.7 can be further simplified, as shown in Fig. 13.8.

13.3.1 GNR Formalism

A typical multicore system consists of processing elements (PEs). The behavior of each PE is captured in C language. In GNR, the PEs are represented by components of type *behavioralPE*. A *behavioralPE* may contain a dedicated hardware, a general-purpose processor, or a custom NISC. The custom NISC is captured by a component of type *NiscArchitecture*. The netlist inside *NiscArchitecture* is either generated automatically [4] or manually. The NISC compiler compiles the application on the given datapath and generates CWs that may be used to fill the content of control memory (Cmem). Of course, the Cmem is sometimes more complex and itself needs to be synthesized, as explained in [4]. Fig. 13.9 shows a simple example of a system with two PEs (BPE1, BPE2), a bus, and an arbiter. BPE2 is implemented by a programmable *NISC* that has a Cmem and a data memory (Dmem).

The key features of GNR are the types and aspects that it adds to components and ports. In GNR, a component x is represented by $(\tau_x, P_x, C_x, L_x, A_x)$, where τ_x is the component's type, P_x is the set of ports, C_x is the set of components inside x, L_x is the set of its internal point-to-point connections, and A_x is the list of aspects

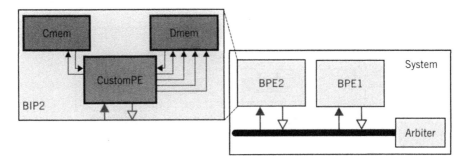

FIGURE 13.9

A sample system in GNR.

that describes the behavior of x for different tools in the toolset. Component type τ_x is defined as follows:

$$\tau_x \in T, T = \{register, register\text{-}file, bus, mux, tri\text{-}state\ buffer, functional\text{-}unit,$$
$$memory\text{-}proxy, controller, NiscArchitecture, behavioralPE, module\}$$

Where, *NiscArchitecture, behavioralPE, module, functional-unit*, and *controller* are hierarchical components (meaning they may contain an internal netlist), while others are basic RTL components with no internal netlist.

Each port p in P_x has a bit-width β_p, and a type θ_p defined as follows:

$$\theta_p \in \{clkPort, ctrlPort, inPort, outPort, cwPort\}$$

Type *clkPort* shows the port is a clock, and type *ctrlPort* shows the port is used to control the component. For example, a register has one port of each type *clkPort, inPort, outPort*, and *ctrlPort* (i.e., load enable). Type *cwPort* means the port is a CW port and is used to drive the control ports of the components in the *NiscArchitecture* (see Section 13.3.6).

The set L_x captures the connections inside component x between its own ports or ports of its internal components. In general, the connections can be defined between the ports' bit-slices as:

$$L_x = \{(\,p1, p2, s1, e1, s2, e2)\,|\,p1, p2 \in \left(P_x \cup \left(\bigcup_{y \in C_x} P_y \right) \right) \text{ and } 0 \le s1 \le e1 < \beta_{p1},$$
$$\text{and } 0 \le s2 \le e2 < \beta_{p2}, \text{ and } e1 - s1 = e2 - s2\}$$

where $s1$ and $s2$ are the start indices of two connected ports $p1$ and $p2$, while $e1$ and $e2$ are their end indices.

A_x is a list of aspects required by different tools for processing component x. The aspects are defined based on component types. Currently, in NISC toolset, each component may have different aspects such as generation (high-level synthesis) aspect

GA_x, compilation aspect CA_x, simulation aspect MA_x, and RTL synthesis aspect NA_x. Generation aspect defines how the internal netlist of that component is generated. Compilation aspect usually captures the relation between the component's behavior and the C-language operations, or application functions. Simulation and synthesis aspects usually contain the description of the component in an HDL, or the information required for generating a hardwired core (e.g., memory, divider, etc.) targeted for a specific platform. For some component types, if an aspect is not specified by the designer, the toolset will generate it automatically. For example, the simulation/synthesis aspects of hierarchical components can either be generated automatically from their internal components, or be explicitly specified by the designer. This feature allows modeling third-party cores or pre-laid-out components that have special technology or manufacturability considerations. The *NiscArchitecture* and *behavioralPE* component types have additional properties as follows:

NiscArchitecture

The *NiscArchitecture* Y represents a NISC processing element that does not have an instruction-set and its CWs are generated by NISC compiler. The compiler aspect of a *NiscArchitecture* is modeled by $CA_Y = (freq, CNST, ORDER, sPt, fPt)$. The *freq* specifies the clock frequency of the *NiscArchitecture* and is used by the compiler to generate the proper control words considering the component delays. A CW contains the control values of components, as well as a set of constant fields *CNST*. The constant fields are used for jump and other operations with a constant operand. Each constant field f in *CNST* has a bit-width or size denoted by β_f. The *ORDER* is a function that defines the ordering of the constant and control fields in the CW. This ordering is used by the compiler to generate the correct CWs. The *sPt* and *fPt* are storage components used for stack pointer and frame pointer. The storage components can be separate registers, or registers in a register file.

behavioralPE

behavioralPE is a component whose behavior is specified in C language, and may be implemented by a general-purpose or custom processor. The compiler aspect of the *behavioralPE* specifies the set of application files (e.g., header files and C files) that execute on that PE. When implemented by NISC, the netlist of *behavioralPE* contains a *NiscArchitecture* and, if necessary, a memory subsystem (Fig. 13.9). The NISC compiler compiles the application C code directly on the datapath of *NiscArchitecture*. The *behavioralPE* may represent instruction-set-based general-purpose, or custom processors as well, where the synthesis aspect is usually a third-party core description that executes the corresponding tools.

13.3.2 GNR Rules

The GNR formal and typed description allows defining rules to validate the correctness of a given netlist. Enforcing such rules significantly improves the productivity of the designer by identifying most of the problems without simulation. Depending on the component type, the rules can restrict the number and types of the ports, instantiated components, and their connectivity. Some of this checking is not possible using HDL-based structural descriptions without logic synthesis. For example, in GNR, if a data port is mistakenly connected to a clock port, or if multiple output ports are connected to one input port of a nonbus component, then it is possible to detect and report the problem. Note that such connections are valid in HDLs but they result in an incorrect design behavior. Using such simple checking in GNR, most architecture problems are quickly determined without simulation. There are two groups of rules that are currently checked on a GNR description: general rules, and NISC-specific rules.

General rules

- Clock ports can only be connected to clock ports:
  $$\forall (p1, p2, \ldots) \in L_x, \tau_{p1} = clkPort \text{ if and only if } \tau_{p2} = clkPort.$$

- Connections in L_x are defined between source ports (i.e., *outPort*) and the destination ports (i.e., *inPort*). For boundary connections (i.e., the connections that involve ports in *Px*), the input ports of *Px* must be the source and its output ports must be the destination.

- Maximum of one connection is allowed to any bit of any destination port. The only exception is for input ports of *bus*-type components, where multiple connections are valid. In digital design, connecting several output ports to a single input port is not valid, unless through tri-state buffers.
  $$\forall (p1, p2, s1, e1, s2, e2), (p3, p4, s3, e3, s4, e4) \in L_x, \text{ if } p2 = p4, \text{ then}$$
  $(p2 \in P_x \text{ and } \tau_x = bus) \text{ or } (s2 > e4) \text{ or } (s4 > e2).$

NISC-specific rules

- Each *NiscArchitecture Y* has one and only one component of type *controller*:
  $\exists! x \in C_Y, \text{ where } \tau_x = controller$

- Only component x with $\tau x = controller$ can have one and only one port of type *cwPort*: $\exists! \; p \in P_x$ and $\theta_p = cwPort$ if and only if $\tau_x = controller.$

- To ensure compilability, each *NiscArchitecture Y* has at least one component of type *register-file*, however, the size of register file can be zero if none of its registers is used: $\exists x \in C_Y, \text{ where } \tau_x = register\text{-}file.$

- In *NiscArchitecture Y*, the bit-width of the *cw* port of the controller component must be equal to the sum of the bit-widths of all control ports, plus the sum of the bit-widths of all control fields in *CNSTY*.
  $$\forall cw \in P_c, \text{ if } \theta_{cw} = cwPort, \text{ then } \beta_{cw} = \sum_{p \in CP_Y} \beta_p + \sum_{f \in CNST_Y} \beta_f$$
  $$\text{where } CP_Y = \{p | p \in \bigcup_{x \in C_Y} P_x \text{ and } \theta_p = ctrlPort\}$$

■ Control connections in *NiscArchitecture Y* are defined between the *cw* port and the control ports of components in *CY*.

$\forall(p1, p2, s1, e1, s2, e2) \in L_x$, if $p2 \in CP_Y$, then $\theta_{p1} = cwPort$ and $s2 = 0$ and $e2 = (e1 - s1) = \beta_{p2} - 1$.

13.3.3 **GNR Syntax**

GNR uses XML language [9] to describe PE models. The GNR syntax is defined in XML Schema [10] to enforce syntax and semantics checking on the given input model. The Schema can also be used for code completion, which further increases the productivity of the designers. Fig. 13.10 shows the partial block diagram of the Schema for modeling a custom PE (*NiscArchitecture*). The PE has several child tags including: <Ports>, <Components>, <Connections>, <CwFields>, <Compiler-aspect>, <Simulation-aspect>, and <Synthesis-aspect>, representing P_Y, C_Y, L_Y, $ORDER_Y$, CA_Y, MA_Y, and NA_Y, respectively. All components in GNR have a <Params> tag that parameterizes that component. For example, the delay or bit-width of the component can be specified as its parameters.

The rest of this section explains different component types in more detail and presents examples on using GNR for capturing components and systems. First we explain how a simple component, namely an ALU, is defined in GNR. Then, we talk about how components are integrated to form a simple IP that can execute a C code. Finally, it explains how the entire system can be captured in GNR.

13.3.4 **Basic Components**

The basic component types include *register, register-file, multiplexer, tri-sate buffer, bus, functional-unit, memory*, and *controller*. Depending on the component type,

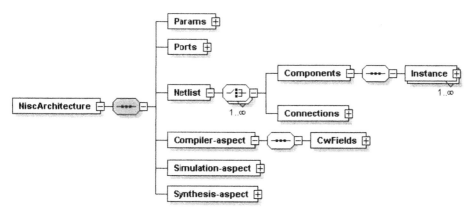

FIGURE 13.10

Block diagram of GNR schema for *NiscArchitecture*.

the compiler aspect of a component may define one or more *machine actions* (MA). The machine actions are the very low-level functionalities of the components. The compiler constructs different behaviors by composing the MAs and scheduling them in proper clock cycles.

There are four types of machine actions: *Read*, *Write*, *Transfer*, and *Execute*. The *Read* and *Write* MAs describe access to a registered storage, while *Transfer* describes data movement from one port to another. The *Execute* MA represents the operations that a component can perform. Each MA may have at most one output that is associated with one output port of the corresponding component. Similarly, an MA may have several inputs, each of which is associated with an input port of the corresponding component. Each MA also defines the timing as well as the control values of each control port of the corresponding component. Finally, if an *Execute* MA specifies the number of pipeline stages, then it is considered as a pipelined operation. Otherwise, if the delay of that MA is longer than the clock period, then it is considered as a multicycle operation. Basic components are divided into four groups: (a) *register* and *register-file* components can only define *Read* and *Write* MAs; (b) *multiplexer*, *tri-sate buffer*, and *bus* components can only define *Transfer* MAs; (c) *functional-unit* and *memory* components can only define *Execute* or *Transfer* MAs; and (d) *controller* components define a set of predefined *Execute* MAs. As an example, assume that the compiler needs to implement an addition between two variables in C code on the datapath of Fig. 13.8. For each operand in the DFG of the program, the compiler must use a *Read* MA from register file RF, and a *Transfer* MA through multiplexer In0 or In1. Then, an *Execute* MA is scheduled on the alu and the result is written back by using a *Transfer* MA through multiplexer Out0 and a *Write* MA on the RF. If the clock period is longer than the sum of all these machine actions, then the ADD operation in DFG takes at most one cycle; otherwise it takes multiple cycles. In addition to *Read* and *Write* MAs, the *register* and *register-file* components specify the variable types that they can store. During compilation, the compiler uses this information to bind local variables to proper register storage. Global variables are bound to the main memory.

For example, Fig. 13.11 shows the GNR description of a small ALU that implements three operations, hence its control port is 2 bits wide (supporting up to four operations). The component has two parameters: BIT_WIDTH and DELAY. The parameters are initialized during the instantiation of a component in a datapath. The simulatable and synthesizable codes of the ALU are described in the <Simulation-aspect> and <Synthesis-aspect> (not shown in the figure). Note that in many cases, if the RTL description of a component is not explicitly specified in its simulation or synthesis aspect, then it can be easily generated from the MA descriptions in its compiler aspect.

13.3.5 Hierarchical Components

Hierarchical components are used to simplify the construction of new components from the available ones. This is done with the *module* component type in

```
<FU typeName="ALU">                              <Operation n="Sub" delay="{@DELAY}">
  <Params>                                         <Output port="o"/>
    <Param n="BIT_WIDTH"/>                         <Input port="i0"/>
    <Param n="DELAY" val="1"/>                     <Input port="i1"/>
  </Params>                                        <Ctrl val="01" port="ctrl"/>
  <Ports>                                        </Operation>
    <InPort n="i0" bitWidth="{@BIT_WIDTH}"/>     <Operation n="Not" delay="{@DELAY}">
    <InPort n="i1" bitWidth="{@BIT_WIDTH}"/>       <Output port="o"/>
    <OutPort n="o" bitWidth="{@BIT_WIDTH}"/>       <Input port="i0"/>
    <CtrlPort n="ctrl" bitWidth="2" default="00"/>  <Ctrl val="10" port="ctrl"/>
  </Ports>                                        </Operation>
  <Compiler-aspect>                            </Operations>
    <Operations>                             </Compiler-aspect>
      <Operation n="Add" delay="{@DELAY}">   </FU>
        <Output port="o"/>
        <Input port="i0"/>
        <Input port="i1"/>
        <Ctrl val="00" port="ctrl"/>
      </Operation>
```

FIGURE 13.11

GNR description of an example ALU.

the NISC architecture model. For example, consider the full GNR description of a two input multiplexer (Mux2) shown in Fig. 13.12. We can repeat this whole code to create a four input multiplexer (Mux4); however, different aspects of the components must be again described for every tool. Another option is to construct a Mux4 from three Mux2 components as shown in Fig. 13.13 and its corresponding GNR shown in Fig. 13.14. At compile time, every time a data transfer is needed from an input port of a Mux4 component to its output port, the compiler schedules several *Transfer* MAs through different internal multiplexers of this module.

For larger multiplexers built using a *module*, the increased number of MAs that the scheduler should schedule for a single data transfer may slow down the compiler. To prevent this negative effect, for frequently used *module* components, we can add compiler aspect that describes the internal behavior of the module and prevents the compiler from going further down the hierarchy for generating the control bits. Fig. 13.15 shows the updated version of Fig. 13.14 with the compiler aspect information. Especially for more complex models, this flexibility of modeling allows the designer to choose where to spend the most time to gain the most productivity. The GNRs in Figs. 13.14 and 13.15 will lead to exactly the same RTL results at the end.

13.3.6 Modeling an NISC Architecture

The *NiscArchitecture* component type is the top module that captures all information about a NISC architecture. The ports of this component are used to connect it to the rest of a system. Fig. 13.16 shows the GNR description of a simple NISC component shown in Fig. 13.8. The datapath of a NISC architecture can have several instances of each component type. A component instance has a unique name

```
<Mux type="Mux2">
  <Params>
    <Param n="BIT_WIDTH"/>
    <Param n="DELAY" val="0"/>
  </Params>
  <Ports>
    <InPort n="i0" bitWidth="{@BIT_WIDTH}"/>
    <InPort n="i1" bitWidth="{@BIT_WIDTH}"/>
    <CtrlPort n="sel" bitWidth="1" default="x"/>
    <OutPort n="o" bitWidth="{@BIT_WIDTH}"/>
  </Ports>
  <Annot_verilog>
    <Synthesis topModuleName="Mux2">
      <VerilogParams>
        <Param n="BIT_WIDTH" val="{@BIT_WIDTH}"/>
      </VerilogParams>
      <VerilogCode>
        <File n="Mux2.v"/>
      </VerilogCode>
    </Synthesis>
    <Simulation topModuleName="Mux2">
      <VerilogParams>
        <Param n="BIT_WIDTH" val="{@BIT_WIDTH}"/>
      </VerilogParams>
      <VerilogCode>
        <File n="Mux2.v"/>
      </VerilogCode>
    </Simulation>
  </Annot_verilog>
  <Annot_compiler>
    <Transfers>
      <DataTransfer inPort="i0" outPort="o" transferDelay="{@DELAY}">
        <Ctrl val="0" port="sel"/>
      </DataTransfer>
      <DataTransfer inPort="i1" outPort="o" transferDelay="{@DELAY}">
        <Ctrl val="1" port="sel"/>
      </DataTransfer>
    </Transfers>
  </Annot_compiler>
</Mux>
```

FIGURE 13.12

GNR description of a Mux2 multiplexer.

and a type name that refer to a component description in the library. The architecture consists of a controller, a register file (RF), a data memory proxy, an ALU, a comparator, and a few multiplexers. The bus-width of the PE is 32 bits. The register file has 32 registers, 2 read ports, and 1 write port. In this PE, suppose that

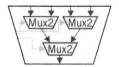

FIGURE 13.13

Construction of Mux4 from several Mux2 components.

a constant field of 10 bits is used for operations with a constant operand. The PE has one clock port, a reset port, and several IO ports for communicating with the data memory unit. The <Netlist> tag shows the components and connections of the PE. For each instantiated component, the proper parameters such as BIT_WIDTH and REG_COUNT are initialized. Thirty four connections are defined for this PE. Each connection determines the source component *src*, source port *sPort*, destination component *dest*, and destination port *dPort*. Twenty one of these connections are shown in Fig. 13.16 and the rest are clock and control connections, which are generated automatically.

In addition to the netlist information of the *NiscArchitecture*, the compiler may also need extra information that is captured in the compiler aspect of the *NiscArchitecture*. The datapath of NISC can be customized and simplified to exactly match the requirements of the application. Therefore, some of the information in the compiler aspect is optional. If such information is not provided in the model, then the corresponding features will be disabled in the compiler. In the rest of this section, we explain the compiler aspect information of NISC architecture.

The compiler aspect of the *NiscArchitecture* component determines one memory component in the datapath as the main memory. However, a datapath can have more than one *memory* component, but the user must directly access the rest using prebinding [11]. All normal memory (load/store) and stack (push/pop) operations are bound to a specified main memory. If the compiler aspect does not specify any memory component as the default main memory, then memory and stack operations will be disabled and the compiler will generate error if these operations are detected in the program. Consequently, the function calls will also be disabled because they depend on stack operations. To support function calls, the Stack Pointer (SP) and the Frame Pointer (FP) registers must also be specified as a register or a register-file element. The FP register always points to the start of the stack of a function, while the SP register points to the end of the stack.

The compiler aspect of the *NiscArchitecture* component also specifies the clock period length in terms of the time unit. This value, along with the operation delays, is used by the compiler to determine if multiple operations can be chained within one cycle, or if the execution of one operation must expand across multiple clock cycles. For example, if the clock period is 1 unit and a multiply operation delay is 2 units, then the multiply is scheduled to last two cycles before the results can be used by another operation, or written back to register file.

```
<Mux type="Mux4">
  <Params>
    <Param n="BIT_WIDTH"/>
    <Param n="DELAY" val="0"/>
  </Params>
  <Ports>
    <InPort n="i0" bitWidth="{@BIT_WIDTH}"/>
    <InPort n="i1" bitWidth="{@BIT_WIDTH}"/>
    <InPort n="i2" bitWidth="{@BIT_WIDTH}"/>
    <InPort n="i3" bitWidth="{@BIT_WIDTH}"/>
    <CtrlPort n="sel0" bitWidth="1" default="x"/>
    <CtrlPort n="sel1" bitWidth="1" default="x"/>
    <OutPort n="o" bitWidth="{@BIT_WIDTH}"/>
  </Ports>
  <Netlist>
    <Components>
      <Instanse n="m0" type="Mux2"/>
      <Instanse n="m1" type="Mux2"/>
      <Instanse n="m2" type="Mux2"/>
    </Components>
    <Connections>
      <Conn src="" srcPort="i0" dest="m0" destPort="i0"/>
      <Conn src="" srcPort="i1" dest="m0" destPort="i1"/>
      <Conn src="" srcPort="i2" dest="m1" destPort="i0"/>
      <Conn src="" srcPort="i3" dest="m1" destPort="i1"/>
      <Conn src="m0" srcPort="o" dest="m2" destPort="i0"/>
      <Conn src="m1" srcPort="o" dest="m2" destPort="i1"/>
      <Conn src="m2" srcPort="o" dest="" destPort="o"/>
      <Conn src="" srcPort="sel0" dest="m0" destPort="sel"/>
      <Conn src="" srcPort="sel0" dest="m2" destPort="sel"/>
      <Conn src="" srcPort="sel1" dest="m3" destPort="sel"/>
    </Connections>
  </Netlist>
</Mux>
```

FIGURE 13.14

GNR of Mux4 built from several Mux2 components.

The structure of the control words is specified in the <CwFields> tag of the compiler aspect of the *NiscArchitecture* component. This structure determines (a) the bit-width and number of constant fields, and (b) the index of control bits (i.e., ordering of the control fields) of each component in the control word. The compiler uses this information to convert the FSM into the corresponding control value stream. Since the control ports are typed, the fields can be added automatically if the user does not specify them. In Fig. 13.16, the total bit-width of the control ports is 35 bits, and the constant width is 10 bits. Therefore, the bit-width of the control words is 45 bits.

```
<Mux type="Mux4">
  <Params>...</Params>
  <Ports>...</Ports>
  <Netlist>...</Netlist>
  <Annot_compiler>
    <Transfers>
      <DataTransfer inPort="i0" outPort="o" transferDelay="{@DELAY}">
        <Ctrl val="0" port="sel1" /><Ctrl val="0" port="sel0" />
      </DataTransfer>
      <DataTransfer inPort="i1" outPort="o" transferDelay="{@DELAY}">
        <Ctrl val="0" port="sel1" /><Ctrl val="1" port="sel0" />
      </DataTransfer>
      <DataTransfer inPort="i2" outPort="o" transferDelay="{@DELAY}">
        <Ctrl val="1" port="sel1" /><Ctrl val="0" port="sel0" />
      </DataTransfer>
      <DataTransfer inPort="i3" outPort="o" transferDelay="{@DELAY}">
        <Ctrl val="1" port="sel1" /><Ctrl val="1" port="sel0" />
      </DataTransfer>
    </Transfers>
  </Annot_compiler>
</Mux>
```

FIGURE 13.15

Extended version of Fig. 13.14 with compiler aspect to speed up compilation.

13.3.7 Communication Modeling

The GNR can also be used for capturing the entire system in pin-accurate level. In a system-level model, the processing elements are instantiated and connected to each other using different communication paradigms. The paradigms include queues, shared memory, buses, and switched networks. As shown in Fig. 13.17, each processing element has one or more communication interfaces (CIs) that should match the communication paradigms it is connected to. Also, the application code of each PE should have proper drivers that match the selected communication interface. The pin-accurate model of the system and the applications may be written manually or may be generated automatically from the Transaction-Level Model (TLM) of the system (shown in Fig. 13.18), where behaviors communicate through abstract channels.

In general-purpose processors, drivers are usually written in assembly. However, custom PEs in general, and NISC in particular, do not have any assembly language because they do not have an instruction-set. An alternative to assembly is to directly write the driver in binary or FSM, which is very inflexible, error-prone, and time consuming. To overcome this issue, the NISC toolset uses prebound functions [8, 11], which expose fine-grained hardware-level control to software developers. The signature of a prebound function defines both the control values for enabling the function, and the ports that are used as inputs and output. The signature is

```
<CustomIP type="simpleIP">
 <Ports>
   <Clock n="clk" bitWidth="1"/>
   <InPort n="reset" bitWidth="1"/>
   <InPort n="dm_r" bitWidth="32"/>
   <OutPort n="dm_addr" bitWidth="32"/>
   <OutPort n="dm_w" bitWidth="32"/>
   <OutPort n="dm_readEn" bitWidth="1"/>
   <OutPort n="dm_writeEn" bitWidth="1"/>
 </Ports>
 <Netlist>
  <Components>
   <Instance n="controller" type="Controller"/>
   <Instance n="RF" type="RF2x1">
    <SetParam n="BIT_WIDTH" val="32"/>
    <SetParam n="REG_COUNT" val="32"/>
   </Instance>
   <Instance n="In0" type="Mux2"/>
   <Instance n="In1" type="Mux2"/>
   <Instance n="Out0" type="Mux4"/>
   <Instance n="comp" type="Comparator"/>
   <Instance n="alu" type="ALU"/>
   <Instance n="mem" type="DataMemProxy"/>
  </Components>
  <Connections>
   <Conn src="controller" sPort="cw" dest="controller" dPort="offset" s="9" e="0"/>
   <Conn src="controller" sPort="cw" dest="In0" dPort="i0" s="9" e="0"/>
   <Conn src="comp" sPort="o" dest="controller" dPort="status" s="0" e="0"/>
   <Conn src="Out0" sPort="o" dest="controller" dPort="address"/>
   <Conn src="Out0" sPort="o" dest="RF" dPort="w0"/>
   <Conn src="RF" sPort="r0" dest="In0" dPort="i1"/>
   <Conn src="RF" sPort="r1" dest="In1" dPort="i0"/>
   <Conn src="In0" sPort="o" dest="comp" dPort="i0"/>
   <Conn src="In1" sPort="o" dest="comp" dPort="i1"/>
   <Conn src="comp" sPort="o" dest="Out0" dPort="i0"/>
   <Conn src="In0" sPort="o" dest="alu" dPort="i0"/>
   <Conn src="In1" sPort="o" dest="alu" dPort="i1"/>
   <Conn src="alu" sPort="o" dest="Out0" dPort="i1"/>
   <Conn src="In0" sPort="o" dest="mem" dPort="addr"/>
   <Conn src="In1" sPort="o" dest="mem" dPort="w"/>
   <Conn src="mem" sPort="r" dest="Out0" dPort="i2"/>
   <Conn src="" sPort="dm_r" dest="mem" dPort="dm_r"/>
   <Conn src="mem" sPort="dm_addr" dest="" dPort="dm_addr"/>
   <Conn src="mem" sPort="dm_w" dest="" dPort="dm_w"/>
   <Conn src="mem" sPort="dm_readEn" dest="" dPort="dm_readEn"/>
   <Conn src="mem" sPort="dm_writeEn" dest="" dPort="dm_writeEn"/>
   <!--GNR parser automatically adds clock and control connections -->
  </Connections>
 </Netlist>
 <Compiler-aspect defaultDMem="mem" clockPeriod="1">
   <CwFields n="cwFields">
     <Field n="const0" bitWidth="10"/>
     <!--GNR parser automatically adds control fields -->
   </CwFields>
   <StackPointer><RegisterFile ref="RF" index="0"/></StackPointer>
   <FramePointer><RegisterFile ref="RF" index="1"/></FramePointer>
 </Compiler-aspect>
</CustomIP>
```

FIGURE 13.16

GNR description of the NISC in Fig. 13.8.

FIGURE 13.17

A system composed of three NISCs communicating using two queues.

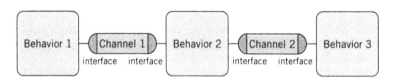

FIGURE 13.18

TLM model of the system in Fig. 13.17.

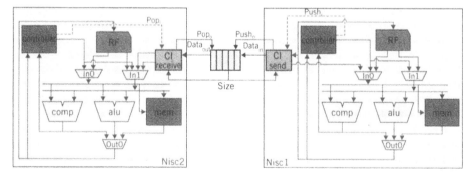

FIGURE 13.19

Block diagram of two NISC PEs communicating using a shared queue.

declared in GNR and is used by the NISC compiler for mapping function calls in a C code to proper RTL operations.

The following example shows how GNR is used for capturing prebound functions and modeling two PEs communicating using a queue. Fig. 13.19 shows the block diagram of the system, where NISC1 pushes the data into the queue and NISC2 pops the data from the queue. The queue has *Push*, *Pop*, *Data$_{in}$*, *Data$_{out}$*, and *Size* ports that are connected to send and receive CIs. When the

```
<Module type="SenderCI">                            </Connections>
 <Ports>                                          </Netlist>
   <Clock n="clk" bitWidth="1"/>                  <Compiler-aspect>
   <ControlPort n="iPush" bitWidth="1"/>           <Functions>
   <OutPort n="oPush" bitWidth="1"/>                <Function n="Push" ...>
   <InPort n="iSize" bitWidth="32"/>                 <Input port="iData"></Input>
   <OutPort n="oSize" bitWidth="32"/>                <Ctrl port="iPush" val="1" />
   <InPort n="iData" bitWidth="32"/>               </Function>
   <OutPort n="oData" bitWidth="32"/>              <Function n="Size"...>
 </Ports>                                            <Output port="oSize"></Output>
 <Netlist>                                          </Function>
   <Components></Components>                       </Functions>
   <Connections>                                  </Compiler-aspect>
   <!--straight through connectivity             </Module>
   removed for brevity-->
```

FIGURE 13.20

GNR of points-to-points queue-based CI for the producer component.

Push signal is "1," one word is pushed into the queue, and when the *Pop* signal is "1," one word is popped from the queue. The *Size* shows the number of words in the queue. In this protocol, the producer pushes the data into the queue and the consumer pops them from the queue. The parties can check queue status (*Size* value) for synchronization between them. The CI components are placeholders in the GNR description of each PE for providing the description of prebound functions, and they act as a proxy for controlling the queue. In general, a proxy component can enable the NISC compiler (hence, a program) to control a component that is outside of the datapath of a NISC component.

Fig. 13.20 shows the GNR description of the queue CI for producer. The CI for consumer is defined similarly. The internal netlist of the CIs is straight through connectivity and is omitted for brevity. In the compiler aspect of the CIs, the prebound functions that they support are specified. For example, SenderCI has *Push* and *Size* functions, while ReceiverCI has *Pop* and *Size* functions.

The names of prebound functions that are exposed to programmers are generated by combining the name of CI instance in the datapath with the aforementioned function names specified in the GNR model of CI. Assuming the instance name of CI is called *ci*, the name of the prebound function for pushing data becomes _$*ci_Push*, where "_$" is a prefix added to avoid name conflicts with other functions in the program. Fig. 13.21(a) and (b) shows the description of send and receive drivers. The send driver first waits until the queue is empty of any previous data (by checking _$*ci_Size* in line 3). Then, it consecutively pushes *N* words into the queue using the _$*ci_Push* function. On the receiver side, the driver receives *N* words in a loop (lines 3-6) by popping from the queue. If the sender has a slower clock frequency than the receiver, then the queue may become empty in the middle of a transaction. In such a case, the receiver must wait (line 4 of Fig. 13.21(b)) until the sender pushes more data into the queue.

```
1 void send(int N, int* data)    1 void receive(int N, int* data)
2 {                              2 {
3   while(__$ci_Size()!=0);      3   for(i=0;i<N; i++){
4   for(i=0;i<N; i++)            4     while(__$ci_Size()==0);
5     __$ci_Push(data[i]);       5     data[i++] = __$ci_Pop();}
6 }                              6 }

         (a)                              (b)
```

FIGURE 13.21

(a) Send, and (b) receive driver code for point-to-point queue-based CIs.

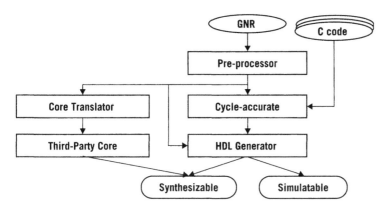

FIGURE 13.22

The flow of an RTL generator.

Similarly, other communication protocols can be captured using this approach. For complex protocols, the time-critical parts are implemented using an FSM inside CIs, and the untimed parts are captured using drivers. In [12], a double hand-shake shared bus was designed for NISC using the above approach. This bus is used in Section 13.4.3 to implement multicore systems for an Mp3 application.

13.3.8 Generating RTL Code from GNR

The RTL Generator of the NISC toolset (Fig. 13.4) generates a hardware description of a system from its pin-accurate model captured in GNR. Fig. 13.22 shows the logical flow for generating HDL from the GNR description of a system and C code of applications for its processing elements. The outputs are synthesizable and simulatable RTL codes. The *Pre-processor* first verifies the syntax of the given GNR file using the GNR Schema. Next, it completes the netlist by (a) resolving the parameters of the components, (b) adding the missing clock and control connections, and

(c) adding the control fields, as explained in Section 13.3.6. The semantic correctness of the completed netlist is validated afterward by applying the rules of Section 13.3.2, and proper warning and error messages are reported by *Pre-processor*. The netlist checker reports unconnected ports, invalid connectivity, and nonexisting referenced component and port names.

After compilation, the compiler generates the contents of data and control memories. The *HDL Generator* uses the GNR and outputs of the compiler to produce the final simulatable and synthesizable HDL codes. The simulatable code is mostly behavioral and simulates much faster than the synthesizable code.

To improve the area, power consumption, and performance of the generated IPs, designers may desire to include hardwired third-party cores. This requires proper support in the ADL and RTL code generator. The RTL synthesis aspect of GNR components allows calling *core-translator* programs to generate proper input files for third-party core generators. For example, for Xilinx FPGAs, the core generator (i.e., LogiCore) requires an XCO file that describes the properties of the core. For other cores, additional information may be required in specific formats. The information of the core is extracted from GNR and passed to core translator programs to generate third-party core files of a specific platform. For example, Fig. 13.23 shows the RTL synthesis aspect of a memory component that contains a command to call an external program, i.e., XilinxcoreGenTranslator.exe. The external translator uses the passed parameters to generate proper XCO and COE (memory content) files that LogiCore needs for generating memory cores.

The *translator* programs usually have different implementations for different target platforms. Using the translators improves productivity by minimizing the manual platform-specific work. In Fig. 13.22, the *Core Translator* generates the input files for third-party core generators by extracting proper information and parameters from the GNR model. The produced cores are combined with the generated HDL code to form the synthesizable code.

```
<Synthesis_Aspect topModuleName="RAM0">
  <Exec command="XilinxCoreGenTranslator.exe">
    <Arg n="UNIQUE_ID" val="RAM0"/>
    <Arg n="CORE_TYPE" val="RAM"/>
    <Arg n="BITWIDTH" val="8"/>
    <Arg n="DEPTH" val="1000"/>
    <Arg n="READ_PORT_COUNT" val="1"/>
    ...
  </Exec>
</Synthesis_Aspect>
```

FIGURE 13.23

Example of GNR code for calling Xilinx Coregen for generating a RAM.

13.4 EXPERIMENTS: DESIGN-SPACE EXPLORATION USING NISC AND GNR

This section presents several experiments using the NISC technology. These experiments show that any NISC architecture can be captured in GNR and the output RTL code can be used for evaluating the quality of the results. This way, the designers can explore different architectures and select the best one that matches their constraints. In the first sets of experiments, several benchmarks are implemented on general-purpose NISCs. The second sets of experiments present designing a custom datapath for a given application, namely a DCT benchmark. The custom DCT experiment also shows the level of control designers can have on the datapath. Unlike high-level synthesis that does not allow designers to modify a generated datapath, NISC gives designers the freedom to adjust the datapath as many times as they want. The last sets of experiments show the use of a GNR and NISC toolset for implementing a simple multicore system.

13.4.1 Designing General-purpose NISCs

Several general-purpose datapaths with different pipeline structures were modeled using GNR and were used to implement different benchmarks by compiling them on these datapaths. Fig. 13.24 shows a basic datapath without pipelining. Fig. 13.25 shows the same datapath with pipeline registers added to the inputs and outputs of the units. Fig. 13.26 extends the datapath of Fig. 13.25 by adding data forwarding paths from the output of the units to their inputs, so that the short-lived variables can be used immediately without being written to RF.

Fig. 13.27 shows the block diagram of a DLX [13] processor that has a NISC controller. We call this processor NISC-style DLX, or NDLX for short. The GNR can also

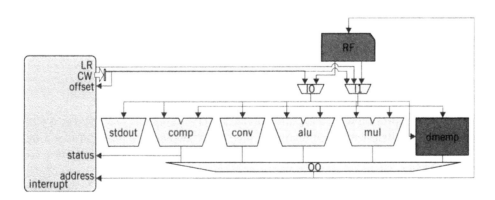

FIGURE 13.24

GN0 with no pipelining or data forwarding.

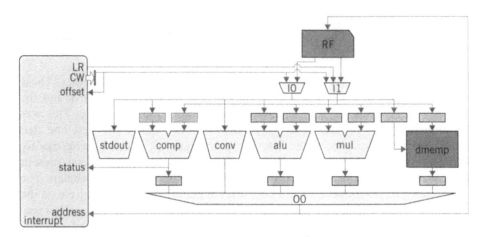

FIGURE 13.25

GN1 with pipelining but no data forwarding.

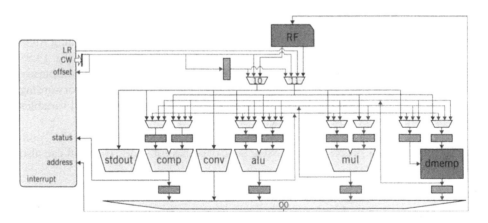

FIGURE 13.26

GN2 with pipelining and data forwarding.

be used for capturing wide datapaths similar to VLIW datapaths [14, 15]. Fig. 13.28 shows a NISC-style wide architecture (called NWA), which is equivalent to a four-slot VLIW machine (1 branch, 2 integer, and 1 memory/div). We captured these six datapaths in GNR and used them in our experiments.

For these experiments, we used three benchmarks from MiBench benchmarks [16] (i.e., adpcm_decoder, crc32, and sha), the Bdist2 kernel from MPEG2, the DCT32 kernel from MP3, and an 8×8 DCT kernel from JPEG. After compiling these benchmarks on the five architectures, Verilog codes for synthesis and simulation

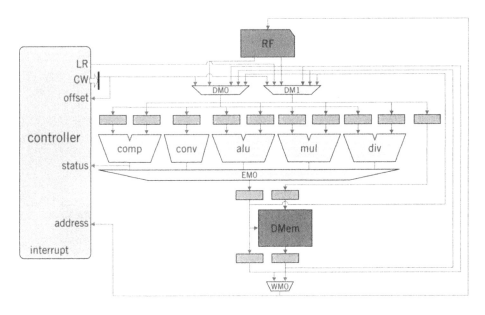

FIGURE 13.27

Block diagram of NDLX.

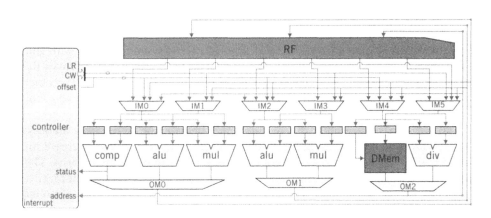

FIGURE 13.28

Block diagram of NWA.

were generated. Using the Xilinx ISE tool, the output Verilog code was synthesized on Virtex4-SX35 FPGA. Table 13.1 shows the number of lines of the GNR code, the clock frequency, and the area of the architectures. All these architectures are described with a few hundred lines of GNR code, which is far less than other ADLs

Table 13.1 Clock frequency of architectures after synthesis.

	GN0	GN1	GN2	NDLX	NWA
Lines of GNR code	170	200	270	250	320
Clock frequency (MHz)	83	120	115	115	75
Area (FPGA slices)	1200	1300	1360	2000	5100

Table 13.2 Execution cycle counts of benchmarks.

	GN0	GN1	GN2	NDLX	NWA
adpcm_decoder	109518	173830	146214	170168	107404
crc32	19083	26109	18081	25106	15068
sha	68874	77622	60522	79133	44156
bdist2	4133	3782	3185	4053	2135
DCT32	961	1018	806	967	494
DCT	12552	15213	12043	22523	12465

that need in the range of thousands of lines of code for architectures with similar capabilities. As the second row of the table shows, pipelining improves the clock frequency significantly while data forwarding slightly reduces it. The clock frequency of NWA architecture is the worst, because of its large area and long wires.

Table 13.2 shows number of cycles for each benchmark on different architectures. While adding pipelining increases the clock frequency, it may increase the cycle counts, especially if there is not enough parallelism in the benchmark. This can be observed by comparing the second and the third columns of the table for GN0 and GN1, respectively. Data forwarding improves the cycle count of the pipelined datapaths. Since GN2, NDLX, and NWA have forwarding, their cycle count is better than GN1. In terms of number of cycles, NWA achieves the best cycle count for almost all benchmarks. The minimum execution cycles of all the benchmarks are shaded in Table 13.2. If the datapaths operate at the same clock frequency (e.g., 50 MHz), then NWA will be the fastest datapath.

Table 13.3 shows the execution delay of the benchmarks when datapaths operate at their maximum clock frequency. The shaded values show the least execution times for each benchmark. When operating at their maximum clock frequency, GN2 yields the minimum execution time for most benchmarks.

Table 13.3 Execution time (ns) of benchmarks at maximum clock frequency.

	GN0	GN1	GN2	NDLX	NWA
adpcm_decoder	1319.5	1448.6	1271.4	1479.7	1432.1
crc32	229.9	217.6	157.2	218.3	200.9
sha	829.8	646.9	526.3	688.1	588.7
bdist2	49.8	31.5	27.7	35.2	28.5
DCT32	11.6	8.5	7.0	8.4	6.6
DCT	151.2	126.8	104.7	195.9	166.2

13.4.2 Custom Datapath Design for DCT

In this section, we include an illustrative example that shows the customization capability of NISC technology. This example explores the design space for different quality metrics such as performance, area, and energy consumption. The step by step details of this customization experiment can be found in [17]. We only include a summary of this experiment here for a quick reference and to show that the NISC cycle-accurate compiler can handle irregular custom datapaths as well. This example also shows how GNR and NISC Technology can enable the designer to directly control the implementation quality and generate customized dedicated hardware. Other ASIP approaches cannot provide this level of customization and control.

The goal is to design a custom pipelined datapath for a DCT algorithm to further improve the performance and power consumption of the design. The definition of Discrete Cosine Transform (DCT) [18] for a 2-D $N \times N$ matrix of pixels is as follows:

$$F[u,v] = \frac{1}{N^2} \sum_{m=0}^{N-1} \sum_{n=0}^{N-1} f[m,n] \cos \frac{(2m+1)u\pi}{2N} \cos \frac{(2n+1)v\pi}{2N}$$

Where u and v are discrete frequency variables ($0 \le u, v \le N - 1$), $f[i,j]$ is the gray level of the pixel at position (i,j), and $F[u,v]$ are coefficients of point (u,v) in spatial frequency. Assuming $N = 8$, matrix C is defined as follows:

$$C[u][n] = \frac{1}{8} \cos \frac{(2n-1)u\pi}{16}$$

Based on matrix C, an integer matrix $C1$ is defined as follows: $C1 = $ round ($factor \times C$). The $C1$ matrix is used in calculation of DCT and IDCT: $F = C1 \times f \times C2$,

(a)

```
for(int i = 0; i < 8; i + +)
    for(int j = 0; j < 8; j + +){
        sum = 0;
        for(int k = 0; k < 8; k + +)
            sum =
                sum + A[i][k]×B[k][j];
        C[i][j] = sum;
    }
```

(b)

```
ij = 0;
do {
    i8 = ij & 0xF8;
    j  = ij & 0x7;
    aL = *(A + (i8|0) );  bL = *(B + (0|j) );  sum  = aL × bL;
    aL = *(A + (i8|1) );  bL = *(B + (8|j) );  sum+ = aL × bL;
    aL = *(A + (i8|2) );  bL = *(B + (16|j) );  sum+ = aL × bL;
    aL = *(A + (i8|3) );  bL = *(B + (24|j) );  sum+ = aL × bL;
    aL = *(A + (i8|4) );  bL = *(B + (32|j) );  sum+ = aL × bL;
    aL = *(A + (i8|5) );  bL = *(B + (40|j) );  sum+ = aL × bL;
    aL = *(A + (i8|6) );  bL = *(B + (48|j) );  sum+ = aL × bL;
    aL = *(A + (i8|7) );  bL = *(B + (56|j) );
    *(C + ij) = sum + (aL × bL);
} while (+ + ij! = 64);
```

FIGURE 13.29

(a) Original and (b) transformed matrix multiplication.

where $C2 = C1^T$. As a result, DCT can be calculated using two consecutive matrix multiplications. Fig. 13.29(a) shows the C code of multiplying two given matrices A and B using three nested loops. As a base for comparison and start point for customizations, a NISC-style implementation of an MIPS M4K datapath [19] (called NMIPS) is chosen. The bus-width of the datapath is 16 bit for a 16-bit DCT precision, and the datapath does not have any integer divider or floating point unit. The clock frequency of 78.3 MHz was achieved after synthesis and Placement-and-Routing (PAR) on a Xilinx Virtex2V250-6 FPGA package. Two synthesis optimizations of retiming and buffer-to-multiplexer conversions were applied to improve the performance.

In general, customization of a design involves both software and hardware transformations. To increase the parallelism in code, the inner-most loop of the matrix multiplication code is unrolled, the two outer loops are merged, and some of the costly operations such as addition and multiplication are converted to OR and AND. In DCT, the operation conversions are possible because of the special values of the constants and variables. The transformed code is shown in Fig. 13.29(b). In the next step, a custom architecture is designed for the transformed DCT code. This architecture is called CDCT1, and is shown in Fig. 13.30(a). Several customizations are applied to this initial custom architecture to improve the performance, area, and energy consumption. These customizations include reducing bit-width of components, removing underutilized resources, and repeatedly adding registers to break the critical path delay. After each customization, the modified C code of Fig. 13.29(b)

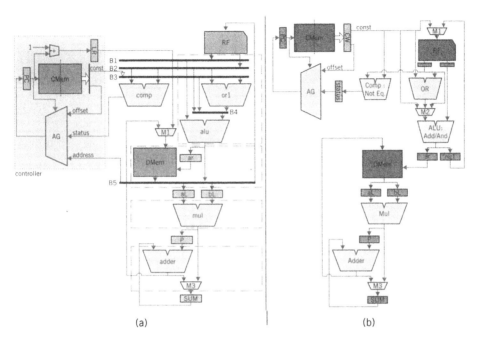

(a)   (b)

FIGURE 13.30

Block diagram of (a) CDCT1 and (b) CDCT7.

is compiled on the refined architecture. In each step, the results are synthesized and analyzed to figure out what part of the datapath can be further customized for more improvements. After seven iterations, the final architecture is called CDCT7, as shown in Fig. 13.30(b). In this last architecture, the multiplier is considered to be a multicycle component because in the target FPGA, this multiplier is mapped to an ASIC unit that cannot be pipelined or optimized.

Table 13.4 shows the summary of the customizations applied to architectures. It also shows the size of corresponding GNR files and the amount of code that has been modified to implement the customizations. The first custom NISC (CDCT1) is derived by modifying 150 lines in NMIPS description. The rest of the architectures need substantially less modifications. The entire DCT experiments took one week, which shows that the GNR and NISC toolset can be used for fast design-space exploration.

Table 13.5 compares the performance, power, energy, and area of all NISC architectures. The third column shows the maximum clock frequency after Placement and Routing. The fourth column shows the total execution time of the DCT algorithm calculated based on the number of cycles and the maximum clock frequencies. Note that although in some cases (such as CDCT4 and CDCT5) the number of cycles

Table 13.4 Summary of customizations and GNR changes.

	Customization	# Lines in GNR	# Modified Lines in GNR
NMIPS	Initial generic architecture	247	–
CDCT1	Custom pipeline design	199	150
CDCT2	Optimizing interconnects	160	50
CDCT3	Removing unused ALU and Comparator operations	160	10
CDCT4	Controller pipelining 1	162	5
CDCT5	Controller pipelining 2	164	5
CDCT6	Bit-width reduction	164	10
CDCT7	Multicycle multiplier, additional pipelined registers at the outputs of the RF	173	15

Table 13.5 Performance, power, energy, and area of the DCT implementations.

	No. of Cycles	Clock Freq.	DCT Exec. Time (us)	Power (mW)	Energy (uJ)	Normalized Area
NMIPS	10772	78.3	137.57	177.33	24.40	1.00
CDCT1	3080	85.7	35.94	120.52	4.33	0.81
CDCT2	2952	90.0	32.80	111.27	3.65	0.71
CDCT3	2952	114.4	25.80	82.82	2.14	0.40
CDCT4	3080	147.0	20.95	125.00	2.62	0.46
CDCT5	3208	169.5	18.93	106.00	2.01	0.43
CDCT6	3208	171.5	18.71	104.00	1.95	0.34
CDCT7	3460	250.0	13.84	137.00	1.90	0.35

increases, the clock frequency improvement compensates for that. As a result, the total execution delay maintains a decreasing trend.

The fifth column shows the average power consumption of the NISC architectures while running the DCT algorithm. All the designs are stimulated with the same data values, and Post-placement and Routing simulation is used to collect

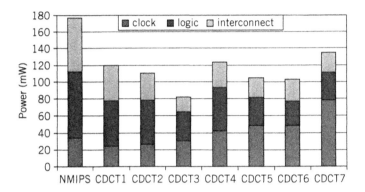

FIGURE 13.31

Power breakdown of the DCT implementations.

the signal activities. To compute the total power consumption, the Xilinx XPower tool is used. The sixth column shows the total energy consumption calculated by multiplying power and execution time. Fig. 13.31 shows the power breakdown of different designs in terms of the clock, logic, and interconnect power.

In these experiments, CDCT1 consumes less power than NMIPS due to its different pipeline structure and better signal gating. CDCT2 consumes less power compared to CDCT1 because of replacing shared bus B5 with short point-to-point connections with lower bus capacitance. The diagram of Fig. 13.31 shows the reduction in interconnect power consumption of CDCT2.

Power consumption of CDCT3 is lower than CDCT2 because of the elimination of unused operations in the ALU and comparator. Elimination of operations reduces the number of fan-outs of the RF output wires. Therefore, reduction in interconnect power, as well as logic power, is achieved. The power breakdown of CDCT3 confirms this fact. Note that as the clock frequency goes up, the clock power gradually increases.

In CDCT4, the power consumption further increases because of: (1) higher clock power due to higher clock frequency and higher number of pipeline registers; and more importantly, because of (2) the power consumption of logic and interconnects added by the retiming algorithm. Since the difference between the delays of the two pipeline stages located before and after the CW register is high, the retiming works aggressively to balance the delay. As a result, it adds extra logic to the circuit. In CDCT5, the status register is added to the output of Comp to reduce the critical path. In this case, the retiming algorithm works less aggressively because the delays of the pipeline stages are less imbalanced. This reduces logic and interconnect power. In CDCT7, the logic and interconnect power remain the same as CDCT5 but the clock power increases due to higher clock frequency.

The last column of Table 13.5 shows the normalized area of different designs calculated based on the number of FPGA slices that each design (including memories)

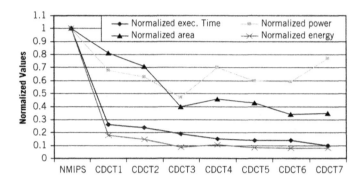

FIGURE 13.32

Comparing different DCT implementations.

occupies. The area trend also confirms the increase in area in CDCT4 followed by a decrease in CDCT5, evidently due to retiming.

Fig. 13.32 shows the performance, power, energy, and area of the designs normalized against NMIPS. The total execution delay of the DCT algorithm has a decreasing trend, while the power consumption decreases up to CDCT3 and then increases. The energy consumption significantly drops at CDCT1 because of the reduction in number of cycles and power consumption. From CDCT1 to CDCT7, the energy decreases gradually in a slow pace.

As shown in Fig. 13.32, CDCT7 is the best design in terms of delay and energy consumption, while CDCT3 is best in terms of power, and CDCT6 in terms of area. As a result, CDCT3, CDCT6, and CDCT7 are considered the pareto-optimal solutions. Note that minimum energy and minimum power are achieved by two different designs: CDCT7 and CDCT3, respectively. Compared to NMIPS, CDCT7 runs 10 times faster, and consumes 1.3 times less power and 12.8 times less energy. It also occupies 2.9 times less area than NMIPS.

We also compared the quality of final design (CDCT7) with a commercial manual design [20]. Their design takes 82 cycles to compute an 8×8 DCT with a 15-bit precision (ours has a 16-bit precision). They have achieved maximum clock frequency of 74 MHz on the FPGA package (we achieved 250 MHz). Therefore, their total execution time of an 8×8 DCT is 1.1 μs. Compared to NMIPS that takes 137.57 μs, the manual design is 125 times faster. This clearly shows two orders of magnitude performance gap between the manual design and software implementation. Compared to CDCT7 that takes 13.84 us to compute DCT, the manual design is 12.5 times faster. These results show that a custom NISC architecture can serve as an intermediate point between software and hardware implementations.

On the other hand, the total area of the manual design is 1365 FPGA slices, while the area of CDCT7 is 220 slices. Note that the low area of CDCT7 allows fitting six of CDCT7 in the same area as that of the manual hardware design. Since the DCT algorithm can usually run on different parts of an image in parallel, the performance of six CDCT7 is almost six times of the performance of one. This makes the CDCT7 only two times slower than the manual hardware design. Note that it took only a couple of days to explore different DCT design alternatives using NISC while it usually takes a significantly longer time to implement and verify manual designs.

To summarize, designing a custom datapath for a given application by properly connecting functional units and pipeline registers is the key to reducing the number of cycles and energy consumption. Also, eliminating the unused logic and interconnects, adjusting the bus-width of the datapath to the application requirement, signal gating, and clock gating are the key to reducing power consumption. The NISC toolset makes these customizations very easy to apply. It allows the designer to modify the component netlist of the datapath and then use the NISC toolset to automatically map the application on the given datapath and generate the results.

13.4.3 **Communicating NISC Components**

In this section, we describe the implementation results of two multi-NISC systems for a fixed-point Mp3 benchmark downloaded from [21]. These NISCs communicate via the shared bus protocol implemented similar to what was described in Section 13.3.7. In general, an Mp3 audio file contains several frames. For a stereo file, each frame has two channels (i.e., left and right). In the Mp3 decoder, the frames go through three main phases, namely, *decode_frame*, *synthesis_frame*, and *output_pcm*. Profiling the Mp3 decoder on the generic NISC architecture of Fig. 13.33 showed that 63% of execution time is spent in *decode_frame*, 25% in *synthesis_frame*, and 11% in the *output_pcm*. We realized that there are two approaches to parallelize the Mp3 application: (a) processing each channel separately, or (b) pipelining the phases. However, the Mp3 decoder was originally targeted for desktop PCs and separating the channels completely requires rewriting most of the code. Alternatively, we decided to separate the *synthesis_frame* phase for each

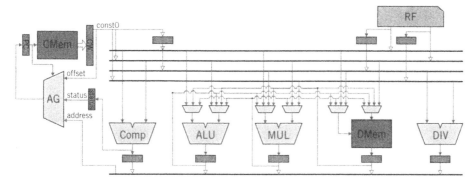

FIGURE 13.33

A generic NISC architecture (GN).

channel because it required minimum code modifications. Such partitioning can reduce the execution time of *synthesis_frame* to half and hence can at most improve the performance by 12.5%. As for the second system, we pipelined the application into two stages where the first pipeline stage implements the *decode_frame* phase and the second stage implements *synthesis_frame* and *output_pcm* phases. In this approach, processing delay of one frame is expected to increase due to the communication overhead. However, since the *decode_frame* of one frame is overlapped with the *synthesis_frame* and *output_pcm* of another frame, the overall performance can be improved by up to 36% (min(63, 25 + 11)).

We implemented the Mp3 decoder on a MicroBlaze, a single GN, and two multicore configurations of GN. Table 13.6 shows the clock speed and area of each architecture as well as their performance for decoding one frame of audio.

Table 13.7 shows the results of implementing the Mp3 decoder in three configurations. The second and fourth columns show the number of cycles for processing one frame, and 25 frames in each configuration and the third and fifth columns show the respective speedups. Fig. 13.34 shows the block diagram of the three implementation configurations. Fig. 13.34(a) shows the *SingleCore* configuration in which the entire Mp3 decoder runs on one GN. Fig. 13.34(b) shows the *Coprocessor* configuration in which the Mp3 decoder runs on two GAs. In this case, one of the GNs acts as a coprocessor for the main GN and runs the *synthesis_frame* phase for the left channel while the main GN runs the same phase for the right channel. The main GN also runs the other phases for both channels. The total performance improvement

Table 13.6 Area and clock frequency of MicroBlaze and GN.

Processors	Clock Freq.(MHz)	Area (Gates)	#Cycles for 1 Frame	Speedup
MicroBlaze	105	39574	8,861,336	1
GN	80	35632	897,452	7.28
Multicore GN	80	73046	–	–

Table 13.7 Throughput of three Mp3 implementations.

Systems	#Cycles for 1 Frame	Speedup for 1 Frame(%)	#Cycles for 25 Frames	Speedup for 25 Frames(%)	Frames/sec
SingleCore	897,452	0.00	22,800,961	0.00	88
Coprocessor	803,357	10.48	20,205,994	11.38	99
Pipelined	917,204	−2.20	16,433,655	27.93	122

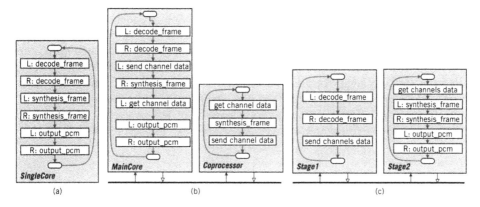

FIGURE 13.34

Implementing Mp3 (a) single core, (b) with coprocessor, and (c) pipelined.

in this case is 10.48%, which is close to the expected 12.5%. For each channel, the main GN sends 1152 words to the coprocessor GN and then receives 1152 words from it. The communication overhead is responsible for the 2% performance loss from the expected upper bound, i.e., 12.5%. Fig. 13.34(c) shows the *Pipelined* configuration, where one GN runs the *decode_ frame* of both channels and sends 2 × 1152 words to the second GN to perform *synthesis_ frame* and *output_ pcm*. In this configuration, the processing time for a single frame is increased by 2%, but the overall throughput of the system is increased by 28%. Similarly, the communication overhead is responsible for the 8% performance loss from the expected upper bound, i.e., 36%. The communication overhead in the *Pipelined* configuration has increased because of the extra synchronization, which was not necessary in the *Coprocessor-Sys* configuration.

According to the Mp3 standard, at least 38 frames must be played per second. MicroBlaze can only run 12 frames per second. The last column of Table 13.7 shows the throughput of these three NISC-based configurations. Clearly, this throughput is much more than the required standard. To save power, the *SingleCore* and *Coprocessor* configurations can run with half their clock frequency. The clock frequency of the *Pipelined* system can be reduced by two-thirds and still meet the throughput constraints of the standard.

13.5 CONCLUSION

This chapter presented Generic Netlist Representation (GNR) for describing general-purpose and custom embedded processors as well as multicore systems using NISC Technology. In NISC Technology the target architecture is a statically scheduled nanocoded architecture that does not have any predefined instruction-set. Therefore,

GNR does not contain any instruction-set or instruction decoder specification, and hence is very concise. The GNR is strongly typed and defines extensible aspects for components so that different tools (such as the compiler and HDL generator) can process them properly. The GNR can also be used to describe multicore systems on a pin-accurate level. Therefore, it can be used for single-core and system-level explorations as well as an intermediate step toward synthesis of TLM to RTL. Several case studies for designing and optimizing single-core and multicore systems using NISC and GNR were presented in this chapter. The NISC toolset is released for public use [6].

REFERENCES

[1] M. Reshadi and D. Gajski. A Cycle-Accurate Compilation Algorithm for Custom Pipelined Data-paths. in *International Symposium on Hardware/Software Codesign and System Synthesis (CODES+ISSS)*, 2005.

[2] A. Agrawala and T. Rauscher. *Foundations of Microprogramming: Architecture, Software, and Applications*. Academic Press, 1976.

[3] B. Gorjiara and D. Gajski. FPGA-friendly Code Compression for Horizontal Microcoded Custom IPs, in *International Symposium on Field-Programmable Gate Arrays (FPGA)*, 2007.

[4] B. Gorjiara. *Synthesis and Optimization of Low-power Custom NISC processors*, PhD dissertation, University of California, Irvine, 2007.

[5] TMS320C62xx CPU and Instruction Set: Reference Guide Texas Instruments, Dallas, 1997.

[6] NISC Technology website: *http://www.cecs.uci.edu/~nisc/*.

[7] B. Gorjiara and D. Gajski. A Novel Profile-driven Technique for Simultaneous Power and Code-size Optimization of Nanocoded IPs. in *International Conference on Computer Design (ICCD)*, October 2007.

[8] M. Reshadi. *No-Instruction-Set-Computer (NISC) Technology Modeling and Compilation*, PhD dissertation, University of California, Irvine, 2007.

[9] XML: *http://www.w3.org/XML/*

[10] XML Schema: *http://www.w3.org/XML/Schema/*

[11] M. Reshadi and D. Gajski. Interrupt and Low-level Programming Support for Expanding the Application Domain of Statically-scheduled Horizontally-microcoded Architectures in Embed-ded Systems. *Design Automation and Test in Europe (DATE)*, April 2007.

[12] B. Gorjiara, M. Reshadi, and D. Gajski. NISC Communication Interface. *Center for Embedded Computer Systems (CECS) Technical Report TR 06-05*, 2006.

[13] J. Hennessy and D. Patterson. *Computer Architecture: A Quantitative Approach*. Morgan Kaufmann Publishers, San Mateo, CA, 1990.

[14] B. Rau, D. Yen, W. Yen, and R. Towle. The Cydra 5 Departmental Supercomputer: Design Philosophies, Decisions and Trade-offs. *IEEE Computers*, 22(1):12–34, January 1989.

[15] R. CodWell, R. Nix, J. O Donnell, D. Papworth, and P. Rodman. A VLIW Architecture for a Trace Scheduling Compiler, *ACM SIGOPS Operating Systems Review*, 21(4), 1987.

[16] MiBench benchmark: *http://www.eecs.umich.edu/mibench/*

[17] B. Gorjiara and D. Gajski. Custom Processor Design Using NISC: A Case-Study on DCT algorithm, in *IEEE Workshop on Embedded Systems for Real-Time Multimedia (ESTIMEDIA)*, 2005.

[18] N. Ahmed, T. Natarajan, and K. R. Rao. Discrete Cosine Transform, in *IEEE Trans. On Computers*, Vol. C–23, 1974.

[19] MIPS32® M4K™ Core: *http://www.mips.com.*

[20] Custom DCT: *http://www.cast-inc.com/cores/dct/cast_dct-x.pdf*, December 2007.

[21] MPEG Audio Decoder: *http://www.underbit.com/products/mad/*, (Last accessed April 10, 2008).

13.6 INDEX TERMS

ADL	GNR	Pipeline
Code size	HDL	Prebinding
Communication	Hierarchical component	Refinement
Compiler	Machine Action	RTL
Compression	Microcode	Synthesis
Customization	Multicore	Third-party core
Cycle-accurate compiler	Multicycle	Verilog
Datapath	Nanocode	VLIW
Formal	NISC Technology	Xilinx
Forwarding path	No-Instruction-Set-	XML
FPGA	Computer	

HMDES, ISDL, and Other Contemporary ADLs

14

Nirmalya Bandyopadhyay, Kanad Basu, and Prabhat Mishra

This chapter describes various contemporary ADLs that are not covered in the earlier chapters. Section 14.1 describes the HMDES ADL, which is used in the Trimaran research infrastructure for compilation and architectural exploration. Section 14.2 describes ISDL, which is a behavioral ADL. This chapter also briefly describes other significant ADLs and associated methodologies including RADL, Sim-nML, UDL/I, Flexware, Valen-C, and TDL.

14.1 HMDES

Machine description language HMDES [1] was developed at University of Illinois at Urbana-Champaign for the IMPACT research compiler. It captures the processor resources and their usage by the instruction set in a programmer and compiler-friendly way. It also supports C-like preprocessing capabilities such as file inclusion, macro expansion, and conditional compilation. The HMDES is used as the specification language for the Trimaran compiler infrastructure [2] that contains IMPACT as well as Elcor research compiler from HP labs. The framework supported by Trimaran is suitable for Explicitly Parallel Instruction Computing (EPIC) architectures. This section describes the HMDES language and Trimaran infrastructure.

14.1.1 HMDES Language

The HMDES (also known as MDES) expresses the machine description in a structured way that is easy to read, write, and modify for both a compiler writer and language user. This language along with the machine description (MD) libraries and tools can be used to write complex machine descriptions. It captures the design in a form of hierarchical specification, where each level in the hierarchy can be placed in a separate MD file. These features are convenient for writing processor description as they allow compiler-specific machine descriptions to be written on top of compiler-independent machine descriptions. It provides a framework where variations of an architecture can be built over the base architecture description [1]. HMDES also

369

supports powerful preprocessing capabilities that facilitates concise representation of structured information.

In HMDES language, structural information is first broken down into sections based on a high-level partitioning of the information. For example, a processor description is divided into three sections. The first section describes the instruction set, the second talks about the resources, and the third one gives information about the registers. Each section contains some entries based on various types of information contained in that section. For example, a section for instruction set (called as *operations* in HMDES) contains one entry for each instruction in the processor. Every entry consists of MD fields based on low-level classification of the information. For example, entries describing processor registers might use one field for register width and another field for compatible registers that can be used as proxy of the first register.

Language syntax

The programming syntax to create a section named *section_name* is as follows:

```
CREATE SECTION section_name
{
}
```

Once a section is created, it can be referred by omitting the CREATE keyword. For example, to create a section *Resource_usage* the programmer needs to write:

```
CREATE SECTION Resource_usage
  // Fields may be declared here
{
  // Entries may be created here
}

//Reference to the existing section Resource_usage
SECTION Resource_usage
  //Additional field declaration may go here
{
  //Entries may be created and/or modified here
}
```

Fields of a section are declared between the section name and the opening "{" bracket while creating a new section or referring to an existing section. The generic form of field declaration looks like,

```
FIELD_TYPE field_name (ELEMENT_TYPE1 ELEMENT_TYPE2 ... ELEMENT_TYPEN);
```

Each field declaration consists of a FIELD_TYPE, name of the field and a declaration of the expected content of the field (ELEMENT_TYPE1 ... ELEMENT_TYPEN). The FIELD_TYPE can be of three types—REQUIRED,

OPTIONAL, and IGNORED. The REQUIRED field implies that usage of this field is mandatory for each entry in the section. Usage of OPTIONAL field is optional, as the name suggests. IGNORED fields are never used. ELEMENT_TYPE may be of INT (integer number), FLOAT (floating point number), STRING (a character string) or LINK. A LINK is the name of an entry from one of the sections specified in the field declaration. In the following example, INT* means that zero or more occurrences of the *element* is possible.

```
CREATE SECTION usage
  REQUIRED time1(INT*);
  OPTIONAL time2(INT*);
{
  ...
}
```

Once a section and its fields are defined, the next task is to enumerate the *entries* inside the section. To create or modify entries for a section directives are placed between the open "{" and close "}" brackets. An entry is created and modified in the following way:

```
entry_name (field1_name(field1_contents) ...
                 fieldN_name(fieldN_contents));
```

In the following example the section *Resource_Usage* enumerates three different kinds of resource usages. Each entry represents a connection between a resource and the time to use that resource.

```
CREATE SECTION Resource_Usage
  REQUIRED use(LINK(RESOURCE));
  REQUIRED time(INT INT*);
{
}
SECTION Resource_usage
{
  RU_slot0                   (use(slot0) time(0));
  RU_integer_alu             (use(integer_alu)) time(0));
  RU_floating_point_alu_s0   (use(floating_point_alu_s0) time(0));
}
```

The MD preprocessor

The MD preprocessor is based on C preprocessor and the Unix shell language. It has a fully recursive implementation allowing the nesting of preprocessor directives. This section briefly describes some of the interesting features of MD preprocessor [1].

The preprocessor supports both the nested C-style comments (/* */) and C++ style single line comments (//). To include a file, the directive will be:

```
$include "name_of_the_file_to_include".
```

The text replacement directive can take two forms:

```
$def def_name def_value
$def def_name {bounded_def_value}
```

In the first form everything between *def_value* and the end of the line with leading and trailing space removed replaces *def_name*. For the second case *def_name* is substituted by everything between "{" and "}". The bounded_def_value can include new lines and matched pairs of curly brackets "{" "}". For example,

```
$def NAME1  name1 text
$def NAME2 { name2 text }
"NAME1"                    "name1 text"
"NAME2"                    "name2 text"
(a) Before preprocessing  (b) After preprocessing
```

HMDES supports expression evaluation.

```
$ = {expr}
```

Like C it also permits conditional inclusion such as,

```
$if(cond){body}
$elif(cond){body}
$else{body}
```

For example,

```
$def CONDVAR 10
$if(CONDVAR == 1) {one}
$else {not one}                 not one
(a) Before Preprocessing    (b) After preprocessing
```

It supports *for loop* that is used for iteration.

```
$for (def_name in value1 value2 ... valueN) {for_body}
```

For each value in the list *value1 value2 .. valueN*, for_body is replicated and def_name replaced by the corresponding value. The values may be specified in one of the following ways:

- A string containing letters, numbers, "-", ".", "+", and "_" (No white space is allowed).
- A quoted string.
- A string bounded by "{" "}".

For example,

```
$for (I in "text 1" {text 2} text 3)      "text 1"
{                                          text 2
   I                                       text
}                                          3
(a) Before Preprocessing                  (b) After Preprocessing
```

The order in which various preprocessing steps take place is given here.

1. Comment removal

2. Text replacement

3. Expression calculation

4. Range expansion

5. Other directives, in the order encountered.

14.1.2 Structural Overview of Machine Description

The HMDES provides a framework to write processor description in a structured way. The framework consists of some components known as *information* in HMDES terminology. Each information is nothing but a logical collection of closely related sections. This consolidated framework provides an abstraction of the machine description in terms of structure (registers, flags etc.) and functionality (operations, latency). Fig. 14.1 gives a pictorial description of this framework. In that figure the *information* components are represented by rectangular boxes. Each information box lists the corresponding *Sections* of that information. The arrows between the sections depict the relation and dependency among the sections.

The remainder of this section describes the six important sections of HMDES that provides structural and behavioral infrastructure required to model an HPL-PD architecture: format information, resource usage information, latency information, operation information, register information, and compiler-specific information. Code fragments of PA-7100 processor written in HMDES are used as illustrative examples.

Format information

Format information in HMDES describes the operands that are supported by different operations (instructions) of a processor. Two MD sections, *Field_Type* and *Operation_Format* are used for this purpose. Entries in the Field_Type section describe the types of operands that can be placed in the operand fields of an assembly operation. Operation_Format entries are used to describe the operation formats supported by the processor. More specifically, it enumerates the field types of operands in operations.

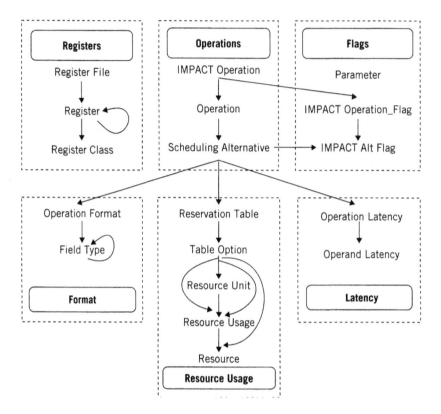

FIGURE 14.1

Structural overview of machine description.

Various operands that may be placed in an assembly operation are enumerated in the Field_Type section. The HMDES language definition for this section looks like:

```
CREATE SECTION Field_Type
    OPTIONAL compatible_with(LINK(Field_Type)*);
{
}
```

The *compatible_with* field is used to specify the flexibility of using one type of operand as a substitute of another. For example, a five-bit literal can be placed in a field big enough to hold a 16-bit literal. In the following example some literal widths of PA-7100 are listed. By convention, the "NULL" field-type specifies that no operand is allowed in that field.

```
SECTION Field_Type
{
    NULL        ();
```

```
REG          ();
Literal5     ();
Literal1     (compatible_with(Literal5));
Label        ();
Literal2     (compatible_with(Literal1 Label1));
}
```

The *Operation_Format* section describes the formats of the assembly level operations in this language. For example, it can specify whether the destination field of an operation should use a register or a 16-bit literal. To define this section HMDES code looks like:

```
CREATE SECTION Operation_Format
    OPTIONAL dest(LINK(Field_Type)*);
    OPTIONAL src(LINK(Field_Type)*);
    OPTIONAL pred(LINK(Field_Type)*);
{
}
```

The "dest", "src" and "pred" list the field types expected for each destination, source and predicate operand, respectively. For example, the Operation_Format section for a code fragment of PA-7100 is shown here:

```
SECTION Operation_Format
{
    Operation_Format_Std1  (dest(REG) src(REG REG));
    Operation_Format_Std2  (dest(REG) src(Literal1 REG));
    Operation_Format_Std3  (dest(REG) sec(REG Literal1));
}
```

Resource Usage Information

Resource usage information describes the processor's resources and the usage of them during execution of the processor. Five HMDES language sections (Resource, Resource_Usage, Resource_Unit, Table_Option, and Reservation_Table) are members of this information. The entries in the Resource section are used to enumerate the resources of the processor.

The Resource_Usage section describes the time when a resource can be used. The Resource_Unit section entries group these resource usages into more meaningful units such as ALU functional units or load-store units. The Table_Option section describes options of available and compatible resources where only one of them needs to be selected. Finally, the Reservation_Table section describes how an operation can use the resources of the processor as it executes. In the *Resource* section the resources of the processor are enumerated. Slots for execution, stages in the integer ALU and floating point ALU are valid examples for entries in the *Resource* section. The HMDES language specification for this section is as follows:

```
CREATE SECTION Resource
{
}
```

To model PA-7100's resources the HMDES description looks like,

```
SECTION Resource
{
    first_slot      (); // Instruction issue slots
    second_slot     ();
    float_issue     (); // Only one floating point operation per cycle
    int_alu         (); // can be issued.
    float_alu_s0    ();
    float_alu_s1    ();
    float_mul_s0    ();
    float_div       ();
    memory          ();
    ...
    ...
}
```

In the aforementioned example, a set of resources are described in the *Resource* section of PA-7100's description. PA-7100 can issue up to two operations per cycle through two issue slots modeled by first_slot and second_slot, but can issue only one floating point instruction and only one integer instruction per cycle. Load and store are considered as integer operations. The limitation of issuing only one integer operation is modeled by one integer ALU (int_alu). There is more than one floating point units (float_alu_s0, float_alu_s1, float_div etc.) though float_issue implies that only a single floating point operation is issued at a time.

Resources are used at particular times, as executions of operations take place. For example, the floating point add operation uses the first stage of the floating point ALU one cycle before it uses the second stage. By convention, the first cycle of an operation is designated as 0th cycle (the time the operation enters the functional unit). From this convention it can be inferred that the floating point add operation uses the first stage of the floating point ALU (float_alu_s0) at time 0 and the second stage (float_alu_s1) at time 1.

Resource_Usage section describes the time resources are used by the operations during execution. HMDES definition of this section is as follows:

```
CREATE SECTION  Resource_Usage
    REQUIRED    use(LINK(resource));
    REQUIRED    time(INT INT*);
{
}
```

Each entry in the Resource_Usage section associates a resource "use" with a list of one or more usage times. As shown here, the first cycle in the execution unit is considered as time 0.

```
SECTION Resource_usage
{
    RU_first_slot          (use(first_slot) time(0));
    RU_second_slot         (use(second_slot) time(0));
    RU_float_issue         (use(float_issue) time(0));

    RU_int_alu             (use(int_alu) time(0));
    RU_float_alu_s0        (use(float_alu_first_slot) time(0));

    RU_float_div_t0_7      (use(float_div) time($0..7));
    RU_float_div_t0_14     (use(float_div) time($0..14));
}
```

The HMDES also provides an option to group a set of resource usage together to form more meaningful units, such as function units or issue units. The grouping is accomplished using the Resource_Unit section.

```
CREATE SECTION Resource_Unit
    REQUIRED use(LINK (Resource_Usage) LINK(Resource_Usage)*);
{
}
```

An example of *Resource_Unit* is to make logical groups of all the resources used in a stage of a pipeline. For example, an assembly of the execution resource usage looks like,

```
SECTION Resource_Unit
{
    first_float_issue_unit    (use(RU_first_slot RU_float_issue));
    second_float_issue_unit   (use(RU_second_slot RU_float_issue));
    float_mul_unit            (use(RU_float_alu_s0 RU_float_alu_s1
                               Ru_float_mul_s1 RU_float_mul_fake_t0));
}
```

The entries first_float_issue_unit and second_float_issue_unit represent two possibilities for issuing floating point operations in first_slot and in second_slot. Both Resource_Unit entries use the float issue unit to model the constraint that only one floating point operation can be issued per cycle.

It is often observed that operations have some flexibility in terms of the resources they use during execution. For example, there may be several functional units capable of executing a particular operation. The flexible use of resources is modeled using Table_Option entries, where a set of Resource_Usage entries and/or Resource_Unit entries are available to choose a favorable option.

```
CREATE SECTION Table_Option
    REQUIRED one_of(LINK(Resource_Unit|Resource_Usage)
                    LINK(Resource_Unit|Resource_Usage)*);
{
}
```

Each table option entry consists of a list of Resource_Unit or Resource_Usage entry names in the "one-of" field. By convention, the first entry in the list is the most desirable option, the second entry is the second most useful option, and so on. For example,

```
SECTION Table_Option
{
    any_int_issue       (one_of(RU_first_slot RU_second_slot));
    any_float_issue     (one_of(first_float_issue_unit
                                second_float_issue_unit));
}
```

The Resource_Usage entries, Resource_Unit entries and the Table_Option entries described so far are used in the "Reservation_Table" section to build reservation tables. Each Reservation_Table entry models how an operation can use resources over its entire execution time. For example, a Reservation_Table entry for a floating point ALU can specify which one of the floating point issue units will be used along with the single ALU. The HMDES declaration for the *Reservation_Table* section is as follows:

```
CREATE SECTION Reservation_Table
    REQUIRED use(LINK(Table_Option|Resource_Unit|Resource_Usage)*);
{
}
```

All the Reservation_Table entries, Table_Option entries, Resource_Unit entries, and/or Resource_Usage entries that are used as the operation executes are listed in the "use" field. To schedule the operation, the requirements specified in the "use" field must be satisfied.

Latency information

The latency information models how to calculate dependence distances between operations. Two HMDES language sections, Operand_Latency and Operation_Latency, are used for this purpose. The Operand_Latency section entries are used to specify the times when operands can be used. Entries in the Operation_Latency section describes when an operation uses its operands.

Operations use their operands at various time points during execution. Registers used as source operands are typically required just before the execution stage and

the destination register operands are effectively written just as the operation exits the execution stage. By considering when operands are used during the execution of the operation, dependence distances can be calculated for the various flow, anti, and output dependences between operations. The times when the operands may be used are enumerated in the *Operand_Latency* section that has the following HMDES definition:

```
CREATE SECTION Operand_Latency
    REQUIRED time(INT*);
{
}
```

For each entries in the Operand_latency section the corresponding "time" field lists the times when that operand can be required. By convention, time 0 is fixed to be the moment just before an operation enters its execution stage (at the very end of the cycle before the execution stage is entered). Thus, time 1 is at the very end of the first cycle in the execution stage.

One example depicting the time of operand usage for the PA-7100 is shown below. The PA-7100 integer ALU operations write their result at time 1 (dest1), loads and most floating point operations write their result at time 2 (dest2), an interlocked floating point multiply writes its result at time 3 (dest3) and the floating point divides write their result at time 8 (dest8) or 15 (dest15), depending on their precision. The names inside the parenthesis indicates the corresponding entries in Operand_Latency section given later. Typically operations read their source operands at time 0 (source0), but the store operation reads the data to store at time 2 (source2).

```
SECTION Operand_Latency
{
    $for (N in 1 2 3 8 15)
    {
        dest${N} (time(N));                    // Destination operands
    }
    $for    (N in 0 2)  {source${N}  (time(N); } // Source operands
}
```

The times enumerated in the *Operand_Latency* section are used by the *Operation_Latency* entries to specify when an operation uses its operands. Following code segment outlines the format of the section Operation_Latency.

```
CREATE SECTION Operation_Latency
    OPTIONAL dest(LINK(Operand_Latency)*);
    OPTIONAL src(LINK(Operand_Latency)*);
    OPTIONAL pred(LINK(Operand_Latency)*);
```

```
        OPTIONAL mem_dest(LINK(Operand_Latency));
        OPTIONAL ctrl_dest(LINK(Operand_Latency));
        OPTIONAL sync_dest(LINK(Operand_Latency));

        OPTIONAL mem_src(LINK(Operand_Latency));
        OPTIONAL ctrl_src(LINK(Operand_Latency));
        OPTIONAL sync_src(LINK(Operand_Latency));
  {
  }
```

The fields "dest", "src", and "pred" associate an Operand_Latency entry with each destination, source, and predicate operands of the operation, respectively. The Operand_Latency entries specify the time(s) when each of those operands are used by the operation. This information, can be used to calculate the dependence distance for all register flow, anti, and output dependences.

The remaining fields in the Operation_Latency section are used to determine the dependence distances for memory, control, and synchronization dependencies. The "ctrl_dest" and "ctrl_src" fields are used to calculate control dependence distances (dependencies preventing operations from moving above or below a branch). The "sync_dest" and "sync_src" fields are used to calculate synchronization dependence.

Operation information

The *Operation* information describes the operations supported by the architecture and how they may be scheduled. Two compiler independent sections *Scheduling_Alternative* and *Operation* are used for this purpose. Additional compiler specific HMDES sections can be required to propagate the operation information to the compiler.

The entries of the section Scheduling_Alternative are used to describe the ways that operations can be scheduled in terms of their operands, the resources they use, and their dependence distance information. The entries of the Operation section specifies the assembly-level operations supported by the architecture and associates scheduling alternatives with each of these operations. A scheduling alternative is a set of three requirements that must be met in order to schedule an operation using that alternative. These three requirements are:

1. The resources required by the alternative (specified by a Reservation Table entry) will be available.

2. The register, control, memory, and synchronization dependence distances for the alternative (specified by a Operation Latency entry) must be satisfied.

3. The operands of an operation must be compatible with alternative allowable operands (specified by one or more Operation Format entry).

The HMDES declaration for the *Scheduling_Alternative* is as follows:

```
CREATE SECTION Scheduling_Alternative
    REQUIRED format (LINK(Operation_format) LINK(Operation_Format)*);
    REQUIRED resv (LINK(Reservation_Table));
    REQUIRED latency (LINK(Operation_Latency));
{
}
```

For example, one way to model the scheduling alternative of PA-7100's floating point multiply and the PA-7100's load operations is shown below.

```
SECTION Scheduling_Alternative
{
    ALT_FMul2      (format(OF_Std1) resv(RT_FPMult2) latency(OL_Lat2));
    ALT_FMul3      (format(OF_Std1) resv(RT_FPMult3) latency(OL_Lat3));

    //Load operations
    ALT_LD_Short   (format(OF_Load1) resv(RT_Load) latency(OL_load));
    ALT_LD_Long    (format(OF_Load2 OF_LOAD2_REG OF_LOAD2_LABEL)
                   RESV(RT_Load) latency(OL_Load));
    . . .
    . . .
}
```

Here floating-point multiply normally takes two cycles, but can interlock and take three cycles when issued six cycles after a single-precision floating-point division. The two-cycle noninterlocking floating-point multiply case is represented by the Scheduling_Alternative entry *ALT_FMul2*. The three-cycle interlocking case is modeled by *ALT_FMul3*. The two and three-cycle latencies are set by the Operation_Latency entry names OL_Lat2 and OL_Lat3, respectively. Every assembly-level operation in the architecture is enumerated in the *Operation* section, which has the following HMDES definition:

```
CREATE SECTION Operation
    REQUIRED alt (LINK(Scheduling_Alternative)
                 LINK(Scheduling_Alternative)*);
{
}
```

For each assembly level operation, one entry is created in the Operation section. As already defined, each Operation entry requires the enumeration of one or more scheduling alternatives for the operation, in the "alt" field. Through these scheduling alternatives, the operation associates with itself the formats it supports, the resources it uses during execution, and its own latency information.

For example, PA-7100's floating point multiply and load operations, in an Operation section is shown here.

```
SECTION Operation
{
    //FP Mul takes 2 cycles unless interlocks due to
    //conflict with divide
    OP_MUL_F    (alt(ALT_FMul2 ALT_FMul3));
    OP_MUL_F2   (alt(ALT_FMul2 ALT_FMul3));

    $for(TYPE in C C2 I F F2)
    {
       OP_LD_SHORT_${TYPE}      (alt(ALT_LD_Short));
       OP_LD_INDEX_${TYPE}      (alt(ALT_LD_Index));
     ...
    }
    ...
}
```

The two operation entries OP_MUL_F and OP_MUL_F2 list the two scheduling
alternatives ALT_Fmul2 and ALT_FMul3. It specifies that ALT_Fmul2 has the higher
scheduling priority.

Register information

The compiler-independent register information describes how the registers overlap. In addition, compiler-specific machine description extensions can be used to
describe how to allocate registers for the architecture. HMDES section *Register* is
used to represent the compiler-independent information. Two more sections *Regis-
ter_Class* and *Register_File* and some additional register section fields can be added
to represent the compiler-specific register allocation information.

The way that processor registers overlap is described by the entries of Register
section. When determining the register dependencies, it is necessary to know how
physical (and virtual) registers overlap so that proper flow, anti- and output dependencies may be created after (before) register allocation. The "Register" section is
used to describe this overlap as shown here:

```
CREATE SECTION Register
    OPTIONAL overlaps (LINK(Register)*);
{
}
```

Compiler-specific information

The compiler-specific information is used to provide any additional information the
compiler needs about the architecture. For example, three IMPACT-specific language
sections are described in this part: Parameter, IMPACT Operation Flag, and IMPACT
Alt Flag. The information in *Parameter* section is used to pass architecture-specific
parameters to the compiler. The entries in *IMPACT Operation Flag* and *IMPACT Alt*

Flag sections are used to enumerate the flags used by an IMPACT-specific extension of the machine description.

14.1.3 Trimaran Infrastructure

Trimaran [2] is an integrated compilation and performance monitoring infrastructure that uses HMDES as its front end specification language. Trimaran is primarily used for HPL-PD [3], a parameterized processor architecture. HPL-PD supports novel features such as predication, control, and data speculation and compiler-controlled management of the memory hierarchy. Trimaran was started as a collaborative effort between the Compiler and Architecture Research (CAR) Group (once a member of Hewlett Packard Laboratories), the IMPACT Group at the University of Illinois, and the ReaCT-ILP Laboratory at New York University (CREST, the Center for Research on Embedded Systems and Technology at the Georgia Institute of Technology). Currently Trimaran is being actively developed and maintained by the CCCP group at the University of Michigan and the Commit Group at MIT [2]. The Trimaran infrastructure comprises of the following components [2].

1. A machine description language HMDES for writing specification of the machine.

2. A parameterized processor HPL-PD that supports Instruction-level parallelism (ILP).

3. A front-end C compiler called IMPACT. It takes care of parsing, type checking, and some high-level optimizations.

4. A compiler back end, called Elcor is also parameterized by HMDES. It performs instruction scheduling, register allocation, and machine-dependent low-level optimizations.

5. An extensible intermediate program representation (IR) equipped with many modern compiler techniques is also used.

6. A cycle accurate simulator is provided configurable by HMDES. It provides run-time information on execution time, branch frequencies, and utilization of resources.

7. An integrated graphical user interface for configuring and running Trimaran system.

Fig. 14.2 shows the overview of the Trimaran system overview. The machine description model (HMDES) allows the user to develop a machine description for the HPL-PD family of processors in a high-level language. This processor description is then translated into a low-level representation Lmdes for efficient use by the compiler. The high-level language allows the specification of detailed execution constraints in a maintainable, retargetable, and easy-to-understand manner. The

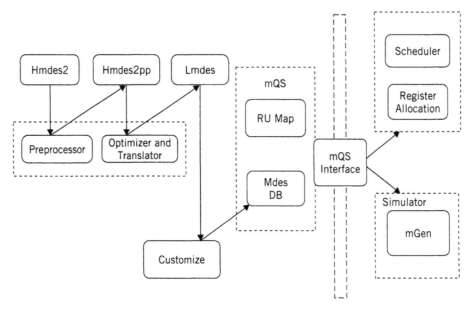

FIGURE 14.2

HMDES in Trimaran.

low-level representation is designed to allow the compiler to check the execution constraints with high space–time efficiency. The target architecture (HPL-PD processor) is described by HMDES [2]. After the macro-preprocessing and compilation of the high-level description takes place, the corresponding low-level specification in Lmdes Version 2 is loaded within the compiler using the *customize* module that reads the textual Lmdes specification and builds the internal data-structures of the HMDES database. The information obtained within the machine description database is made available to various modules of the Trimaran infrastructure through a query interface called mQS. The queries available to mQS are used in the simulator and compiler of the Trimaran. The details of the Trimaran infrastructure is available at *http://www.trimaran.org* [2].

14.2 ISDL

Instruction Set Description Language (ISDL), is a behavioral machine description language. The ISDL was developed at MIT and used by the Aviv compiler [4] and Gensim simulator generator [5]. Fig. 14.3 shows the overview of ISDL-based processor design. The compiler front end takes a source program written in C or C++, and then parses it to produce an intermediate format description in SUIF. The compiler back end takes this SUIF and ISDL description as inputs to produce assembly code specific to the target processor.

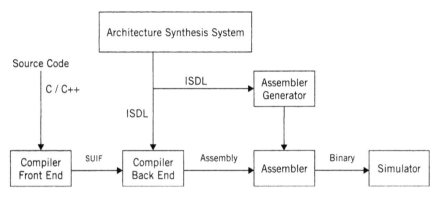

FIGURE 14.3

High-level view of system.

14.2.1 ISDL Language

This section describes the six primary components of ISDL: instruction word format, global definitions, storage resources, assembly syntax, constraints, and optimization information.

Instruction word format

The instruction word format defines the instruction word, which is divided into fields. Each of these fields are divided into different subfields, the bitwidth of which are provided as well. The instruction word is assembled by concatenating all the subfields in a specified order. The following example describes how a 24-bit instruction word is divided into 3 subfields of 8 bits each namely, DBM.OP, DBM.MODE and MAIN.OP.

```
DBM = OP[8],MODE[8];
Main= OP[8];
```

Global definitions

The Global Definitions section defines four types: Tokens, Non-terminals, Split Functions, and Macro definitions. Tokens represent the primitive operands of an instruction like register and memory bank names, immediate constants, etc. Syntactically related tokens are grouped together. The tokens differentiate between elements in a group by returing a value which corresponds to a particular element. An example of token definition of a binary operand is shown here:

```
Token X[0..1] X_R ival {yylval.ival = yytext[1] - '0';}
```

In this example, X[0..1] is the assmebly format of the operand. X_R is the symbolic name of the token used for reference. The *ival* is used to describe the type of value returned by the token. The last field describes the computation of the value. In this case, the values returned may be 0 or 1.

Nonterminals serve a number of useful purposes. They are used to group syntactically unrelated tokens. A number of possible alternatives in an instruction are factored out into nonterminals. The following example demonstrates the fact:

```
MOVE X:(addressing mode) Y:(addressing mode)
```

If there are 6 addressing modes, then this would require 36 rules to describe all possible alternatives if there were no nonterminals. However, with nonterminals, it requires two rules: One for the instruction and the other for the addressing mode. Every nonterminal has an action portion that allows it to include arbitary C code to be executed with every rule. Nonterminals also help in defining new sets of rules. The following code segment defines a nonterminal named XYSRC:

```
Non_Terminal ival XYSRC:X_D{$$ = 0;}|Y_D{$$ = Y_D + 1;};
```

Each Nonterminal definition consists of the type of returned value, a symbolic name as it appears in the assembly, an action that describes the possible token combinations and the return value associated with each of them.

Split function takes long bitfields as input and splits them up into existing subfields of the instruction word. As shown here, the right-hand side exhibits the concatenation of DBM.OP, DBM.MODE, and Main.OP. Thus, this section generates a function that splits a long input into shorter bitfields, as specified by the right-hand side.

```
Split.ADR  DBM.OP+DBM.MODE+Main.OP
```

The Macro definition type defines standard text replacement macros. A self-explanatory example is shown here:

```
# define ADD 56
```

Storage resources

The storage resources are the only structural information modeled by ISDL. The storage section lists all the storage resources visible to the programmer. It lists the names and sizes of the memory, register file, special registers and stack, and program counter. This information is used by the compiler to determine available resources, and how they should be used. An example of a storage section is shown here:

```
Memory X    = 0x1FF x 0x24
RegFile AGU_R   = 0x8 x 0x24
Register SP = 0x8
Register X0 = 0x24
CRegister SR    = 0x10
```

```
ProgramCounter  = 0x10
Stack SS    = 0xf x 0x20
. . . .
```

The is example shows a storage section which contains instantiations of memory, register files, single registers, control register, program counter, and stacks.

Assembly syntax

The assembly syntax section is separated into fields corresponding to the different operations that can be performed in parallel with a single instruction. For each field, a list of possible alternative operations are listed. Each operation consists of a name, a list of tokens or nonterminals as parameters, a set of commands that manipulate the bitfields, RTL description, timing details, and costs. The RTL description is used by the compiler to decide on operation selection. Multiple costs are allowed including operation execution time, code size, and costs due to resource conflicts. The timing information of ISDL describes when the various effects of the operation takes place. An example is shown here:

```
Field Main:
ADC XYSRC,AC        {Main.OP = 0x21|(AC<<3)|(XYSRC<<4);}
            {AC<- AC + XYSRC + CCR[0];}
            {cycle = 2 + dbm; size = 1 + dbm;}
            {latency = 1;}
. . . .
```

Constraints

The constraint section uses a set of Boolean rules to present to the compiler the set of operations that cannot be executed in parallel. Conflicts in instructions issued at different times can be shown by time shifting the constraints. Constraints may be datapath, bitfield or syntactic in nature. In the first two, two parallel operations can try to utilize the same datapath and bitfield respectively. The last type is artifacts of the assembler syntax. This following example constraint denotes that any REP instruction is illegal when fetched consecutively after the DO instruction. The "[1]" indicates a time shift of 1 for the REP instruction.

```
~(DO *) & ([1] REP *,*)
```

Optimization information

The Optimization information section of the ISDL description enables the compiler to generate better, optimized codes. Some uses of these optimizations are in delay slot instructions and in branch prediction hints.

14.2.2 ISDL-driven Methodologies

The ISDL is used to enable various design automation tasks including assembler generation, simulator and hardware generation, and compiler generation for exploration and rapid prototyping.

Assembler generator

The ISDL can be used to automatically generate an assembler, which led to the decoupling of the compiler from the simulator. As shown in Fig. 14.3, the simulator takes a binary file as input. The presence of the assembler allows to write and test assembly programs, even in the absence of compilers. The output of the compiler is an assembly file, which is not only more comprehendible than binary, but also is easy for debugging purposes. An assembler generator was designed by Hadjiyiannis et al. [6], which receives an ISDL description as input and produces an assembler that is used to assemble the compiler's output to a binary file. The Lex and Yacc files generated by the assembler generator, when compiled produces an executable. This executable parses the assembly and generates the instruction words. In addition to the parser, the assembler consists of a shell script. The latter processes the input files through the C preprocessor before parsing them, which allows for macro substitutions and file inclusions. File inclusions allow to include a common kernel into the code which acts as an operating system in the compiled code. Initialization operations such as setting up trap and exception vectors and enabling interrupts are performed by the kernel. The Lex file contains the lexical analyzer for the token and a symbol table which defines the following tokens:

- Predefined types such as integers, floats, strings, etc.
- All the operation names.
- The tokens defined in the global definition section of the ISDL description.
- Lables used in the symbol table to represent symbolic address.

The contents of the YACC file are the following:

- Main Driver: It performs initializations for file I/O, and also some clean-up in the case of a syntax error.

- Parser: It contains a top-level rule that performs two passes, which enable forward and reverse references in the labels. The first pass is used to fill in the symbol table and produce a listing file for the sake of debugging. The second pass composes an instruction from the fields in the assembly syntax section of the ISDL description.

Each field in the assembly syntax pattern is a rule, which consists of a list of possible patterns, one for each operation that is described in the specification for that field. A rule is also added for each nonterminal specified in the global definitions section.

The unambiguity of the assembly token is guranteed by the usage of Lex and Yacc for the assemblers. Certain syntax restrictions are imposed since the assembler preprocesses the input file with the C preprocessor. Thus, although the system may force the use of a modified assembly syntax, this is not considered as a problem since the input will be coming from the compiler, which will create this modified syntax. However, the output binary file created by assembling this modified syntax is identical to that of a commercial assembler.

GENSIM simulator generator

GENSIM [5] is a tool that automatically generates an instruction-level simulator from an ISDL description of the candidate architecture. This simulator can be used to measure the performance index of a program. This simulator is also known as XSim simulator. The XSim simulator consists of six parts: user interface and file I/O, scheduler, state monitors, state, disassembler, and processing core. The total simulator code, except for the graphical user interface, is written in C. Among the different parts of the simulator, the user interface, state monitors and scheduler codes are common to all architectures and hence, are implemented as a library. The other parts are architecture-specific and hence are generated as C source code from the ISDL description. The C code is now compiled and linked with the common library to create an architecture-specific executable program for the simulator. In order to generate the state data structures, sufficient memory should be allocated to each storage element defined in the ISDL description. The disassembler is generated by reversing the assembly function provided by the bitfield assignments. The processing core contains routines that correspond to the RTL statements in the description. These are translated to C functions that can perform the operations described in the RTL.

HGEN: hardware synthesis

The ISDL also utilizes the architecture synthesis system for hardware synthesis. The output of the architecture synthesis system is an ISDL description, which drives the ISDL to hardware compiler, also known as HGEN compiler. The HGEN compiler outputs synthesizable Verilog, which is then used to create the hardware implementation.

AVIV Compiler

The AVIV compiler is a retargetable code generator. The compiler accepts the application program and machine description of the target processor as inputs. The machine description is written in ISDL. The AVIV produces optimized code that is able to run on target processor. The main influence of AVIV is on architectures exhibiting ILP including VLIW processors. Instruction selection, resource allocation, and scheduling are important issues in case of retargetable code optimizations. The AVIV deals with all these issues concurrently. As shown in Fig. 14.3, the front end of AVIV converts a source program into an intermediate format, which comprises of expression DAGs and control flow informations. The starting point of AVIV is therefore, some basic block DAGs connected through control information.

The next task of AVIV is to extract machine dependent parallelism using the ISDL description of target processor. Conversion of basic block DAGs into split node DAGs aid in this. The instruction set information contained in the ISDL machine description language helps to create several databases that are required to create split node DAGs. The database may store various informations like correlation between target processor operations and SUIF operations, all possible data transfers explicitly stated in the target machine description. The machine description in ISDL is also used to know the constraints on the instructions in the target processor.

14.3 RADL

The language RADL [7] was developed at Rockwell Inc. as an extension to LISA language [8]. The motivation to develop RADL was to go beyond Instruction set architecture (ISA) and generate a cycle and phase accurate simulator of pipelined digital signal processors (DSP). The RADL has explicit support for pipeline modeling, including delay slots, interrupts, hardware loops, hazards, and multiple interacting pipelines. The main contribution of this language is to capture the behavior of pipeline, inter-pipeline control, and data communication with ease and flexibility. The RADL demonstrates that the description of pipelines supporting tool retargetability can be achieved in a straightforward and intuitive manner. Typical pipeline aspects, such as control signals, stall strategies, and latch registers are made explicit in RADL. Many of the tedious issues such as copying unchanged latch fields to the next latch register and resetting control signals are handled automatically.

14.4 SIM-nML

Sim-nML [9] was developed in Indian Institute of Technology, Kanpur as a performance estimator of a system in the paradigm of hardware–software codesign. Sim-nML is an extension to the nML language [10]. nML is powerful enough to express the behavior of the processor at instruction level. It is an extensible formalism developed for specifying the processor at higher level of abstraction. nML formalism models the instruction set architecture of a machine, but does not provide any control flow constructs. Sim-nML tries to complement nML by capturing control flow and inter-instruction dependencies by extending nML formalism.

To estimate the performance of the hardware–software co-partitioned systems, Sim-nML extends nML by abstracting out the control flow with the help of *resource usage model*. In resource usage model, a resource is an abstraction of a piece of hardware such as registers, ALUs, a functional block etc., for which instructions contend and control resolves the conflict due to contention. For example, consider the simple processor described here:

```
resource fetch_unit, execution_unit, retire_unit
reg AC [1, card(8) ]
reg PC [1, card (32)]
reg temp [1, card (8)]
op plus ()
syntax = "add"
image = "000000"
action = {AC = AC + tmp; }
uses = execution_unit #1
op multiply ()
syntax = "mult"
image = "000001"
action = { AC = AC * tmp ; }
uses = execution_unit # 3
op binaction = plus | multiply
op instruction (x: binaction, data:card (8))
syntax = format ("%s %d", x.syntax, data)
image = format ("11%6d %8b", x.image, data)
action = [tmp = data; x.action; }
preact = { PC = PC +2}
uses = fetch_unit : preact&#1, x.uses.
                   retire_unit #1 : action
```

In a resource usage model, a set of resources are obtained by an instruction and the resources are held till the next set of resources are available. Therefore, a specification of sequence of resources used by an instruction results in an abstract specification of the control flow. For example, "*fetch_unit, execution_unit, retire_unit*" means that fetch_unit is acquired first if it is free and is held till the execution_unit becomes free. Once the execution_unit is free, the unit is acquired and fetch_unit becomes free.

To control the flow of instruction, it is necessary to specify the minimum amount of time for which each resource is held. For example, fetch_unit #1, execution_unit #1, retire_unit #1 means that at first the fetch_unit is acquired. Although execution_unit is available immediately, the instruction waits in fetch_unit for one time unit before acquiring the execution_unit. In a similar manner, before completion it acquires retire_unit and holds it for one time unit. In this way Sim-nML provides an environment for hardware–software codesign, where complex issues such as branch prediction, hierarchical memory etc. can be modeled. The tool Sim-HS is capable of generating behavioral and structural models from Sim-nML specification. Behavioral Sim-HS can produce behavioral Verilog model from a Sim-nML specification to enable fast functional simulation. Similarly, structural Sim-HS can generate synthesizable Verilog models (implementation) from the Sim-nML specification.

14.5 UDL/I

Unified design language UDL/I [11] was developed as a hardware description language for compiler generation in COACH ASIP design environment at Kyushu University, Japan. It is used for describing processors at an RT-level on a per-cycle basis. The instruction-set is extracted automatically from the UDL/I description [12], and then it is used for generation of a compiler and a simulator. COACH works on simple RISC processors and does not explicitly support ILP or processor pipelines. The processor description is synthesizable with the UDL/I synthesis system [13]. The major advantage of the COACH system is that it requires a single description for synthesis, simulation, and compilation. Designer needs to provide hints to locate important machine states such as program counter and register files. Due to complexity in instruction-set extraction (ISE), ISE is not supported for VLIW and superscalar architectures.

14.6 FLEXWARE

FlexWare is a CAD system for DSP or ASIP design [14]. The FlexWare system includes the CodeSyn code generator and the Insulin simulator. The target processor description captures both structural and behavioral aspects. The machine description for CodeSyn contains three components: instruction set, available resources (and their classification), and an interconnect graph representing the datapath structure. The instruction set description is a list of generic processor macro instructions to execute each target processor instruction. The simulator uses a VHDL model of a generic parameterizable machine. The parameters include bit-width, number of registers, ALUs, and so on. The application is translated from the user-defined target instruction set to the instruction set of the generic machine. Then, the code is executed on the generic machine.

14.7 VALEN-C

Valen-C is an embedded software programming language proposed by Kyushu University, Japan [15, 16]. Valen-C is an extended C language which supports explicit and exact bit-width for integer type declarations. A retargetable compiler (called valen-CC) has been developed that accepts C or Valen-C programs as an input and generates the optimized assembly code. Although valen-CC assumes simple RISC architectures, it has retargetability to some extent. The most interesting feature of valen-CC is that the processor can have any datapath bit-width (e.g., 14 bits.).

The valen-C system aims at optimization of datapath width. The target processor description for valen-CC includes the instruction set consisting of behavior and assembly syntax of each instruction, as well as the processor datapath width.

14.8 TDL

Target Description Language (TDL) [17] was developed at Saarland University, Germany. The language is used in a retargetable postpass assembly-based code optimization system called PROPAN [18]. A TDL description consists of four sections: resource, instruction set, constraints, and assembly format. TDL offers a set of predefined resource types whose properties can be described by a predefined set of attributes. The predefined resource types comprise functional units, register sets, memories, and caches. Attributes are available to describe the bit-width of registers, their default data type, the size of a memory, its access width, and alignment restrictions. The designer can extend the domain of the predefined attributes and declare user-defined attributes if additional properties have to be taken into account. Similar to behavioral languages, the instruction-set description of TDL is based on attribute grammar [19]. The TDL can also be used for VLIW architectures, so it differentiates operation and instruction. The instruction-set section also contains definition of operation classes that groups operations for the ease of reference. TDL provides a nonterminal construct to model common components among operations. Similar to ISDL, TDL uses Boolean expressions for constraint modeling. A constraint definition includes a premise part followed by a rule part, separated by a colon. The following code segment describes constraints in TDL [17]:

```
op in {C0}: op.dst1 = op.src1;
op in {C1} & op2 in {C2}: !(op1 && op2);
```

The first rule requires the first source operand to be identical to the destination operand for all operations of the operation class C0. The second rule prohibits any operation of operation class C1 to be executed in parallel with an operation of operation class C2. The assembly section deals with syntactic details of the assembly language such as instruction or operation delimiters, assembly directives, and assembly expressions. The TDL is assembly-oriented and provides a generic modeling of irregular hardware constraints. The TDL provides a well-organized formalism for VLIW DSP assembly code generation.

14.9 CONCLUSIONS

This chapter described various contemporary ADLs and their associated methodologies for design automation of embedded processors. The ADLs covered in this chapter includes HMDES, ISDL, RADL, Sim-nML, UDL/I, FlexWare, Valen-C and TDL.

REFERENCES

[1] J. Gyllenhaal, B. Rau, and W. Hwu. HMDES Version 2.0 specification. Technical Report IMPACT-96-3, IMPACT Research Group, University of Illinois, Urbana, IL, 1996.

[2] http://www.trimaran.org. *Trimaran*.

[3] http://trimaran.org. *HPL-PD Architecture Specification: Version 1.1*, 2000.

[4] S. Hanono and S. Devadas. Instruction selection, resource allocation, and scheduling in the AVIV retargetable code generator. In *Proc. of Design Automation Conference (DAC)*, pages 510–515, 1998.

[5] G. Hadjiyiannis, P. Russo, and S. Devadas. A methodology for accurate performance evaluation in architecture exploration. In *Proc. of Design Automation Conference (DAC)*, pages 927–932, 1999.

[6] G. Hadjiyiannis, S. Hanono, and S. Devadas. ISDL: An instruction set description language for retargetability. In *Proc. of Design Automation Conference (DAC)*, pages 299–302, 1997.

[7] C. Siska. A processor description language supporting retargetable multipipeline DSP program development tools. In *Proc. of International Symposium on System Synthesis (ISSS)*, pages 31–36, 1998.

[8] V. Zivojnovic, S. Pees, and H. Meyr. LISA—machine description language and generic machine model for HW/SW co-design. In *IEEE Workshop on VLSI Signal Processing*, pages 127–136, 1996.

[9] V. Rajesh and R. Moona. Processor modeling for hardware software codesign. In *Proc. of International Conference on VLSI Design*, pages 132–137, 1999.

[10] M. Freericks. The nML machine description formalism. Technical Report TR SM-IMP/DIST/08, TU Berlin CS Dept., 1993.

[11] H. Akaboshi. *A Study on Design Support for Computer Architecture Design*. PhD thesis, Dept. of Information Systems, Kyushu University, Japan, January 1996.

[12] H. Akaboshi and H. Yasuura. Behavior extraction of MPU from HDL description. In *Proc. of Asia Pacific Conference on Hardware Description Languages (APCHDL)*, 1994.

[13] http://pjro.metsa.astem.or.jp/udli. *UDL/I Simulation/Synthesis Environment*, last accessed December, 1997.

[14] P. Paulin, C. Liem, T. May, and S. Sutarwala. FlexWare: A flexible firmware development environment for embedded systems. In *Prof. of Dagstuhl Workshop on Code Generation for Embedded Processors*, pages 67–84, 1994.

[15] A. Inoue, H. Tomiyama, F. Eko, H. Kanbara, and H. Yasuura. A programming language for processor based embedded systems. In *Proc. of Asia Pacific Conference on Hardware Description Languages (APCHDL)*, pages 89–94, 1998.

[16] A. Inoue, H. Tomiyama, H. Okuma, H. Kanbara, and H. Yasuura. Language and compiler for optimizing datapath widths of embedded systems. *IEICE Trans. Fundamentals*, E81-A(12): 2595–2604, 1998.

[17] D. Kastner. TDL: A hardware and assembly description languages. Technical Report TDL 1.4, Saarland University, Germany, 2000.

[18] D. Kastner. Retargetable Postpass Optimization by Integer Linear Programming. PhD thesis, Saarland University, Germany, 2000.

[19] J. Paakki. Attribute grammar paradigms—a high level methodology in language implementation. ACM Computing Surveys, 27(2):196–256, June 1995.

Index

395